PRAISE FOR
DATA STRUCTURES AND ALGORITHMS IN JAVASCRIPT

"An incredibly useful resource for those of us who did not have a formal computer science background or simply never had the time to properly learn formal aspects of programming in general. It's a tough book that will make you work, but worth every bit of effort."

—RAYMOND CAMDEN, SENIOR DEVELOPER
EVANGELIST FOR ADOBE

"Finally a book about algorithms and data structures that uses JavaScript! Read this if you want to learn important foundational techniques for writing and analyzing code—how to sort data, how to structure data for efficient lookup, and more. While similar to academic textbooks in its comprehensiveness, it's much easier to understand."

—AXEL RAUSCHMAYER, JAVASCRIPT
SPECIALIST, AUTHOR OF THE *2ality*
BLOG AND NUMEROUS JAVASCRIPT BOOKS

DATA STRUCTURES AND ALGORITHMS IN JAVASCRIPT

DATA STRUCTURES AND ALGORITHMS IN JAVASCRIPT

Optimizing Performance and Solving Programming Challenges

by Federico Kereki

no starch press®

San Francisco

Printed in the United States of America

First printing

28 27 26 25 24   1 2 3 4 5

ISBN-13: 978-1-7185-0262-8 (print)
ISBN-13: 978-1-7185-0263-5 (ebook)

Published by No Starch Press®, Inc.
245 8th Street, San Francisco, CA 94103
phone: +1.415.863.9900
www.nostarch.com; info@nostarch.com

Publisher: William Pollock
Managing Editor: Jill Franklin
Production Manager: Sabrina Plomitallo-González
Production Editor: Rachel Monaghan
Developmental Editor: Jill Franklin
Cover Illustrator: Gina Redman
Interior Design: Octopod Studios
Technical Reviewer: Daniel Zingaro
Copyeditor: Lisa McCoy
Proofreader: Scout Festa

A catalog record of this book is available from the Library of Congress.

For customer service inquiries, please contact info@nostarch.com. For information on distribution, bulk sales, corporate sales, or translations: sales@nostarch.com. For permission to translate this work: rights@nostarch.com. To report counterfeit copies or piracy: counterfeit@nostarch.com.

To Sylvia Tosar, my wife and constant companion and rock throughout this endeavor. Your unwavering support and love fueled every step of this journey, and I'm endlessly grateful for your presence and companionship.

And to my father, Eugenio Kereki, who is no longer here to witness this new book, but whose influence echoes through these pages. I dedicate this book to his memory, knowing his spirit resides in the love and guidance he shared.

About the Author

Federico Kereki is a Uruguayan systems engineer with a master's degree in education and over 30 years of experience as a consultant, system developer, and writer. Kereki is a subject matter expert at Globant, where he gets to use a variety of development frameworks, programming tools, and operating systems. He has taught several computer science courses at Universidad de la República; Universidad ORT, Uruguay; and Universidad de la Empresa. He specializes in efficiency and optimization, with a particular focus on algorithm and data structure design and usage. He's also the author of several books: *Mastering JavaScript Functional Programming*, 3rd edition (Packt, 2023), *Modern JavaScript Web Development Cookbook* (Packt, 2018), and *Essential GWT* (Pearson, 2010), as well as multiple articles for *Linux Journal, Linux Magazine, IBM DeveloperWorks*, the *OpenReplay* and *Globant* blogs, and others.

About the Technical Reviewer

Dr. Daniel Zingaro is an award-winning associate professor of mathematical and computational sciences at the University of Toronto, Mississauga. He is well known for his uniquely interactive approach to teaching and internationally recognized for his expertise in active learning. He is also the author of *Learn to Code by Solving Problems* and *Algorithmic Thinking* (both from No Starch Press).

BRIEF CONTENTS

CONTENTS IN DETAIL

3
ABSTRACT DATA TYPES

4
ANALYZING ALGORITHMS

PART II: ALGORITHMS

5
DESIGNING ALGORITHMS

PART III: DATA STRUCTURES   175

10
LISTS   177

11
BAGS, SETS, AND MAPS   203

18
IMMUTABILITY AND FUNCTIONAL DATA STRUCTURES   469

ANSWER KEY   487

BIBLIOGRAPHY   545

INDEX   549

PREFACE

Ever since I started learning about computers, back in the past millennium, I always had a soft spot for algorithms and data structures and for trying to get the best possible, most efficient, fastest solution to computing problems.

Over the years I've seen that developers are always taught about the new programming languages and frameworks, but I've also noticed that the focus on algorithm development and data structure design has been somehow minimized, which means programmers don't have all the needed tools to face more complex challenges.

I've also seen that many job interviews routinely include questions that are easily answered if you have a background in algorithms, but are a big obstacle otherwise. Similarly, coding challenges often require some specific data structure to optimally solve problems, and that's another place where knowledge about performance and order of algorithms is necessary.

I wanted this book to cover as much ground as possible, and you'll find many topics to explore here. Obviously, with the speed at which all areas of computer science advance, it's totally impossible to cover every single algorithm and describe every possible data structure, so I focused on the most important ones, providing a basis for eventual adaptations and enhancements.

I hope you, the reader, will derive several benefits from the book: more experience with the JavaScript language, a feeling for the performance aspects of algorithms, and the ability to adapt and enhance data structures to provide optimum solutions for your coding problems. Thanks for reading!

ACKNOWLEDGMENTS

Getting a book from the idea stage to an actual printed tome is a group effort, and several people were involved therein. First of all, I would like to thank Jill Franklin, managing and development editor, for starting me on this project and providing support and help throughout the process. Special thanks are due to three people: Daniel Zingaro, the technical reviewer who had to deal with way too many data structures and algorithms; Lisa McCoy, who had to copyedit all my descriptions and explanations; and Scout Festa, who proofread everything in the book—if any errors or issues are left, they are entirely my fault, not this trio's! Finally, many thanks must also go out to the rest of the No Starch Press team, without whom the book wouldn't exist: Bill Pollock, publisher; Sabrina Plomitallo-González, production manager; and Rachel Monaghan, production editor.

INTRODUCTION

This book deals with two basic concepts in computer science: data structures and algorithms. It follows a structure similar to university curricula and adds examples taken from coding challenges and interview questions, using them to discuss the relative advantages and disadvantages of specific algorithms and data structures.

All examples are fully coded in JavaScript, with particular attention given to modern language features that simplify coding. Performance is also considered, from both a theoretical point of view (order of algorithms) and a practical one (scaffolding, measuring). Each chapter ends with a series of questions that amplify the concepts covered in the chapter and provide further examples for the reader to apply. The answers to the questions are provided at the end of the book.

Who Should Read This Book?

This book is geared toward three groups of readers. The first and main group is JavaScript frontend (web) and backend (Node.js) developers, as it explores how we can apply data structures and algorithms to solve and optimize complex problems.

The second group of readers is computer science (CS) students, as the book covers topics that appear in most CS courses. These students should be familiar with several programming languages, so the JavaScript focus shouldn't be a hindrance. The algorithms don't heavily depend on aspects of the language and can be translated into other languages with little difficulty.

Finally, the third group of readers includes programmers who are preparing for coding interviews or are interested in competitive programming. These readers will profit from actual implementation of algorithms and data structures and from seeing examples of the kinds of questions they'll encounter.

What's the Book's Approach?

The book always takes a practical approach to real-life use cases. It considers common problems and discusses appropriate algorithms and data structures. We'll explore multiple versions and optimizations and develop several implementation variants to provide a deeper understanding of alternative possible solutions.

All of the algorithms are programmed using JavaScript, as it's a widely available language that could be applied for both frontend and backend work. JavaScript is also well known and commonly used today, and it should be applicable to all kinds of problems.

What's in the Book?

The book is structured in three parts. Part I covers the basics and highlights important aspects of JavaScript that are used throughout the rest of the book. We'll explore functional programming to understand some design considerations that are applied in later chapters. We'll also consider abstract data types (ADTs), which is a concept that involves data structures and algorithms. Finally, we'll study the topic of performance as it relates to algorithms, which will be applied often in the rest of the book. The following chapters are in Part I:

Chapter 1: Using JavaScript   In this chapter we'll cover important features of JavaScript that are used in the rest of the book, but we'll just consider the highlights, as it is assumed you are already familiar with the language. Topics will include the current JavaScript version, transpilation, typing, arrow functions, spreading, destructuring, modules, and more. I'll also introduce some of the many tools that are available to help you develop JavaScript code.

Chapter 2: Functional Programming in JavaScript   Here we'll consider functional programming to highlight some design features that are used in the rest of the book. Topics will include what functional programming is, why you should use it, whether JavaScript is a functional programming language, the declarative programming style, side effects, and higher-order functions, among others.

Chapter 3: Abstract Data Types   In this chapter I'll introduce the concept of abstract data types as a basis for considering data structures and their associated operations. In later chapters, all structures will be considered as ADTs to highlight their pros and cons as well as their performance. The key topics will include what ADTs are and how to implement them in JavaScript.

Chapter 4: Analyzing Algorithms   Here we'll consider the performance aspects of algorithms in terms of both space and speed. We'll discuss the concept of complexity classes and how (and when) it applies to the design of algorithms and data structures. The topics we'll explore include what an algorithm's performance is; the big O notation; complexity classes; the differences among best, average, worst, and amortized cases; how to measure performance; and time versus space trade-offs.

Part II of the book focuses on algorithms, and it's concerned with strategies for algorithm design. In particular, we'll consider searching, sorting, shuffling, and sampling—all of which have well-known algorithms. The chapters in this part are as follows:

Chapter 5: Designing Algorithms   In this chapter we'll consider strategies for the design of algorithms and look at examples of each case. We'll discuss general practices, recursion, brute-force search, greedy algorithms, divide-and-conquer algorithms, backtracking, dynamic programming, branch-and-bound, transform-and-conquer, and problem reduction.

Chapter 6: Sorting   Here we'll discuss several common and important sorting algorithms to produce an ordered sequence of data out of unordered data. Some algorithms (such as heapsort) will receive only a brief mention, because they will be further analyzed in later chapters where the corresponding data structure is described. Topics include a description of the sorting problem, internal versus external sorting, JavaScript's own sort function, comparison-based algorithms (such as bubblesort, selection sort, insertion sort, quicksort, and mergesort, among others), and comparison-less sorting algorithms (like bitmap sort, counting sort, and radix sort).

Chapter 7: Selecting   This chapter will show algorithms for just finding the kth smallest value in a list or array, as opposed to sorting where we wanted to order the complete set. We'll discuss the selection problem in general, using JavaScript's minimum and maximum functions; doing selection by sorting (or partial sorting); and several other algorithms like quickselect, Floyd-Rivest, median of medians, and sorting by selection.

Chapter 8: Shuffling and Sampling   This chapter can be considered a complement to Chapter 6. In this case we want to produce a random disordered sequence of data, instead of a totally ordered one, as might be needed for a computer card game or statistical sampling. We'll look at the shuffling problem, how to do random sorting, the Fisher-Yates algorithm, random key sorting, and random sampling algorithms.

Chapter 9: Searching   Here we'll consider several common searching algorithms meant to quickly answer whether a specific value is or isn't included in some set of data. Some algorithms will just be introduced here, but we'll explore them more thoroughly in later chapters where the corresponding data structures are described and analyzed. Topics in this chapter include a description of the searching problem, JavaScript's own search functions, linear searching (with or without sentinels), jump searching, binary searching, and interpolation searching.

Part III of the book is devoted to data structures, and it considers several data structure types, starting with simple, linear ones and finishing with more complex nonlinear structures. The following chapters are included in this part:

Chapter 10: Lists   This chapter discusses the simplest structure, a linked list, which also includes several variants. We'll delve into lists in detail (what they are; their various types; their ADT; single-, double-, and circular-linked lists), stacks (what they are and several implementations), queues (what queues are and what they're used for, their ADT, and many implementations), and deques (their objective, ADT, and implementation).

Chapter 11: Bags, Sets, and Maps   Here we'll discuss structures that let you represent sets (no repeated elements) and bags (repeated elements allowed), with maps (key/value pairs) as a special important case. We'll see what bags and sets are and their implementation (including JavaScript's own versions as well as array- and list-based versions), finishing with hashing and bitmaps.

Chapter 12: Binary Trees   This chapter considers binary trees, and in particular binary search trees (BSTs), which are the basis for many algorithms. We'll discuss what a tree is, tree traversal (pre-, in-, and postorder algorithms), and using binary search trees for searching (including splay trees, balanced search trees like AVL and red-black trees, and randomized binary search trees).

Chapter 13: Trees and Forests   Here we'll study more general variants of trees, including forests (sets of trees). Topics include what trees and forests are, how to represent them in several ways, traversal algorithms (breadth- and depth-first algorithms), B-trees and variants that are oriented for searching, and red-black trees as a variant of BSTs.

Chapter 14: Heaps   In this chapter we'll consider heaps, a variant of binary trees that are stored without the need for dynamic memory and allow easy implementation of priority queues and sorting. We'll discuss what heaps are, binary heaps and variants (ternary or d-ary heaps), heapsort (a heap-based sorting algorithm), heap-based sampling algorithms, and treaps, a heap-related BST.

Chapter 15: Extended Heaps This chapter expands on the concept of heaps, considering variants that allow extra operations such as changing (altering the value of a key) and melding or merging (joining two or more heaps into one). Topics include binomial heaps, lazy binomial heaps, Fibonacci heaps, and pairing heaps.

Chapter 16: Digital Search Trees Here we'll consider trees specially designed to search for strings, as in a common "dictionary" where we'd look up words. We'll include tries, radix tries, ternary tries, and other variants of these structures.

Chapter 17: Graphs In this chapter we'll consider graphs, which are currently used in many applications, such as Google Maps or for calculating dependencies in a software project. Topics include what graphs are, different ways of representing them (such as adjacency lists or adjacency matrices), graph traversals and path finding (including shortest-path algorithms), and topological sorting.

Chapter 18: Immutability and Functional Data Structures This final chapter will discuss the immutability aspect and explore how algorithms can be altered to avoid modifying the input structures, producing a new one instead. We'll see what functional data structures are, what immutability means, object freezing, algorithms needed to avoid modifying data structures, and some examples of specific functional data structures such as lists, queues, and trees.

The book finishes with the answers to the questions found at the end of each chapter; sometimes answers are given in full, and other times hints or links to solutions are provided.

NOTE *All of the source code for this book is available at* https://github.com/fkereki /data-structures-and-algorithms-book.

PART I

THE BASICS

In this first part of the book, we'll start by looking at important JavaScript features used throughout the book, and then go on to explore functional programming (FP) for design insights, abstract data types (ADTs) as they relate to data structures, and the concept of algorithm performance, which plays a crucial role when designing for efficiency.

1

USING JAVASCRIPT

JavaScript has evolved and added significant functionality since its original 1995 version. The addition of *classes* to the language aided with object-oriented programming so that you no longer need to work with complex prototypes. *Destructuring* and *spread* operators have simplified working with objects and arrays, and they allow you to manage multiple assignments at once. The introduction of *arrow functions* lets you work in a more succinct, expressive way, enhancing JavaScript's functional programming capability. Finally, the concept of *modules* has simplified code organization and lets you partition and group your code in logical ways. This chapter briefly explores these modern features of the language that help you write better, shorter, more understandable code.

The JavaScript language isn't the only thing that has evolved, however, so this chapter also will introduce some of the many tools that are now available to help you develop JavaScript code. Environments like Visual Studio Code with special fonts provide better code readability. Other tools help produce documented, well-formatted code, and validation utilities can detect static or type-related errors. In addition, many online tools exist to help deal with incompatibilities among browsers and servers.

Modern JavaScript Features

We'll start with an exploration of some modern JavaScript features that will simplify coding: arrow functions, classes, spreading values, destructuring, and modules. This list isn't exhaustive, and we'll look at other features in later chapters, including functional programming, map/reduce and similar array methods, functions as first-class objects, recursion, and more. We certainly can't cover all of the language's features, but here the focus is on the most important and newer features that are used throughout the book.

Arrow Functions

JavaScript provides many ways to specify a function, such as:

- Named functions, which are the most common: `function alpha() {...}`
- Nameless function expressions: `const bravo = function () {...}`
- Named function expressions: `const charlie = function something() {...}`
- Function constructors: `const delta = new Function()`
- Arrow functions: `const echo = () => {...}`

All of those definitions work basically the same way, but arrow functions—JavaScript's new kids on the block—have these important differences:

- They may return a value even without including a `return` statement.
- They cannot be used as constructors or generators.
- They don't bind the `this` value.
- They don't have an `arguments` object or a `prototype` property.

In particular, the first characteristic in the previous list is used a lot in this book; being able to omit the `return` keyword will make for shorter, more succinct code. For example, in Chapter 12, you will see the following function:

```
const _getHeight = (tree) => (isEmpty(tree) ? 0 : tree.height);
```

Given a `tree` argument, this function returns 0 if the tree is empty; otherwise, it returns the tree object's `height` attribute.

The following example uses return and is an equivalent (but longer) way to write the same function:

```
const _getHeight = (tree) => {
  return isEmpty(tree) ? 0 : tree.height;
};
```

The longer version isn't necessary: shorter code is good.

If you use the shortened version and want to return an object, you need to enclose it in parentheses. Here's another arrow function example from Chapter 12:

```
const newNode = (key) => ({
  key,
  left: null,
  right: null,
  height: 1
});
```

Given a key, this function returns a node (an object, in fact) with that key as an attribute, plus null left and right links and a height attribute set to 1.

Another common feature of the arrow function is providing default values for missing parameters:

```
const print = (tree, s = "") => {
  if (tree !== null) {
    console.log(s, tree.key);
    print(tree.left, `${s}  L:`);
    print(tree.right, `${s}  R:`);
  }
};
```

You will see what this code does in Chapter 12, but the interesting part is that the recursive function, if not provided with a value for s, will initialize it with the empty string.

Classes

Although we won't use classes much in this book, modern JavaScript has come far from its beginnings, and instead of having to deal with the prototype and adding tangled code to implement inheritance, now you can achieve inheritance easily. In the past you could use classes and subclasses, different constructors, and all of that, but implementing inheritance wasn't easy. JavaScript classes now make it much more straightforward. (See *https://developer.mozilla .org/en-US/docs/Learn/JavaScript/Objects/Inheritance* if you want to learn how to do inheritance in old-style JavaScript.)

Take a look at a partial, slightly modified example from Chapter 13 that shows an actual class and how to define it:

```
❶ class Tree {
❷   _children = [];

❸   constructor(rootKey) {
      this._key = rootKey;
    }

    isEmpty() {
      return this._key === undefined;
    }

❹   get key() {
      this._throwIfEmpty();
      return this._key;
    }

❺   set key(v) {
      this._key = v;
    }
}
```

You can either define a simple class, as is the case here ❶, or extend an existing one. For instance, you could have another class BinaryTree extends Tree to define a class based on Tree. You can define attributes outside a constructor ❷; you don't need to do it inside a constructor ❸. Constructors are available if you need more complex object instance initialization.

Getters ❹ and setters ❺ are other powerful features. They bind an object's property to functions that are invoked whenever we try to modify or access that property.

Other features not used in this example are static properties and methods; such attributes aren't part of the class instances, but rather belong to the class itself.

NOTE *Starting with ECMAScript 2022, JavaScript also includes private properties: fields, methods, getters, setters, and so on.*

The Spread Operator

The spread operator (...) allows you to, well, *spread* an array, string, or object into separate values in a single operation, providing some interesting array and object usages.

Arrays are applied like this:

```
  const myArray = [3, 1, 4, 1, 5, 9, 2, 6];
❶ const arrayMax = Math.max(...myArray);
❷ const newArray = [...myArray];
```

Entering ...myArray is the same as entering 3, 1, 4, 1, 5, 9, 2, 6, so the first usage of ...myArray in this example produces 9 ❶, and the second provides a new array with exactly the same elements of myArray ❷.

You can also use the spread operator to build a copy of an object, which you can then modify independently:

```
   const myObject = { last: "Darwin", year: 1809 };
❶ const newObject = { ...myObject, first: "Charles", year: 1882 };
   // same as: { last: "Darwin", first: "Charles", year: 1882 };
```

In this case, newObject ❶ first gets a copy of the attributes of myObject, and then the year attribute is overwritten. You could do this "the old way" with many individual assignments, but using the spread operator allows for shorter and clearer code.

A third usage of the spread operator is for functions that need to deal with an undefined number of parameters. Earlier versions of JavaScript had the arguments array-like object to handle this situation. The arguments object is "array-like," because .length is the only array property it provides. The arguments object doesn't include any other properties that arrays have.

For example, you could write your own Math.max() version like this:

```
const myMax = (...nums) => {
  let max = nums[0];
  for (let i = 1; i < nums.length; i++) {
    if (max < nums[i]) max = nums[i];
  }
  return max;
};
```

You could now use myMax() like you'd use Math.max(), but there's no reason to reinvent that function. This example shows how you can imitate the features of existing functions—in this case, the ability to pass many arguments to a function.

The Destructuring Statement

The *destructuring* statement is related to the spread operator. It allows you to assign several variables at the same time, which means you can combine several independent assignments into one and write shorter code. For example:

```
[first, last] = ["Abraham", "Lincoln"];
```

In this case, you assign "Abraham" to the first variable and "Lincoln" to the last variable.

You also can mix destructuring and spreading:

```
[first, last, ...years] = ["Abraham", "Lincoln", 1809, 1865];
```

Assign the initial elements in the array to first and last, as in the previous example, and assign all of the rest of the elements (the two numbers) to the years array. This combination lets you write code more succinctly, using a single statement, where previously several would have been required.

In addition, you can use default values when variables on the left side have no corresponding values on the right:

```
let [first, last, role = "President", party] = ["Abraham", "Lincoln"];
```

In this example, the destructuring statement assigns a default value to role and leaves party undefined.

You can also swap or rotate variables, which is a technique used frequently later in this book. Consider this line from code in Chapter 14:

```
[heap[p], heap[i]] = [heap[i], heap[p]];
```

This directly swaps the values of heap[p] and heap[i] without using an auxiliary variable. You also could write something like [d, e, f] = [e, f, d] to rotate the values of three variables, again without requiring more variables.

Finally, another pattern that we'll often use is to return two or more values at once from a function. For example, you could write a function to return two values in order:

```
const order2 = (a, b) => {
  if (a < b) {
    return [a, b];
  } else {
    return [b, a];
  }
};

let [smaller, bigger] = order2(22, 9); // smaller==9, bigger==22
```

The other way of returning many values at once is with an object. You still can do that, but returning an array and using destructuring is more compact.

Modules

Modules allow you to split code into pieces you can import when needed, providing a way to package functionality that is easier to understand and maintain. Each module should be an aggregation of related functions and classes, providing a set of features. A standard practice related to using modules is *high cohesion*, which means elements you put together should truly belong together, as unrelated functionalities should not be mixed in the same module. A related concept called *low coupling* means that distinct modules should be interdependent as little as possible. JavaScript lets you package functions in modules to provide a well-structured design, with greater readability and maintainability.

Modules come in two formats: *CommonJS modules* (an earlier format, used mostly in Node.js) and *ECMAScript modules* (the latest format, generally used by browsers).

CommonJS Modules

With CommonJS modules, write the code in the style of this (abridged) example from Chapter 16:

```
// file: radix_tree.js - in CommonJS style

❶ const EOW = "■";
  const newRadixTree = () => null;
❷ const newNode = () => ({ links: {} });
  const isEmpty = (rt) => !rt; // null or undefined
  const print = (trie, s = "") => { ... }
  const printWords = (trie, s = "") => { ... }
  const find = (trie, wordToFind) => { ... }
  const add = (trie, wordToAdd, dataToAdd) => { ... }
  const remove = (trie, wordToRemove) => { ... }

❸ module.exports = {
    add,
    find,
    isEmpty,
    newRadixTree,
    print,
    printWords,
    remove
  };
```

The `module.exports` assignment at the end ❸ defines what parts of the module will be visible from the outside; whatever is not included ❶❷ won't be accessible for the rest of the system. This way of writing code is well aligned with the *black box* software concept. Users of a module shouldn't need to learn or even know about its internal details to allow for higher maintainability. As long as the module keeps providing the same functionality, its developers are free to refactor or improve it without impacting any users.

If you wanted to import a pair of the functions that the module exports, for example, you'd use the following code style, which employs destructuring, to specify what you want:

```
const { newRadixTree, add } = require("radix_tree.js");
```

This allows access (via destructuring) to the `newRadixTree()` and `add()` functions, out of all the functions exported by the `radix_tree` module. If you want to add something to the Radix tree, you can call `add()` directly; similarly, you can call `newRadixTree()` to create a new tree.

Of course, you can also do this:

```
const RadixTree = require("radix_tree.js");
```

In order to add something to a tree or create a new one, you have to call `RadixTree.add()` and `RadixTree.newRadixTree()` instead. This usage makes for longer code, but it also lets you access all the functions in the `radix_tree` module. I prefer the first style that employs destructuring, because it makes clear what I am using, but it's really up to you.

ECMAScript Modules

The more modern ECMAScript style of defining modules also works with separate files, but instead of creating a `module.exports` object, you rewrite the module you just saw in the previous section as follows:

```
// file: radix_tree.js - in modern style

const EOW = "■";
❶ export const newRadixTree = () => null;
const newNode = () => ({ links: {} });
❷ export const isEmpty = (rt) => !rt; // null or undefined
const print = (trie, s = "") => { ... }
const printWords = (trie, s = "") => { ... }
const find = (trie, wordToFind) => { ... }
const add = (trie, wordToAdd, dataToAdd) => { ... }
const remove = (trie, wordToRemove) => { ... }

❸ export {
    add,
    find,
    print,
    printWords,
    remove
};
```

You can export something directly wherever you define it ❶❷ or postpone doing so until the end ❸. Both methods work (and I don't think anybody would really use *both* styles, as I did for this example), but most people prefer having all export statements together at the end. It's really your choice.

NOTE   *You can also use ECMAScript* import *and* export *statements in Node.js, but only if you use the* .mjs *extension instead of the* .js *extension, which is reserved for CommonJS modules.*

You can import functions from an ECMAScript module in the following way, which is a different usage in comparison with the CommonJS modules, although the end result is exactly the same:

```
import { newRadixTree, add } from "radix_tree.js";
```

If you want to import everything, use the following code instead; this will give you access to an object, including all the functions exported by the module, as earlier:

```
import * as RadixTree from "radix_tree.js";
```

All the exports you've seen so far are *named* exports; you can have as many of them as you want, and you can also have a single unnamed default export. In a given file, instead of defining what you want to export, as described earlier, you include something like this instead:

```
// file: my_module.js
export default something = ... // whatever you want to export
```

Then, in other parts of the code, you can do the following to import something:

```
import whatever from "my_module.js";
```

You can name what you imported whatever you like (whatever isn't a good name) instead of using the name the module creator intended. That isn't usual practice, but sometimes name conflicts arise when using modules by different authors.

Closures and Immediately Invoked Function Expressions

Closures and immediately invoked function expressions aren't actually new, but understanding them will be useful when following the examples in this book. A *closure* is the combination of a function plus its encompassing scope to which the function has access. It allows you to have private variables, which in turn allows you to create the equivalent of classes and modules. For instance, consider the following function:

```
function createPerson(firstN, lastN) {
  let first = firstN;
  let last = lastN;
  return {
    getFirst: function () {
      return first;
    },

    getLast: function () {
      return last;
    },

    fullName: function () {
      return first + " " + last;
    },

    setName: function (firstN, lastN) {
      first = firstN;
      last = lastN;
    }
  };
}
```

The returned value (an object) will have access to the first and last variables in the scope of the function. For example, consider the following:

```
const me = createPerson("Federico", "Kereki");
console.log(me.getFirst()); // Federico
console.log(me.getLast());  // Kereki
console.log(me.fullName()); // Federico Kereki

me.setName("John", "Doe");
console.log(me.fullName()); // John Doe
```

Those variables aren't accessible anywhere else. If you try to access me.first or me.last, you get undefined. Those variables are in the closure, but there's no way to access them, because they work as private values.

Using closures also allows you to simulate modules. For that, you'll need an *immediately invoked function expression (IIFE)*, pronounced "iffy," which is a function defined and executed as soon as it's defined.

Say you want a module to work with taxes. Without using the new modules, you could work in a similar way as with the createPerson(...) function:

```
const tax = (function (basicTax) {
  let vat = basicTax;
  /*
    ...many more tax-related variables
  */

  return {
    setVat: function (newVat) {
      vat = newVat;
    },
    getVat: function () {
      return vat;
    },
    addVat: function (value) {
      return value * (1 + vat / 100);
    }
    /*
      ...many more tax-related functions
    */
  };
})(6);
```

You create a (nameless) function and call it immediately, and the result works like a module. You can pass initial values to the IIFE, such as 6 percent for the default value-added tax (VAT). The vat variable, and others you may declare, are internal and cannot be accessed directly. However, the provided functions, addVat(...) and any others you may want, can work with all the internal variables.

Use the IIFE-based module as follows:

```
console.log(tax.getVat());    // 6: the initial default
tax.setVat(8);
console.log(tax.getVat());    // 8
console.log(tax.addVat(200)); // 216
```

Modules can provide the same basic functionality, but you will see cases when you'll want to use closures and IIFEs—for example, in Chapter 5 where memoizing and precomputing an array of values are discussed.

JAVASCRIPT STANDARDS AND VERSIONS

JavaScript's frequent updates have caused incompatibilities among versions. This book works with the latest version of the language, which is formally known as ECMAScript. ECMA originally stood for European Computer Manufacturers Association, but it's now considered to be a noun rather than an acronym. See the ECMA website (*https://www.ecma-international.org*) for more detailed information about JavaScript, including a very full specification of the whole language.

The Mozilla Developer Network (MDN; *https://developer.mozilla.org/en-US/docs/Web/JavaScript*) is also an invaluable resource for information on features like functions, classes, and modules.

All web browsers, as well as Node.js, may not implement the latest version of the language, but if you encounter an issue, you can use Babel to *transpile* (transpile is a portmanteau of *translate* and *compile*) your code into equivalent, but older (yet compatible) code, so you can still run it. See *https://babeljs.io* for more details on Babel as well as installation and configuration instructions. I ran all the code in the book in the current version of Node.js and didn't encounter any problems. To install Node.js, see *https://nodejs.org/en/download*.

JavaScript Development Tools

Let's turn our attention to some tools to add to your arsenal to help write better-looking code, check for common defects, and more. You won't use all of them in this book, but they are helpful and usually the first things I install whenever I start a new project.

Visual Studio Code

An integrated development environment (IDE) will help you write code quickly and easily. This book uses the Visual Studio Code (VSC) IDE. Other popular IDEs include Atom, Eclipse, Microsoft Visual Studio, NetBeans, Sublime, and Webstorm, and you could work with any of those as well.

Why use an IDE? Although a simple text editor, such as Notepad or vi, might be all you need, an IDE like VSC provides more functionality. With a text editor, you have to do more work yourself, constantly switching between tools and entering commands repeatedly. Using VSC (or any IDE) is a time-saver that allows you to work in an integrated fashion and with many tools in a single click.

VSC is open source, free, and updated monthly, with new features added frequently. You can use it for JavaScript and many other languages. Frontend developers use VSC for basic configuration and recognition ("IntelliSense") for HTML, CSS, JSON, and more. You can expand it via a vast catalog of extensions as well.

NOTE *Visual Studio Code, despite the similar name, is not related to Microsoft's other IDE, Visual Studio. You can use Visual Studio Code in Windows, Linux, and macOS, because it was developed in JavaScript and packaged for the desktop using the Electron framework.*

VSC also provides good performance, integrated debugging, an integrated terminal (to launch processes or run commands without having to leave VSC), and integration with source code management (typically Git). Figure 1-1 shows some of my own work in VSC with the code for this book.

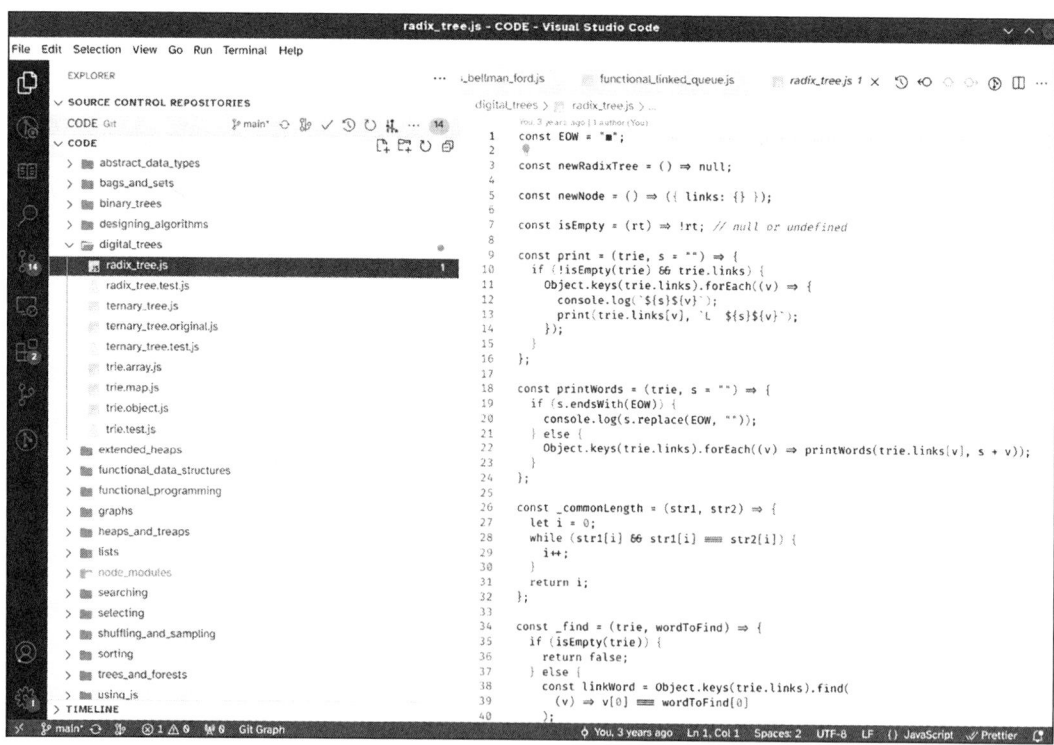

Figure 1-1: Using Visual Studio Code

Go to *https://code.visualstudio.com* to download the proper version for your environment and follow the installation instructions. If you like to live on the edge, install the Insiders' Version to gain access to new features, but be aware that you risk suffering from some bugs. For some Linux distributions, instead of downloading and installing the package yourself, you can use your package manager to handle installation and updating.

Fira Code Font

A quick way to start a (possibly heated) argument among developers is to mention that a given font is the best one for programming. Dozens of monospaced fonts for programming exist, but few include *ligatures*, which is when two or more characters are joined together. JavaScript code is a good candidate for ligatures, because otherwise you need to enter common symbols (such as ≥ or ≠) as two or three separate characters, which just doesn't look as good.

NOTE *The & character originally was a ligature of* E *and* t *to spell* et, *which means "and" in Latin. Another ligature in English is* æ *(as in* encyclopædia *or* Cæsar*) combining the letters* a *and* e. *Many other languages include ligatures; German joins two* s *characters together in* ß, *like in* Fußball *(football).*

The Fira Code font (*https://github.com/tonsky/FiraCode*) provides many ligatures and enhances the look of your code. Figure 1-2 shows all the possible ligatures for JavaScript. Fira Code includes ligatures for other languages as well.

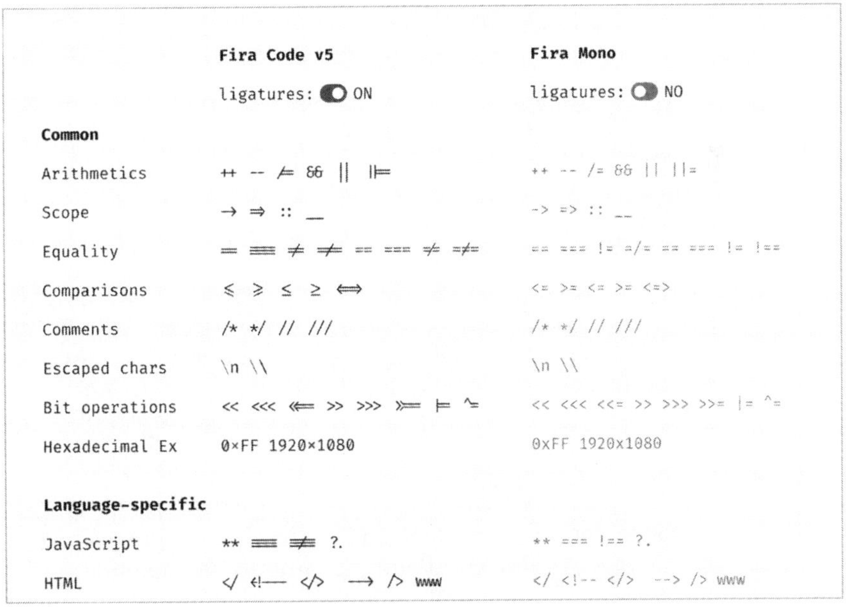

	Fira Code v5	Fira Mono
	ligatures: ◉ ON	ligatures: ◉ NO
Common		
Arithmetics	++ -- ≠ && ‖ ‖≡	++ -- /= && ‖ ‖=
Scope	→ ⇒ :: __	-> => :: __
Equality	= ≡ ≠ ≠≡ == === ≠ ≠≡	== === != =/= == === != !==
Comparisons	≤ ≥ ≤ ≥ ⟺	<= >= <= >= <=>
Comments	/* */ // ///	/* */ // ///
Escaped chars	\n \\	\n \\
Bit operations	<< <<< ⇐ >> >>> >≡ ⊨ ^≡	<< <<< <<= >> >>> >>= |= ^=
Hexadecimal Ex	0×FF 1920×1080	0xFF 1920x1080
Language-specific		
JavaScript	** ≡ ≠≡ ?.	** === !== ?.
HTML	</ ⇐— </> ⟶ /> www	</ <!-- </> --> /> www

Figure 1-2: A sample of the many ligatures Fira Code font provides (cropped from the Fira Code website)

After downloading and installing the font, if you are using Visual Studio Code, follow the instructions at *https://github.com/tonsky/FiraCode/wiki/VS-Code -Instructions* to integrate the font with the IDE.

Prettier Formatting

How to format source code can be another source of disagreements. Every developer you work with will likely have their own take on this issue, asserting that their standard is best. If you work with a team of developers, you may be familiar with the situation shown in the "How Standards Proliferate" xkcd comic (Figure 1-3).

Figure 1-3: "How Standards Proliferate" (courtesy of xkcd, https://xkcd.com/927)

Prettier is an "opinionated" source code formatter that reformats code according to its own set of rules and a few parameters you can set. Prettier's website states, "By far the biggest reason for adopting Prettier is to stop all the on-going debates over styles." All the source code examples in this book are formatted with Prettier.

Installing Prettier is simple; follow the instructions at *https://prettier.io,* and if you use Visual Studio Code, also install the Prettier extension from *https://marketplace.visualstudio.com/items?itemName=esbenp.prettier-vscode.* Be sure to tweak VSC's settings to enable the `editor.formatOnSave` option so all code will be reformatted upon saving. Consult the documentation on the Prettier website to learn more about configuring Prettier to your liking.

JSDoc Documentation

Documenting your source code is development best practice. JSDoc (*https:// jsdoc.app*) is a tool that helps you produce documentation for your code by aggregating specifically formatted comments. If you add comments preceding your functions, methods, classes, and so on, JSDoc will use them to produce documentation for your code. We don't use JSDoc in this book, because the text explains the code. However, for normal work, using JSDoc helps developers understand all of a system's pieces.

Here's a code snippet that adds a key to a heap from Chapter 14 to show how JSDoc produces documentation:

```
/**
 * Add a new key to a heap.
 *
 * @author F.Kereki
 * @version 1.0
 * @param {pointer} heap - Heap to which the key is added
 * @param {string} keyToAdd - Key to be added
 * @return Updated heap
 */
const add = (heap, keyToAdd) => {
  heap.push(keyToAdd);
  _bubbleUp(heap, heap.length - 1);
  return heap;
};
```

JSDoc comments start with the /** combination, which is like the usual comment format but with one extra asterisk. The @author, @version, @param, and @return tags describe specific information about the code; the names are self-explanatory. Other tags you can use include @class, @constructor, @deprecated, @exports, @property, and @throws (or @exception). See *https://jsdoc .app/index.html* for a complete list.

After installing JSDoc according to the instructions at *https://github.com/ jsdoc/jsdoc*, I processed this example file, which produced the results shown in Figure 1-4.

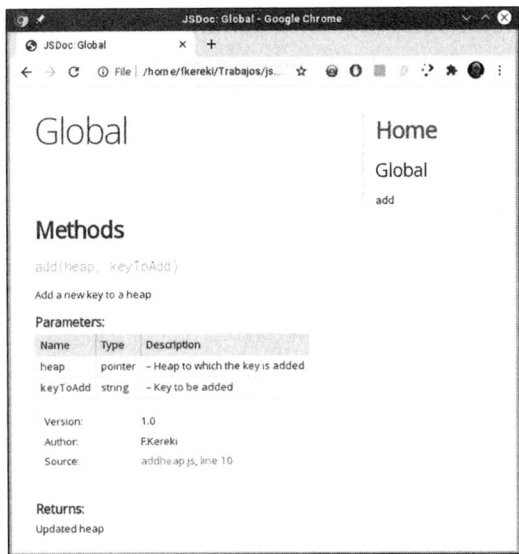

Figure 1-4: A sample documentation web page automatically generated by JSDoc

Of course, this is a simple example with only a single file. For a complete system, you would get a home page with links to every page of documentation.

ESLint

JavaScript presents many possibilities for misuse and misunderstanding. Consider this simple example: if you use the == operator instead of ===, you may find cases in which x==y and y==z, but x!=z, no matter what the transitive law may say. (Try x=[], y=0, and z="0".) Another tricky case is if you accidentally enter (x=y) instead of (x==y), which would be an assignment rather than a comparison; it's not very likely you want the former.

A *linter* is a tool that analyzes code and produces warning or error messages about any doubtful or error-prone features you might be using. In some cases, a linter may even fix your code properly. You also can use linters in conjunction with source code versioning tools. Linters can keep you from posting code that doesn't pass all checks. If you use Git, go to *https://git-scm.com/book/en/v2/Customizing-Git-Git-Hooks* to read about precommit hooks.

ESLint works well for linting in JavaScript. It was created in 2013 and is still going strong. Go to *https://www.npmjs.com/package/eslint* to download and install, and then configure it. Be sure to carefully read the rules at *https://eslint.org/docs/rules/*, because you can set many different rules, but you shouldn't turn them all on unless you want to start some linting wars.

Finally, don't forget the VSC extension at *https://marketplace.visualstudio.com/items?itemName=dbaeumer.vscode-eslint*, so you can see whatever errors ESLint detects.

NOTE *With ESLint, the eqeqeq rule (see* https://eslint.org/docs/rules/eqeqeq*) would have detected the problem with the type-unsafe == operator and even would have fixed it by substituting === instead. In addition, the no-cond-assign rule would have warned about the unexpected assignment.*

Flow and TypeScript

For large-scale coding, consider using Flow and TypeScript, which let you add information about data types to JavaScript. Flow adds comments that describe what data types are expected for function inputs and outputs, variables, and so forth. TypeScript is actually is a superset of JavaScript that is transpiled into it.

Figure 1-5 (shamelessly based on an example from the TypeScript home page) shows the kinds of errors you can detect with type information.

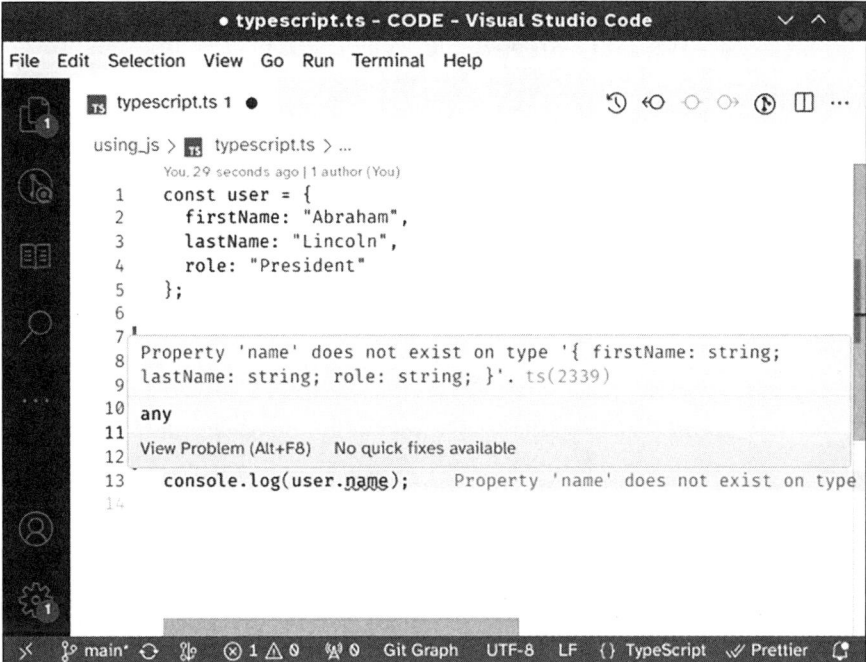

Figure 1-5: A type error in TypeScript code caught on the fly by ESLint

In this example, I'm trying to access an attribute that doesn't exist (`user.name`) according to the type data deduced from earlier lines of code. (Note that I'm using ESLint, which is why I can see the error in real time.)

We won't use either of these two tools in this book, but for big projects that involve many classes, methods, functions, types, and so on, consider adding them to your repertoire.

Online Feature Availability Resources

If you're working server-side with the latest version of Node.js, you probably won't need to worry about any specific feature being available. However, if you're doing frontend work, a given function may not be available, such as Internet Explorer support. If that happens, you'll need to transpile with something like Babel, as mentioned earlier in the chapter.

The Kangax website (*https://compat-table.github.io/compat-table/es2016plus/*) provides information on multiple platforms, detailing whether a function is fully, partially, or not available. Kangax provides a listing of all the JavaScript language features, with examples for each, and you'll find a table on the website that shows what features are available for each different JavaScript engine, such as features found on browsers and Node.js. Generally speaking, when you open it with a browser, green "Yes" boxes mean you can use the feature safely; boxes in different colors or text imply the feature is partially available or not available.

The *Can I Use?* website at *https://www.caniuse.com/* lets you search by function and shows the available support in different browsers. For instance, if you search for arrow functions, the website will tell you which browsers support it, since what date, and the percentage of global users with direct access to that feature without polyfills or transpiling.

NOTE   *If you are hazy on the term* polyfill, *see "What Is a Polyfill" at* https://remysharp .com/2010/10/08/what-is-a-polyfill *by Remy Sharp (creator of the concept). A polyfill is a way to "replicate an API . . . if the browser doesn't have it natively." The MDN website often provides polyfills for new features, which is helpful if you need to deal with older browsers that don't provide them or need details on how something works.*

Figure 1-6 shows information on the availability of arrow functions across browsers; hovering with the mouse provides more data, such as when the feature was first available.

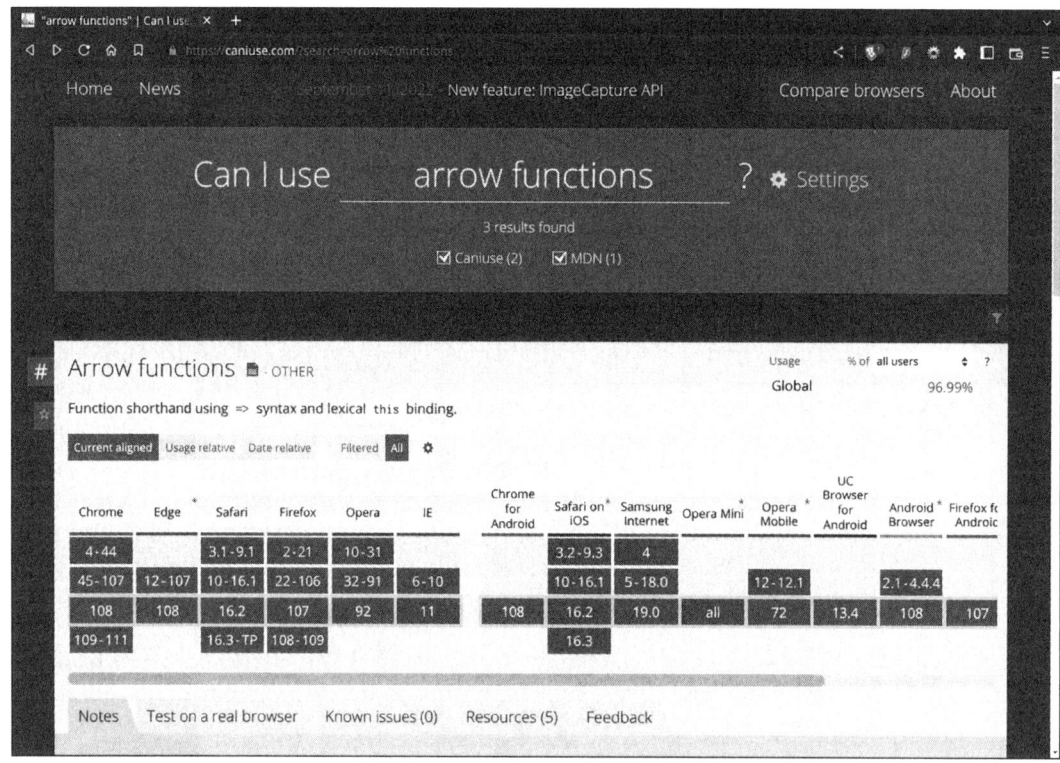

Figure 1-6: The Can I Use? website shows whether a given feature is available in browsers.

The Can I Use? site provides information only about browsers; it doesn't include server-side tools like Node.js, but you'll likely find a need for it at some time.

Summary

In this chapter, we looked at some of JavaScript's new and important modern features, including spreading, destructuring, arrow functions, classes, and modules. We also considered some additional tools you might want to include for your development work, such as the VSC IDE, Fira Code font for neater screen displays, Prettier for source code formatting, JSDoc to generate documentation, ESLint to check for defects or bad practices, and Flow or TypeScript to add data type checking. Finally, in order to ensure that you're not using unavailable functions, two online resources were presented: Kangax, and Can I Use?, both of which will help you avoid unimplemented or only partially implemented JavaScript features.

In the next chapter, we'll go deeper into JavaScript and explore its functional programming aspects, providing a starting point for the examples in the rest of this book.

2

FUNCTIONAL PROGRAMMING IN JAVASCRIPT

Functional programming (FP) is a programming paradigm based on functions, which are the only building blocks for your code. Using FP improves your code's modularity and helps you write more understandable, testable, and easy-to-maintain code while avoiding bugs; it's a win all around.

Some people claim that FP is an enlightened approach to programming that leaves *object-oriented programming (OOP)* and other paradigms far behind. Others say that it is only theoretical, is not applicable "in the real world," and causes more problems than it solves. As in most domains, the differences aren't black or white, but a shade of gray. It's not a matter of using only FP or never using FP. We'll use it in this book, but not dogmatically, and in this chapter, you'll see how JavaScript lets you work functionally, how and when to use FP, and instances when *not* to use it. Wherever FP applies and is more appropriate, we'll use it, but if OOP is better suited, it will be used instead.

NOTE *FP is not a modern fad. The second-oldest programming language still in common use (after FORTRAN, which appeared in 1957) is LISP (which appeared one year later), and it's completely based on FP. Many other FP languages have appeared in the more than 60 years since, and even nonfunctional languages, such as JavaScript, also provide the same kind of functionality.*

Why Use Functional Programming?

Consider what features are most important to you when programming and then ask whether FP provides them. Most programmers usually want the code they write to be:

Understandable   Code is written once but read many times, and users should be able to "get" your functions and their relationships without great effort. FP tends to produce shorter and cleaner code, which is easier to understand.

Maintainable   Your code most likely will need maintenance sometime in the future, and it should be easy to do. The same features that make FP code easier to understand also allow for easier maintenance. You also don't need to worry about breaking other things when modifying any given function.

Testable   Unit testing is a common part of development work, allowing you to verify the behavior of each component of your code. Unit tests also serve as a sort of documentation, providing examples of how to use any given function to people reading your code. If your programming style doesn't favor writing easily testable code, you'll have problems. FP always produces functions you can test on your own.

Modular   You should divide your code's features into independent modules, each pertaining to one specific aspect of your program, so that if you make changes in one module, they won't affect the rest of your code. The goal in FP is to write independent functions you can refactor or modify without impacting other functions. Writing independent functions helps you achieve *separation of concerns* (distinct parts of a program have little overlap). Also, modules tend to be *highly cohesive*, which means they include functions that really belong together, and they are *loosely coupled*, so changes in a function are not likely to require changes in other functions.

Reusable   Reusing code saves time and money. Because functions stand on their own in FP, you can use a well-written set of functions anywhere.

Of course, object-oriented code can do all of these things as well. No one is ever going to say that FP is a silver bullet that solves all your software development problems. I always suggest taking the middle road; a well-thought-out, balanced mix is usually the best solution.

JavaScript as a Functional Language

Is JavaScript functional? When people discuss functional languages, they typically mention Haskell, Erlang, Scala, and the like; no one usually includes JavaScript. That might be because no precise definition exists for what constitutes a functional language, what features a functional language should provide, or how it should work. For the purposes of this book, we'll consider a language to be functional if (and only if) it supports common FP features; you'll see how JavaScript compares. We'll take advantage of items like functions as first-class objects, array functions, pure functions, higher-order functions, and recursion and avoid side effects (or *impurities*, in FP lexicon) using (local or global) state, mutating objects or arguments, and so on.

Functions as First-Class Objects

In JavaScript, to say that functions are *first-class objects* means you can do anything with them that you'd be able to do with other objects. You can store functions in variables, pass functions as arguments, or return a function as a result from some other function.

Consider an example of doing an application programming interface (API) call. If you work with something like Axios (or SuperAgent, or other similar libraries that simplify the process of doing an async call to a remote server), you've likely seen code like this:

```
axios.get("your/url/api").then((response) => {
  // ...do something with the response
})
```

The parameter to the `.then()` method is a function, and it's passed in the same way you pass a number or an array.

In Chapter 12, we'll be doing that as well, and we'll be also able to assign default values to function parameters in order to change the way the function performs:

```
const preOrder = (tree, visit = (x) => console.log(x)) => {
  if (tree !== null) {
    visit(tree.key);
    preOrder(tree.left, visit);
    preOrder(tree.right, visit);
  }
};
```

The `preOrder()` function takes two arguments: a tree (you'll learn about trees later) and a `visit` function; if you don't provide the `visit` function, the default value will be a simple function that just logs whatever you pass to it.

Taking one or more functions as parameters makes the function a *higher-order function*; the other identifying characteristic for such a function is returning a function as a result. Common functions (those that do not receive or return functions) are called *first-order functions*.

Working with the same example, you could also write the following:

```
❶ const myVisit = (x) => {
    // ...do some interesting things with x
  }

  let myTree = newTree();
  // ...set up myTree, add to it, etc.

❷ preOrder(myTree, myVisit);
```

The function is stored in a variable ❶, as functions are just another type of value that can go in a variable, and then the variable's contents are passed to a function ❷ in exactly the same way that you pass myTree, another variable with a different kind of value.

You'll see more examples throughout this chapter, including functions that return functions as results.

Declarative-Style Programming

FP works in a higher-level, *declarative* style instead of the imperative style used in procedural, "common" programming. With declarative programming, you specify what you want, but not the individual detailed steps necessary to accomplish it, as you would do in procedural coding. The best example of declarative coding involves arrays. Working with arrays usually entails loops, which you can do "by hand" (think of using a while loop) or with the preferred for statement, but JavaScript lets you work declaratively with some specific array functions. In fact, we'll look at some *methods* for this, but the same concept applies; a method is just a function after all.

The following list details some of the available array functions that let you search or select elements from an array:

.filter()   Picks elements that satisfy some condition out of an array

.find() and .findIndex()   Search an array to find an element that satisfies some condition

.some()   Lets you know whether at least one element of an array satisfies a condition

.every()   Lets you test whether all elements of an array satisfy a condition

Other functions transform an array into a new array or a single result:

.map()   Lets you transform one array into another by applying a given function to its elements

.reduce()   Applies a given operation to the whole array from left to right, reducing it to a single result

.reduceRight()   Works like .reduce() but from right to left

This list isn't meant to be complete: there are more functions that transform arrays, like .flat() or .flatMap(), but you won't see them here.

Some people say that these functions are slower than the corresponding hand-written loops and that you'll suffer some performance hit for using them. Even if those things were true, they are irrelevant. Unless you are having some real performance problems and after analyzing your code you reach the conclusion that the array functions are to blame, writing longer, more bug-prone code doesn't really make much sense.

Filtering an Array

Let's take a look at a common task: going through an array, selecting elements that satisfy some condition, and dropping the rest. The .filter() method does exactly that: you provide a *predicate* (a function that produces a boolean result in terms of its arguments), and a new array is produced, with only the elements of the original array for which the predicate returned true.

For example, to select all values under 21, the following predicate would be useful:

```
const under21 = (value) => value < 21;
```

The under21() function gets a value and returns true if the value is less than 21. Now you can write the code as follows:

```
let myArray = [22, 9, 60, 12, 4, 56];
let newArray = myArray.filter(under21);
console.log(newArray); // [9, 12, 4]
```

This specifies that you want to apply the "under 21" check to filter the original array, and the result is a new array with only the values that satisfied the given test. You don't need to write anything to control a loop, initialize an output array, or anything else. The code is much shorter and truly declarative; it specifies only *what* you wanted to get, not *how* to get it.

Searching an Array

Other methods let you search an array for some element that satisfies a predicate, returning either the element or its position in the array:

find() Goes through the array from beginning to end, testing for the given predicate; if an element of the array satisfies it, the element is returned; if no elements satisfy the predicate, undefined is returned. A recent new method, findLast(), does the same, but searches from the end to the beginning.

findIndex() Is similar to find(), but it returns the position of the first element satisfying the predicate or -1 if no elements satisfy it. The new findLastIndex() returns the position of the last element that satisfies the predicate.

These methods are practical, because they provide all the needed search code in a single line. There is no limit to the complexity of a search predicate. You aren't limited to looking only for a value, for example, and can test any condition, as you did with under21() previously.

Here are examples of using both of these searching methods at once, since they are related and similar:

```
let myArray = [22, 9, 60, 12, 4, 56];
const under21 = (value) => value < 21;
❶ console.log(myArray.find(under21));        // 9
❷ console.log(myArray.findIndex(under21)); // 1

const equal21 = (value) => value === 21;
❸ console.log(myArray.find(equal21));        // undefined
❹ console.log(myArray.findIndex(equal21)); // -1
```

The first element of myArray that satisfies the under21 predicate is 9 ❶, and it's at position 1 of the array ❷. If you redo the calls using an equal21 predicate that checks whether a value is 21, you get undefined ❸ and -1 ❹, because no such element is found in the array.

NOTE   *If you are searching for a specific value, .includes(), .index(), and .lastIndexOf() are quite useful, though less flexible than the FP-oriented methods described earlier, because they just let you search for a value and not test for any possible condition.*

Testing an Array

Related methods for searching are .some() and .every(). The first checks whether *any* element of an array satisfies some predicate, and the second checks whether *all* elements satisfy it:

```
let myArray = [22, 9, 60, 12, 4, 56];
const under21 = (value) => value < 21;
console.log(myArray.some(under21));  // true
console.log(myArray.every(under21)); // false

const equal21 = (value) => value === 21;
console.log(myArray.some(equal21));  // false
console.log(myArray.every(equal21)); // false
```

If you test whether some elements of the array are under 21, the answer is true, but not all elements are true. If you repeat these tests for equality to 21, both answers are false.

You could omit these functions if you have the methods in the previous section (see question 2.7 at the end of the chapter).

Transforming an Array

Algorithms often need to go through a set of elements (such as an array) and apply some operation to each of them to create a new set. For instance, in a web application, you could have a list of strings, which could represent numbers, and you'd want to transform that list into a list of the corresponding numeric values. Setting up a loop to go through all elements of an array systematically, processing elements one by one, and producing a new array is a common procedure and usually taught early on to developers. This kind of transformation is also key in FP, and JavaScript calls it the .map() function.

Consider an introductory example that just produces a new array, with all values multiplied by 10:

```
let myArray = [22, 9, 60, 12, 4, 56];
console.log(myArray.map((x) => 10*x));
// [220, 90, 600, 120, 40, 560]
```

Start with the same array you've been using. If you map it with a function that multiplies its argument by 10, you get a new array whose values are 10 times the original ones.

Using .map() instead of a common loop means clearer code, and mapping is a well-known pattern in FP. The code is also shorter, which means fewer chances of introducing bugs. And finally, the code produces a new array instead of modifying the original one, so the function is *pure* (you'll learn what that means later in this chapter). Whenever possible, use .map(), but be aware that some of its features could bite you; see question 2.4 at the end of this chapter for an example.

Reducing an Array to a Single Value

Here is another common task: writing loops that go through a complete array, performing some kind of operation and ending up with a single computed value. (A typical example is an array with a list of numbers and adding them up.) You can implement this kind of task in a functional way using either the .reduce() or, less frequently, the .reduceRight() function. The following sums up a complete array:

```
const myArray = [22, 9, 60, 12, 4, 56];
const mySum = myArray.reduce((a, v) => a + v, 0); // 163
```

This logic does all that's needed to apply a function that receives two values and returns their sum to the complete array, starting with 0. In other words, it sums all the elements of the array. The a argument stands for *accumulator* (initially 0), and v stands for *value* (each of the elements of the array). You don't need to provide the initial value for .reduce(), but it's safer to do so. If you try to reduce an empty array without an initial value, you'll get a runtime error.

To show the power of .reduce(), look at another case where you calculate more than one result: averaging the array values. For that, you need not only their sum, which you already know how to get, but also their count (forget that you could use `myArray.length` to provide the latter; this is just an example):

```
const myArray = [22, 9, 60, 12, 4, 56];
myArray.reduce((a, v) => ({ s: a.s + v, c: a.c + 1 }), { s: 0, c: 0 });
// { s: 163, c: 6 }
```

The object has two fields (s for the sum and c for the count), and the reducing function recalculates those values at each step; the final result matches the previous example.

If you need to process an array from right to left, `.reduceRight()` works the same way, but it starts at the end of the array and goes back to the first element. Of course, you also could reverse the original array by first using `.reverse()` and then `.reduce()`, but that would cause a side effect. The array would be reversed in place (see "Mutating Arguments" on page 33 for more about this).

Looping Through Arrays

The `.forEach()` array function takes care of looping through an array, invoking a callback for each element, so you just have to declare what kind of work you want to do and nothing else. You can redo the array summing logic using this function:

```
const myArray = [22, 9, 60, 12, 4, 56];
❶ let sum = 0;
❷ myArray.forEach((v) => {
    sum += v;
});
❸ console.log(sum); // 163, as earlier
```

First set the sum variable to 0; this variable will get the sum of all elements in the array ❶. Then go through the array ❷ and specify what you want to do with each v element. In this case, add it to sum, and the final result ❸ is exactly the same as before.

Higher-Order Functions

As mentioned previously, functions that receive other functions as arguments or that return functions as results are called *higher-order functions*. This means all functions that work with callbacks are higher-order functions, and so are all the array methods just discussed. Some of these functions allow you to work in a more declarative style (like the ones you saw earlier), and others allow you to extend a function and modify what it does—for example, adding logging as an aid for debugging or memoizing for better performance.

Consider one of the uses for higher-order functions: returning a new function, with an example of wrapped behavior. Wrapping produces a new function that keeps its original functionality but adds some extra behavior. Imagine you want to add logging to a function for debugging purposes. You could modify the function of course, but doing so is risky, because you could accidentally touch something you shouldn't. You also could use a debugger, but a wrapper function provides more flexibility.

The original function could be something like this:

```
const myFunction = (arg1, arg2) => {
  // Do something with arg1 and arg2
  // and eventually return something.
}
```

You then could modify it to add logging:

```
const myFunction = (arg1, arg2) => {
  console.log("Entering myFunction with ", arg1, arg2);
  // Do something with arg1 and arg2
  // and calculate something to return.
  const toReturn = something;
  console.log("Exiting myFunction, returning ", toReturn);
  return toReturn;
}
```

If the function had several return statements, however, you'd need to modify them all.

Using a higher-order logging function is better:

```
const addLogging = (fn) => (...args) => {
❶ console.log("Entering ", fn.name, " with ", ...args);
❷ const toReturn = fn(...args);
❸ console.log("Exiting ", fn.name, " returning ", toReturn);
❹ return toReturn;
}
```

The addLogging() function receives a function as argument and returns a new function ❹. First, this new function logs the original function's name ❶ and the arguments it received. Then it actually calls the original function to calculate whatever the function calculated ❷. After that, the new function logs the result ❸ and finally returns that value.

Here's a simple example:

```
❶ const sum2 = (a, b) => {
   console.log("Calculating...");
   return a+b;
 }

❷ addLogging(sum2)(22, 9);
 // Entering sum2 with 22 9
 // Calculating...
 // Exiting sum2 returning 31
```

The sum2() function logs something and returns the sum of its arguments ❶. When you pass sum2 as a parameter to addLogging and call the resulting function ❷, you get the extra logging you wanted without touching the original function.

Side Effects

When a function depends only on the parameters it receives and doesn't produce any side effects, it is called a *pure function*. The concept of pure functions is closely related to mathematical functions: given an $f(x)$ function, all it does when given a value for x is calculate a new value.

Using pure functions is advantageous because they don't produce any side effects, such as changing the program state, modifying variables, mutating objects, and so on. This means that when you call a pure function, you don't need to worry about any possible changes anywhere in your code or the possibility of something else getting broken; you can concentrate on which arguments you pass to the function, knowing no "surprises" will happen. When given the same arguments, the function will always return the same result. That result won't depend on any "outside" variables or state, which could change and then cause the function to produce a different result. On the other hand, a pure function cannot depend on random numbers, the time of day, the result of input/output (I/O) functions, and so on; it depends only on its input.

Using Global State

The most common reason for side effects is using nonlocal variables that are shared with other parts of the code. Since a pure function always produces the same output given the same inputs, if a function depends somehow on variables outside it, it automatically becomes impure.

The problem goes deeper; debugging a function that depends on global state is more difficult, because in order to understand why a function returned a given value, you must also understand how the program state was reached, and that itself requires understanding all the previous history of the running code. The injunction against using global variables is a good one, even if you aren't specifically following FP tenets.

Keeping Inner State

You can extend the practice of not using external variables to include avoiding internal variables, which keep state between calls. Even if no global variables are present, internal variables may cause future calls to the function to return different output, even when provided with the same input arguments.

Using internal state is why functional programmers don't like working with objects. OOP requires data to be stored in an object to use for calculations, which automatically enables the possibility of impure code, because some methods could calculate results depending not only on their arguments but on the internal attributes as well.

Mutating Arguments

We've considered working with (and possibly modifying) external or internal variables, but there is one more "sin" you might commit: modifying the actual arguments to the function. In JavaScript, arguments are passed by value, unless they are objects or arrays, which are passed by reference. The latter implies that if the function modifies the arguments, it will actually be modifying the original object or array, which is certainly a side effect. You saw a possible case earlier in the chapter when emulating .reduceRight() by first applying .reverse() to reverse an array—an unexpected side effect (see the section "Reducing an Array to a Single Value" on page 29).

Detecting this kind of error can be difficult, because JavaScript itself provides several functions and methods that modify their input by definition. For example, if you decided to sort an input array, doing myArray.sort() would actually modify the original array, which is something the caller of your function might not be aware of. Other array methods, such as pop() or splice(), also affect the involved array; many such *mutator* methods exist. (Note, however, that ECMAScript recently added the toSorted(), toReversed(), and toSpliced() methods that do not affect the original array.)

Returning Impure Functions

Some functions are inherently impure. For example, if you call an API to get the daily news, you expect it to return new results every time. If you're programming a game and need to use Math.random() to generate randomness, you'll want a different result every time; it won't be useful if it always returns the same number. Similarly, any function that deals with the current date or time will be impure, because its results depend on an outside condition (the time), which could be considered a part of global state.

For I/O-related functions, the returned result could vary for other reasons. I/O errors can happen unexpectedly; for example, an external service could crash or access rights to some filesystem could change. This means that at any time, the function might throw an exception instead of returning data. Even a seemingly safe function like console.log(), which doesn't produce internal changes, is impure, because the user will see a change in console output.

Impure Functions

Doing away with all impurities likely isn't feasible, so the next option is to consider how to reduce the problem size. One solution is to avoid the usage of state, and another solution is to use the *injection* pattern to control impurity.

Avoiding State

With regard to setting global state, fortunately, there is a well-known solution. If a function needs to *get* global state, just provide the function whatever state elements it needs as arguments. This method solves that problem, because the function won't then need to access global state directly. On the other hand, if a function needs to *set* global state, it shouldn't do so directly. The function should produce an updated state and return it, and the caller should be responsible for updating the global state. If there's a need to update state, at least it will be done at a higher level; whatever provides the state data to the function also will update the state.

These two rules also simplify testing. Instead of having to set some global state, provide the function with the initial state and then check whether the returned new state is correct.

Using Injection

So the problem of working with state is solved, but what about when you really need some impure function—for instance, for I/O or for random numbers? The technique shown here provides more flexible code, simplifies unit testing, and allows easier maintenance.

Suppose a function needed to call an API and did it directly by using Axios, as shown earlier in this chapter (see the section "Functions as First-Class Objects" on page 25):

```
doSomething(a, b, c);
...
function doSomething(x, y, z) {
  // ...
  axios.get("/some/url");
  // ...
}
```

Instead of calling the API directly, provide (or inject) a function to do it:

```
❶ const getData = (url) => axios.get(url);

❷ doSomething(a, b, c, getData);

function doSomething(x, y, z, getter) {
  // ...
  getter("/some/url");
  // ...
}
```

You define a new getData() function that actually calls the API ❶ and passes it to the (modified) doSomething() function as a new extra argument ❷. You haven't actually avoided using impurities (such as doing I/O), but now the caller is in control by injecting a relevant function.

This solution is the same as what was used for avoiding global state, since using axios.get() directly is in fact using a method of a global object, and what you are doing to avoid this situation is providing an extra parameter to be used by the function, so it won't have to access anything global directly. The whole code still will be doing its I/O as required, but now the lower doSomething() function is pure, and for testing purposes, you can provide a mock function.

Summary

In this chapter, we've described features of FP, how it's supported by JavaScript, and a few examples of its usage. Working in an FP-oriented way allows for clearer, more maintainable code, and we'll be using that style throughout the book whenever it makes sense. In later chapters, we'll aim to apply FP and use functions for algorithms and data structures. Chapter 18 takes a totally functional route and explores how the concept of FP extends to functional data structures.

Questions

2.1 Pure or Impure?

Consider the following function that calculates the circumference of a circle given its radius but accesses a global variable to do it. Is it pure or impure?

```
const PI = 3.14159265358979;
const circumference = (r) => 2 * PI * r;
```

2.2 Prepare for Failure

The addLogging() function doesn't take into account the possibility of the original function throwing an exception. Can you modify it to produce proper results in that case as well?

2.3 You Got Time?

Write an addTiming() higher-order function that will take a function as a parameter and produce an equivalent new function, but will log timing data on the console. The kind of solution you want is along the lines of the addLogging() function mentioned in this chapter; take care to account for exceptions as well.

2.4 Parsing Problem

If you attempt to do ["1", "2", "4", "8", "16", "32"].map(parseInt), a weird [1, NaN, NaN, NaN, 1, 17] result is produced; can you explain why? Hint: check what parameters are passed by map() to your function.

2.5 Deny Everything

Write a negate() higher-order function that, given a predicate, will produce a complementary predicate that returns the opposite result. For example, if you had an isAdult() function that checked whether its argument was 21 or more, negate(isAdult) would check whether its argument was *not* 21 or more. (Tip: you may find this function useful for the next two questions.)

2.6 Every, Some . . . None?

Create a .none() method that will check whether no element of an array satisfies a given predicate.

2.7 No Some, No Every

Write the equivalents of .some() and .every() based on .find() or .findIndex().

2.8 What Does It Do?

Explain what the following code produces and why:

```
["James Bond", 0, 0, 7].map(Boolean)
```

3

ABSTRACT DATA TYPES

An *abstract data type (ADT)* is defined by the *operations* it supports and the *behavior* it provides. Throughout this book, we'll study data structures insofar as they allow the implementation of specific ADTs; in a very practical sense, you could say that an ADT specifies needs and requirements in general. This book won't study data structures just for the sake of it; we'll always see them in the context of an ADT and the operations we need the data structure (and associated algorithms) to support. In this chapter, you'll learn more about what ADTs are and how to implement them in JavaScript.

An ADT may be implemented in many ways, possibly with varying performance (a topic we'll discuss in the next chapter) by using alternative data structures and algorithms. For example, you could implement a set with an array, or with a list, or with a tree, but the performance won't be the same in all cases. An actual implementation (meaning some data structure plus the algorithms that work with it) may be called a *concrete data type (CDT)*, but you won't see that term here.

How a data type is implemented is not abstract; it's a concrete aspect that affects the developer. The definition of a data type requires no coding, but the implementation certainly does. Let's first review some basic concepts about data types, abstraction, and operations, and then we'll move to defining ADTs in detail.

The Theory

What are data types, and how do we work with them? Can they be defined abstractly, or must we always resort to actual implementations? What can we do with data types, what operations do they provide, and what effects do they have? Before starting with ADTs, let's take a closer look at some basic software concepts that motivate the focus for this chapter.

Data Types

Programming languages originally included only a few built-in data types, such as characters, strings, booleans, and numbers (either integer or floating point), and developers couldn't add any new ones; the given options were all they had to work with. After concepts like *classes* were added to programming languages, developers were able to add new, more complex data types. A data type in general (both those provided by the language and any you create) is defined by the set of possible values it may represent and the operations that can be performed on it; for example, it's possible to concatenate two strings, perform logical operations with booleans, do arithmetic with integer numbers, or compare floating point numbers.

When using a data type, the details of its internal representation don't usually matter—only what you can do with it and how you can use it to get results. Input and output are all that matter. The basic idea of an ADT is specifying the operations that can be done, leaving aside the internal aspects. (If languages provide bit operations or some low-level features, you might need to learn internal representation details, but for most programming tasks, you won't need to do that.)

Modern languages, JavaScript included, allow users to define their own data types. At first, developers had only simple records (such as representing a date with three numeric fields for day, month, and year), but now you can go further and use classes to hide implementation details, so users need to care only about using the newly defined data type and nothing else.

NOTE *ADT can also stand for* algebraic data type, *which is a different concept representing a type formed by combining other types.*

Abstraction

We have been bandying about the concept of abstraction, so now let's consider more specifically what that term means. Basically, *abstraction* implies hiding or omitting details and reaching instead for an overarching higher-level idea. When we talk about abstraction, we are purposefully ignoring implementation aspects, at least for the time being, to concentrate on our needs, no matter how we'll get around to code solutions for them. For example, do you need to store and retrieve strings? A dictionary ADT would be your solution; you'll see how to implement it later, but no matter how you do it, that's the data type you need.

Software engineering has three similar and related concepts:

Encapsulation   Designing modules as if they had a "shell" or "capsule" around them, so only the module is responsible for handling its data. The idea is to wrap together data and the methods that work on that data in a single place for a more coherent, cohesive design.

Data hiding   Hiding inner details of a module's implementation from the rest of the system, ensuring that they can be changed without affecting any other parts of the code. This mechanism ensures that no one can access internal details from the outside. In other words, encapsulation brings everything together, and data hiding ensures that nobody can mess with internals from "outside."

Modularity   Dividing a system into separate modules that can be designed and developed independently from the rest of the system. Using modules correctly provides both encapsulation and data hiding.

An ADT defines only what operations it can perform; it doesn't go into detail about how those operations will be implemented. In other words, with an ADT, you describe what you can do "in the abstract" rather than going into concrete detail. Let's consider some different types of operations we can perform on an ADT.

Operations and Mutations

A common way to classify data types is by *mutable* versus *immutable* values. For example, in JavaScript, objects and arrays are mutable. After creating an object or array, you can modify its values without creating a new object or array. On the other hand, numbers and strings are immutable; if you apply an operation to either of those data types, a new, different, and distinct value is produced.

When designing a new date type (such as an object with three separate integer values, like the date example mentioned earlier in this chapter), you could opt to provide operations to set the day, month, or year, which would mean that date objects are mutable. On the other hand, if those operations returned a new date object instead of modifying the existing one, date objects would be immutable.

React Redux developers are well aware of immutability and what it requires. If you want to modify the state of a React application that uses Redux, you cannot just modify it directly; you must generate a new state with whatever changes you want. Redux assumes that you manage your state data in an immutable way. (We'll discuss immutability further in Chapter 18.)

The following list shows the categories of operations that apply to an ADT:

Creators   Functions that produce a new object of the given type, possibly taking some values as arguments. Using the date ADT example, a creator could build a new date out of day, month, and year values.

Observers   Functions that take objects of a given type and produce some values of a different type. For the date ADT, a getMonth() operation might produce the month as an integer, or an isSunday() predicate could determine whether the given date falls on a Sunday.

Producers   Functions that take an object of a given type, and possibly some extra arguments, and produce a new object of the given type. With the date ADT, you could have a function that added an integer number of days to a date, producing a new date.

Mutators   Functions that directly modify an object of a given type. A setMonth() method could modify an object (change its month) instead of producing a new one.

With an immutable data type, only the first three types of operations apply; for mutable data types, mutators also apply.

Implementing an ADT

Consider a situation where you want to implement a *bag* or *multiset*, which is a container like a set, but it allows for repeated elements. (Sets cannot have repeated elements by definition.) We'll also add an extra operation ("greatest") to make it more interesting. Table 3-1 provides an example of how ADTs are described throughout the book.

Table 3-1: Operations on Bags

Operation	Signature	Description
Create	→ bag	Create a new bag.
Empty?	bag → boolean	Given a bag, determine whether it's empty.
Add	bag × value → bag	Given a new value, add it to the bag.
Remove	bag × value → bag	Given a value, remove it from the bag.
Find	bag × value → boolean	Given a value, check whether it exists in the bag.
Greatest	bag → value | undefined	Given a bag, find the greatest value in it.

Ignore the middle column for now and focus on the other two. The Operation column names each operation that is provided, and the Description column provides a simple explanation of what the operation is supposed to achieve. You want to be able to create a new (empty) bag and also test whether the bag is empty. You need to be able to add new values to the bag and remove previously entered values from it, and both of those operations will change the bag's contents. Finally, you want to be able to find whether a given value is in the bag and also determine the greatest value in the bag.

You could also have a column specifying the type of the operation—creator, observer, producer, and so on—but that's usually understood from the operation's description and not explicitly included.

What's the operation's Signature, the middle column in Table 3-1? Unless using TypeScript or Flow (as mentioned in Chapter 1), JavaScript doesn't let developers specify types for functions and variables, but adding that information (even if only in comments or a table like this one) helps users better understand what the function expects and returns.

Specifying a function's parameters and the returned result is called a *signature*, and it's based on a *type system* called *Hindley-Milner*. You start with the types of the function's parameters, in order, separated by ×, followed by an arrow, and then the types of the function's results.

Let's consider some examples. Table 3-1 shows that the create() function doesn't take any parameters and returns a bag-type result. Similarly, add() takes two parameters, a bag and a value, and it returns a bag as a result. Finally, the greatest() function takes a bag parameter and returns either a value or undefined.

The complete Hindley-Milner system includes several more details, such as constraints on the types, generic types, undetermined number of parameters, class methods, and so on, but for our needs, the definitions shown in Table 3-1 will suffice.

Implementing ADTs Using Classes

Let's use a class to start implementing a bag ADT. The objects will have two attributes: count, which counts how many elements are in the bag, and data, which is an object with a key for each element and a value that represents how many times that key appears in the bag. Keep in mind, we're not looking for an especially performant way to implement a bag (we'll get to that in Chapter 11). For now, we're just looking at an example of using classes.

For instance, if you add the strings HOME, SWEET, and HOME to a bag, the object would look like the following:

```
{
  count: 3,
  data: {
    HOME: 2,
    SWEET: 1,
  },
};
```

The count attribute has a value of 3 to reflect that three strings were added to the bag. The data part includes a HOME attribute with a value of 2 (since HOME was added twice) and a SWEET attribute with a value of 1.

Listing 3-1 shows the complete Bag class.

```
class Bag {
❶ count = 0;
  data = {};

❷ isEmpty() {
    return this.count === 0;
  }

❸ find(value) {
    return value in this.data;
  }

❹ greatest() {
    return this.isEmpty() ? undefined : Object.keys(this.data).sort().pop();
  }

  add(value) {
❺   this.count++;
❻   if (this.find(value)) {
      this.data[value]++;
    } else {
      this.data[value] = 1;
    }
  }

  remove(value) {
❼   if (this.find(value)) {
❽     this.count--;
❾     if (this.data[value] > 1) {
        this.data[value]--;
      } else {
        delete this.data[value];
      }
    }
  }
}
```

Listing 3-1: A possible implementation for the bag ADT

A new object is initialized with a zero count and an empty set of values ❶. You can tell whether the object is empty by checking whether the count is zero ❷. To see whether the bag contains a given key ❸, check whether it appears in the data object with the in operator. Finding the greatest key ❹ is not hard because of JavaScript's functionality. You first get an array with all the keys in data (all the values that were added to the bag), and after sorting it, you pop() its last element, which will be the greatest key in the bag.

To add a key to the bag, increment the count by 1 ❺ and then check whether the key is already in the bag ❻; if it is, increment its count; if it isn't, add it with a count of 1.

To remove a key from the bag, first verify that the key actually is in the bag ❼. If it isn't, don't do anything at all. If you find the key, decrement the count ❽ and then see how many times the key appears in the bag ❾. If its count is greater than 1, decrement it by 1. If it's exactly 1, just remove the key from the data object.

How can you use this object? Taking a few words from the song "Home, Sweet Home" (the original song from 1823, not the newer one by Mötley Crüe), you can do something like the code shown in Listing 3-2 that adds part of the lyrics to the bag.

```
   const b = new Bag();
❶ console.log(b.isEmpty());    // true

❷ b.add("HOME");
   b.add("HOME");
   b.add("SWEET");
   b.add("SWEET");
   b.add("HOME");

   b.add("THERE'S");
   b.add("NO");
   b.add("PLACE");
   b.add("LIKE");
   b.add("HOME");

❸ console.log(b.isEmpty());    // false

❹ console.log(b.find("YES")); // false
   console.log(b.find("NO"));  // true

❺ console.log(b.greatest());  // THERE'S
❻ b.remove("THERE'S");
   console.log(b.greatest());  // SWEET
```

Listing 3-2: A test for the bag implementation

The newly created bag is empty ❶, as expected. You can add several keys to it ❷, and the bag will obviously no longer be empty ❸. (See question 3.1 for a more compact way of chaining similar operations.) The find operation ❹ works as expected; "YES" isn't in the bag, but "NO" is. Finally, the greatest key in the bag is "THERE'S" ❺, but after removing it ❻, "SWEET" is the new greatest value.

Implementing ADTs Using Functions (Mutable Version)

Now that you've created a concrete implementation of an ADT, how would it change if you were using functions instead of classes? Listing 3-3 uses the same representation based on an object with count and data attributes. The differences essentially will be syntactical, like passing the bag object as an argument to a function, instead of referring to it as this in a method.

```
const newBag = () => ({ count: 0, data: {} });

const isEmpty = (bag) => bag.count === 0;

const find = (bag, value) => value in bag.data;

const greatest = (bag) =>
  isEmpty(bag)
    ? undefined
    : Object.keys(bag.data).sort().pop();

const add = (bag, value) => {
  bag.count++;
  if (find(bag, value)) {
    bag.data[value]++;
  } else {
    bag.data[value] = 1;
  }
  return bag;
};

const remove = (bag, value) => {
  if (find(bag, value)) {
    bag.count--;
    if (bag.data[value] > 1) {
      bag.data[value]--;
    } else {
      delete bag.data[value];
    }
  }
  return bag;
};
```

Listing 3-3: An alternative (mutable) implementation of the bag ADT

The code in Listing 3-3 is similar to Listing 3-1 that used classes. The newBag() function returns an object with the count and data fields, like the constructor in the Bag class did. For the other five functions (isEmpty, find, greatest, add, and remove), there are only two differences in comparison with the classes-based code: you access the object using the bag parameters instead of using this, and you expressly return bag at the end of the add() and remove() mutator methods. In this case, however, you don't really need to do this, because you are actually modifying the bag parameter, which was passed by reference to the functions. (That's the standard way JavaScript passes objects as arguments.) However, if you were to implement this ADT in some other way that didn't use an object, returning the new concrete data type would be mandatory. Since you don't want external dependencies on internal aspects of an implementation, the simplest (and safest) way to work is by always returning the new updated object, whatever its type.

The code to use this ADT implementation, shown in Listing 3-4, is quite similar to the class-based version in Listing 3-2.

```
❶ let b = newBag();
❷ console.log(isEmpty(b));     // true

❸ b = add(b, "HOME");
  b = add(b, "HOME");
  b = add(b, "SWEET");
  b = add(b, "SWEET");
  b = add(b, "HOME");

  b = add(b, "THERE'S");
  b = add(b, "NO");
  b = add(b, "PLACE");
  b = add(b, "LIKE");
  b = add(b, "HOME");

  console.log(isEmpty(b));     // false

❹ console.log(greatest(b));    // THERE'S
❺ console.log(find(b, "YES")); // false
  console.log(find(b, "NO"));  // true

❻ b = remove(b, "THERE'S");
  console.log(greatest(b));    // SWEET
```

Listing 3-4: A test for the mutable implementation of bags

The simple differences are in object creation ❶, testing whether the bag is empty ❷, adding ❸ and removing ❻ elements, getting the greatest value ❹, and finding whether a value is in the bag ❺. Instead of writing b.something(...), you would write something(b, ...).

Implementing ADTs Using Functions (Immutable Version)

Finally, let's consider an immutable implementation of our ADT. (In Chapter 18, we'll see immutable data structures in more detail, with several more cases.) There's no particular reason here for immutability, other than wanting to work in a more functional way and avoiding side effects, as described in Chapter 2.

In this situation, as you want to develop an immutable bag, you may not modify the bag object directly, so you need to change the implementation of the mutator methods; the rest will stay the same. The solution just requires creating and returning a new object if the bag needs any changes. To add a new value, use the following code:

```
const add = (bag, value) => {
❶ bag = { count: bag.count - 1, data: { ...bag.data } };
  if (find(bag, value)) {
    bag.data[value]++;
  } else {
```

```
      bag.data[value] = 1;
    }
    return bag;
};
```

Since adding a new value to a bag can never fail, you always need to produce a new object, so you actually do that ❶.

To remove a value from a bag, first check whether the value to remove is in it before proceeding to remove it:

```
const remove = (bag, value) => {
❶ if (find(bag, value)) {
    ❷ bag = { count: bag.count - 1, data: { ...bag.data } };
      if (bag.data[value] > 1) {
        bag.data[value]--;
      } else {
        delete bag.data[value];
      }
    }
    return bag;
};
```

As before, start by checking whether the value is in the bag ❶; if it is ❷, create a new object, which you'll return.

In this case, the code modifications are minimal, but with more complex data structures (as we'll see later in this book), creating a new copy of an existing structure may not be so easy or quick, and you'll need to do extra processing or structuring.

Summary

In this chapter, we introduced the concept of abstract data types, which you'll see in the rest of the book when analyzing the pros and cons of competing data structures and algorithms. Defining an ADT is the first step when deciding what structure should be used and how algorithms should be implemented. Understanding the concept of ADTs will help you get the best possible performance for your code.

In the next chapter, we'll study a complementary concept: How can we compare concrete implementations of ADTs, or in other words, how can we tell whether one algorithm is actually better or worse than another? We'll also introduce analysis of algorithms and concepts related to classes of performance.

Questions

3.1 Chaining Calls

Modify the Bag methods so you can chain additions in the following fashion:

```
const b = new Bag();
b.add("HOME").add("HOME");
b.add("SWEET").add("SWEET").add("HOME");
```

You should also be able to chain removals and other operations, such as the following, that would remove two values and then test whether the bag becomes empty:

```
b.remove("NO").remove("HOME").isEmpty();
```

3.2 Arrays, Not Objects

Can you implement the bag ADT using arrays instead of objects? You could represent the bag with an ordered array to make the greatest() function implementation really speedy. Of course, add() should take care of maintaining the order of the array.

3.3 Extra Operations

Only a few extra operations for a bag were described in this chapter, but for some applications, you might need added or changed operations; can you think of any?

3.4 Wrong Operations

When defining an ADT, how could you specify error results, such as possibly throwing an exception or returning some kind of special value?

3.5 Ready, Set . . .

In this chapter, we discussed a bag, but in later chapters, we'll work with sets, which don't allow repeated values. Can you think ahead and whip up an appropriate ADT?

4

ANALYZING ALGORITHMS

In the previous chapter, we discussed abstract data types, and later in this book we'll consider many more with alternative implementations and algorithms. When facing several possible ways of implementing the same abstract data types, consider the efficiency of each concrete implementation, which requires an analysis of the involved algorithms. We'll study the basics of such analysis in this chapter to help us make better decisions. What data structure should you pick? What algorithm should you implement? Knowing objectively how to analyze their performance will produce the right answers.

Performance

When measuring the efficiency of a given algorithm, the key is to consider the resources (such as time or random access memory [RAM]) the algorithm needs, and then you can compare different algorithms based on the needed amount. (This method doesn't really apply to small problems. For instance, if you have a dictionary with just a dozen keys, no matter how it's structured or what algorithm you apply for searching, the results will be fast.)

We always want to minimize resource usage (faster processing time, less needed RAM), but we cannot really directly compare time complexity (speed) to space complexity (memory). Often, faster-performing algorithms require larger amounts of memory, and vice versa; smaller, simpler structures may imply slower algorithms. (You'll see an example later in this chapter.) All of these considerations are moot, however, if an algorithm takes way too much time or requires more RAM than available.

In all the cases in this book, we'll see that the space complexity of algorithms is fairly stable. It grows in direct proportion to the number of input elements, so there really may be no grounds to select one algorithm over another. On the other hand, we'll see that time complexity results in many variations, providing a solid basis for choosing which data structure to use and which algorithms to implement.

Accordingly, whenever the book refers to the complexity of any given algorithm, it's always referring to time complexity, or how long the algorithm takes to perform its function in relation to the size of its input data.

Complexity

All data structures always have some basic parameter upon which the efficiency of all algorithms depends. For instance, if you are searching in a dictionary, the number of keys in the dictionary will probably impact the searching speed; more keys equal more time. If sorting a set of values, having more values means a slower sort; for example, ordering the 5 cards in a poker hand can be done really quickly, but ordering a whole deck of 52 cards takes longer. In all cases, we'll call that input parameter n, and you'll express the algorithm's time complexity as a function of that input; this is *analysis of algorithms*. An algorithm will be more efficient when that function's values are small, or at least, it will grow slowly in comparison to the growth of the input size.

In some cases, an algorithm's performance may be directly linked to the data itself; for example, sorting an almost-in-order sequence is likely faster than sorting a completely disordered, random sequence of values. This means we'll be considering best- or worst-case performance, as well as average performance. If nothing is specified, we'll aim for an upper bound on the algorithm's complexity, so in this book, we'll be looking at worst-case complexity unless otherwise noted.

In general, we won't try (or won't be able) to get a precise expression for the complexity function. We'll look at how it compares with common mathematical functions, such as n or n^2 or $n \log n$, and consider in which class an algorithm is in to compare it with others on an equal basis. Algorithms in the same class don't perform at the same speed, but roughly speaking, all algorithms in the same class will perform in the same way for larger inputs, growing at the same rate and keeping the same relationship among them. In other words, an algorithm that's 10 times speedier will most likely keep being thus; it won't become 100 times faster or half as much slower than others in its class.

Notations for Complexity

To express a given function's behavior when its argument grows, we use a family of notations called *asymptotic notations*. This family includes five different notations, including the most often used: *big O* notation. The *O* stands for "order"—or, more accurately, the German word *Ordnung*. (You'll see the other four notations soon.)

Big *O* notation groups functions according to how they behave for growing values of their n parameter. Depending on what algorithm or data structure we're studying, n could be the number of values to sort, the size of a set to be searched, or how many keys are added to a tree. This is made clear on a case-by-case basis when discussing performance.

Describing a function in terms of its big *O* behavior implies an upper bound on how the function grows. Without diving in to mathematical functions too deeply, if the behavior of a function $f(n)$ is $O(g(n))$, that means that when n grows, both functions grow in the same proportion. (A complete definition also specifies that this relationship need not occur for all values of n, but only for large enough ones. For small values of n, the relationship may not apply.) In other words, saying that the behavior of a given algorithm is *O(some function)* already implies how the needed time will grow for larger values of n.

Let's get back to the five notations (Table 4-1).

Table 4-1: The Five Asymptotic Notations

Notation	Name	Description
$f(n) = o(g(n))$	Small o	$g(n)$ grows much faster than $f(n)$; the growth rate of $g(n)$ is strictly greater than that of $f(n)$.
$f(n) = O(g(n))$	Big O	$g(n)$ is an upper bound for $f(n)$; the growth rate of $g(n)$ is greater than or equal to that of $f(n)$.
$f(n) = \Theta(g(n))$	Big Theta	$g(n)$ is a bound from above and below for $f(n)$; both $g(n)$ and $f(n)$ grow at the same rate.
$f(n) = \Omega(g(n))$	Big Omega	$g(n)$ is a lower bound for $f(n)$; the growth rate of $g(n)$ is less than or equal to that of $f(n)$.
$f(n) = \omega(g(n))$	Small omega	$g(n)$ grows much slower than $f(n)$; the growth rate of $g(n)$ is strictly less than that of $f(n)$.

We'll mainly be using the big *O* notation; the others are included for completeness. Big theta is more accurate than big *O*, which is really a bound, but you are aiming for a good, close one that doesn't behave too differently from the original function. Getting a precise, exact expression for the behavior of any algorithm is quite complex (and there still are many algorithms for which the precise order isn't yet known), so working with orders is appropriate. For example, if your personal debt is a few dollars or a few millions, actual numbers aren't really needed to know that in the former case you're doing very well and in the latter you're in serious trouble.

NOTE   *Donald Knuth, renowned computer scientist, author of* The Art of Computer Programming *books, and expert on analysis of algorithms, once suggested that the big* O *should be a big omicron, another Greek character that looks exactly like an uppercase* O, *but it didn't pan out. See* https://danluu.com/knuth-big-o.pdf *for the full story.*

Another (rough) interpretation is that the big *O* bound represents a worst case, while the big omega bound represents the best case, or the smallest amount of time some algorithm could take. In that sense, the big theta case implies an algorithm with a stable performance, because both the worst and best cases grow at the same rate. With this interpretation, the small *o* notation means an even worse upper limit, and the small omega would be a worse lower limit in the sense that actual behavior is greatly separated from these two bounds, with quite different growth rates.

Complexity Classes

Most often we find that algorithms involve only a few common orders. Table 4-2 shows the orders you'll see in this chapter.

Table 4-2: Common Orders

Order	Name	Example
$O(1)$	Constant	Accessing the first element of a list and popping the top of a stack (Chapter 10)
$O(\log n)$	Logarithmic	Searching an ordered array with binary search (Chapter 9) and average height of a binary tree (Chapter 12)
$O(n)$	Linear	Searching an unordered array (Chapter 9) and inorder traversal of a tree (Chapter 12)
$O(n \log n)$	Log-linear	Sorting an array with heapsort and average behavior of quicksort (Chapter 6)
$O(n^2)$	Quadratic	Sorting an array with bubble sort and worst case for quicksort (Chapter 6)
$O(k^n)$	Exponential	Testing whether a binary formula is a tautology ($k = 2$) and a naive implementation of the Fibonacci series ($k = 1.618$)
$O(n!)$	Factorial	Finding the optimum traveling salesman solution and sorting by random permutations (in Chapter 6)

The last two orders are algorithms that are so slow, you won't use them in real life; their time complexity grows way too fast to be usable.

Figure 4-1 is a simple chart showing how the seven functions from Table 4-2 behave. Clearly an $O(\log n)$ algorithm would be preferred instead of an $O(n^2)$ algorithm.

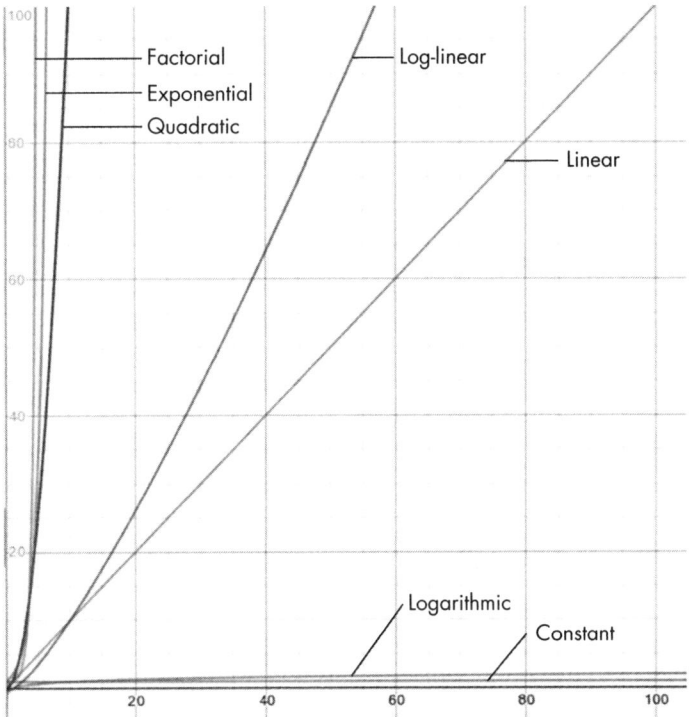

Figure 4-1: This chart (drawn using Desmos, https://www.desmos.com/calculator) shows the seven functions from Table 4-2.

The two first orders at the bottom of the chart (constant and logarithmic) are excellently well behaved. When considering linear (the diagonal line from bottom left to upper right) and log-linear orders (the closest curve to the diagonal), growth starts to be important. The next order, quadratic, goes off the chart for $x = 10$ with the value $x^2 = 100$. Finally, the exponential and factorial orders are even worse behaved; their growth makes them impossible to use.

You can look at this behavior another way by answering a simple question: What happens with a given algorithm if the input size is 10 times bigger? If the algorithm is $O(1)$, the amount of time will stay the same, with no growth. With an $O(\log n)$ algorithm, the required time would grow, but by a fixed amount. An $O(n)$ algorithm would (nearly) multiply its time by 10, and an $O(n^2)$ algorithm would be around 100 times longer. An $O(n \log n)$ algorithm would be in between those two. The difference is clear, but note for future reference that it's much closer to $O(n)$.

The results from the previous paragraph are also the reason why $O(n)$ is used rather than $O(9n)$ or $O(22n)$. The ratio between these three algorithms is constant, so if n grows, they will grow at the same rate. On the other hand, an $O(n^2)$ algorithm will grow so much faster, it's really is a class by itself. Constant values are meaningless when comparing classes: $O(n^2)$ will always grow faster (and also become larger) than an $O(n)$ algorithm, even throwing in some constant factor, if n is large enough.

Performance Measurements

When measuring an algorithm's performance, the best-case performance is an algorithm's behavior in ideal conditions; for instance, in the previous section we mentioned doing a search and finding the desired element at the first position of an array. You can't ever assume you'll always get this optimum performance, but it's a baseline to compare other performances.

The complementary case is *worst-case* performance, which means you try to measure how an algorithm will perform in the slowest possible way. For instance, later in the book we'll see algorithms that usually have $O(n \log n)$ performance that may degenerate to $O(n^2)$ performance for specific input data ordering. The worst-case analysis is important, because you should always assume that possibility will happen; it's the safest (but most pessimistic) analysis.

A third possibility is *average-case* performance, which means determining how an algorithm will behave with typical or random input. In Chapter 6 you'll see that quicksort's average performance is $O(n \log n)$ despite cases when performance is much worse.

The fourth possibility is amortized time. Some algorithms often take a short time to perform, but periodically require more time. If you look at one individual operation, the result may be poor, but if you consider the average performance over a long series of operations, you may find that, overall, the amortized time is much better than the worst case, letting you predict the result of sequences of operations.

Let's consider a simple example: adding elements to a fixed-size array. If every time you want to add a new element you need to copy the current array to a new (and longer) array, the cost of each addition would be $O(n)$. However, if the array is full, an alternative strategy is to copy it to a new double-sized array, leaving empty space to wait for future insertions.

Let's look at how this strategy works. Consider a situation with an array that is almost fully occupied (cells in gray), with just one empty space (cell in white) at the end, as shown in Figure 4-2.

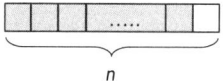

n

Figure 4-2: An array with only one empty space

When adding a new element (an $O(1)$ operation with constant time), the array is full but you don't need to worry yet (Figure 4-3).

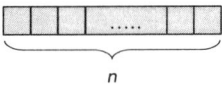

Figure 4-3: Now the array is full.

However, if you need to add another element, there's no place for it, so you copy the array to a new double-sized one and then add the new value, as shown in Figure 4-4.

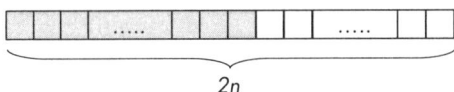

Figure 4-4: A new double-sized array provides space for the new value and more.

After this process, which is an $O(n)$ operation, you now have n free cells and may rest easily, knowing that upcoming insertions won't need any copying and will be $O(1)$. The next time the array becomes full, the process will be repeated: a lengthy single duplication followed by many fast additions. Averaging the cost of many insertions, the costlier (infrequent) doubling will be compensated for by the inexpensive (frequent) simple additions, and the amortized performance will be $O(1)$.

NOTE *The following section is much more mathematically minded than the rest of the book. If you wish, you can skip the demonstrations and study only the results. The rest of the book won't delve into so much math. This section is simply to give you a taste of what complete, formal proofs look like.*

Analysis of Algorithms in Practice

Let's consider examples of actual orders. Suppose you want to search an ordered array of length n for a given key. The worst case, linear search, is going sequentially through the whole array (because you haven't yet learned the better algorithms described later in the book) without finding the key. In this case, the linear search performance is $O(n)$ because you have to go through the whole array: n steps and n (failed) tests. The best-case performance is finding the key you wanted on the first attempt: $\Omega(1)$.

The average requires a bit of algebra. You need to consider all cases: you could find the given element at the first place, the second place, and so on, all the way up the nth, which means n possibilities in all. On average, you have to test $(1 + 2 + \ldots + n)/n$ elements. The sum of numbers from 1 to n equals $n(n + 1)/2$, so the average needed (dividing by n) ends up $(n + 1)/2$. This expression is clearly proportional to n, so the average behavior of the algorithm is indeed $O(n)$. If you had to consider using this algorithm, you'd think in terms of $O(n)$, assuming the worst; hoping for the best case isn't realistic.

There's another way to look at this calculation. The search could succeed at the first element or could take up to the n*th; on average,* (n + 1)/2. *Or, it could succeed either at the second element or the* (n − 1)*th; on average again,* (n + 1)/2. *The same reasoning applies for the third, fourth, and subsequent elements. For each case in which the search finishes in a few steps, a complementary case drives the average number of steps up to* (n + 1)/2. *Since in every case the average is the same, you can conclude that's the result. You arrive at the same result with a bit more "hand waving" but less algebra.*

Let's discuss another way of searching an ordered array, a binary search, which you'll see in Chapter 9. Instead of starting at the beginning of an array and going through all its elements, you start at the *middle* of the array. If you find the key you want, you're done. If not, you can discard half the array (if the key you want is less than the middle element, you know it can't be in the higher part of the array) and recursively search in the other part. You search in that new part by picking its middle element, comparing, and so on.

Consider an array with the numbers 4, 9, 12, 22, 34, 56, and 60. If you wanted to check whether 12 was in it, first you'd look at the middle element: 22. That's not what you want, so you can discard the second half of the array (34, 56, and 60), because you know that 12, if present, must be in the first half. Now look for 12 in the array that is now 4, 9, and 12. Start by looking at its middle element (9) and then discard it and the first half of the array (4). The last step of the search looks at an array with a single element (12). Its middle (and only) element is what you were looking for, so the search succeeded. If you were looking for 13 instead, the search would fail at this point, since no more pieces of the array exist.

To see how this algorithm performs, count how many times you need to test an element; assume that the array's length n is 2^{k-1} for some $k > 0$, so all halves of the array always have an odd number. (This is just to simplify calculations; see question 4.9.) In one case the element is found on the first attempt. In two cases the key is found on the second try—namely, the middle elements of the chosen halves. In four cases, the third try is successful, and in eight cases, the fourth try succeeds. Figure 4-5 shows this for an array with 15 elements.

4	3	4	2	4	3	4	1	4	3	4	2	4	3	4

Figure 4-5: Starting in the middle of an array with 15 elements

For a general array, the total number of comparisons is $S = (1 \times 1 + 2 \times 2 + 3 \times 4 + 4 \times 8 + \ldots + k \times 2^{k-1})$, which you must divide by the number of elements in the array to get the average. To calculate S do a math trick and first write a more general formula. Write $S = 1 \times 2^0 + 2 \times 2^1 + 3 \times 2^2 + \ldots + k \times 2^{k-1}$ and then define $f(x) = 1x^0 + 2x^1 + 3x^2 + \ldots + kx^{k-1}$; note that $S = f(2)$. It follows from calculus that $f(x)$ is the derivative of $g(x) = 1 + x + x^2 + x^3 + \ldots + x^k$. Since a well-known result says that $g(x) = (x^{k+1} - 1)/(x - 1)$, by deriving you find the following:

$$f(x) = \frac{(k+1)x^k(x-1) - (x^{k+1} - 1)}{(x-1)^2}$$

Now undo the generalization you just made by setting $x = 2$ and remembering that $n = 2^{k-1}$. You can write $S = (k + 1)(n + 1) - (2n + 1)$. Dividing by n you find that the average number of comparisons is $(k - 1) + k/n$. You can write $k = \log n$ (taking logarithms in base 2 and rounding upward) so the average performance of the algorithm is $\Theta(\log n)$. Whew! The worst case (a failed search) requires k tests, so again you are justified in saying that binary search is an $O(\log n)$ algorithm.

Using big O notation is "safer" and provides better "cover." Of course, you could also say that the binary search is $o(n)$ or, even worse, $o(n^2)$ because those functions behave in a worse way, growing faster. The small o and small omega bounds are good for a rough estimate, but you want to be more precise and aim for closer bounds whenever possible.

Most analysis of algorithms involves recurrences as shown here, and some studies are even more complex mathematically than what you just saw. Recurrences usually take a few well-known forms, such as $P(n) = aP(n - 1) + f(n)$ or $Q(n) = aQ(n/b) + f(n)$, among practically infinite possibilities. There are several tricks to help find expressions for $M(n)$ in each case (in particular, the "master theorem" quickly provides a solution to the $Q(n)$ recurrence style, but a whole book could be written on that).

Time and Space Complexity Trade-offs

Earlier in the chapter, we mentioned we'd look at time performance, because from the point of view of the storage requirements, algorithms are usually well behaved. Let's explore a simple problem and see how time and space trade-offs apply.

Say we have a (long) array of numbers and frequently require finding the sum of the values in a range of positions, from i to j, both inclusive, with $i < j$. (This problem has to do with breaking a long string of text into justified lines.) A first solution just needs a couple of auxiliary variables, so extra memory requirements are $O(1)$, but finding the sums themselves requires $O(n)$ time:

```
const sumRange = (arr, from, to) => {
  let sum = 0;
  for (let i = from; i <= to; i++) {
    sum += arr[i];
  }
  return sum;
};
```

This function is clear and correct—it consists of just a loop summing all the values between from and to inclusive in sum—but its performance will impact the process negatively, because you are calling it frequently. For a function that will be called many times, a better performance is preferred.

You also can apply a concept of dynamic programming (which we'll study in more detail in Chapter 5) and work by tabulation, precomputing the sums from position 0 to all other positions:

```
const precomputeSums = (arr) => {
  let partial = Array(arr.length);
  partial[0] = arr[0];
  for (i=1; i<arr.length; i++) {
    partial[i] = partial[i-1] + arr[i];
  }
}
```

With this array of partial sums, if you need the sum of elements 0 to q, you already have those, and for the sum of elements $p > 0$ to q, just calculate the sum of elements from 0 to q minus the sum of elements from 0 to $p - 1$:

```
const sumRange2 = (partial, from, to) =>
  from === 0 ? partial[to] : partial[to] - partial[from-1];
```

This solution implies further $O(n)$ processing to compute the partial sums, which is only done once, and $O(n)$ extra memory, but it provides $O(1)$ sums for ranges, so you can see the trade-off: use more memory to apply faster algorithms or save memory by accepting a slower performance.

Which version should you choose? That depends on the problem and whether the current performance is acceptable, and even possibly whether enough memory is available!

Summary

In this chapter, we've discussed the definitions that are relevant to studying how algorithms perform in terms of either operations or memory requirements. We've seen several classes of algorithms that help decide how to implement a given solution to a problem by comparing efficiency in response to larger inputs. In the next chapter, we'll switch gears and study ways to create algorithms in preparation for the rest of the book, where we'll consider many varied data structures and the algorithms that perform them.

Questions

The questions in this chapter are visibly different from all other questions in the book because they are more mathematically oriented. Feel free to skip ahead.

4.1 How Fast Did You Say?

An analyst has just completed a study of a new algorithm and concludes that its running time, depending on its input size n, is exactly $17n \log n - 2n^2 + 48$. What do you say about that result?

4.2 Weird Bound?

Is it valid to say that n is $O(n^2)$? What about $o(n^2)$? Other orders?

4.3 Of Big Os and Omegas

What can you deduce if a certain function is both $f(n) = O(g(n))$ and $f(n) = \Omega(g(n))$?

4.4 Transitivity?

If $f(n) = O(g(n))$ and $g(n) = O(h(n))$, how are $f(n)$ and $h(n)$ related? What if instead of big O, you were looking at other orders: small o, big theta, and so on?

4.5 A Bit of Reflection

It seems clear that for any function $f(n)$, you have $f(n) = \Theta(f(n))$. What would you say if working with other orders instead of big theta?

4.6 Going at It Backward

If $f(n) = O(g(n))$, what is the order of $g(n)$ relative to $f(n)$? What if $f(n) = o(g(n))$?

4.7 One After the Other

Suppose you have a process that consists of two steps: the first is an $O(n \log n)$ algorithm and the second is an $O(n^2)$ algorithm. What's the order of the whole process? Can you give a general rule?

4.8 Loop the Loop

A different but related question: suppose your process consists of an $O(n)$ loop that does an $O(n^2)$ process at each step. What's the order of the whole? Again, can you provide a general rule?

4.9 Almost a Power . . .

When analyzing binary search, you learned that if the array's length is 2^{k-1} for some $k > 0$, the initial array and all subsequent arrays would have an odd length. Can you prove this?

4.10 It Was the Best of Times; It Was the Worst of Times

What happens if the best-case running time of an algorithm is $\Omega(f(n))$ and the worst case is $O(f(n))$?

PART II

ALGORITHMS

In this second part of the book, we'll shift our discussion away from the theoretical topics of functional programming, abstract data types, and analysis of algorithms and focus on how to design algorithms, considering several strategies for discovering solutions to specific problems.

5

DESIGNING ALGORITHMS

This chapter covers several techniques for designing algorithms. We'll start with recursion, which solves a problem by breaking it up into one or more simpler cases of the same problem. We'll also look at *dynamic programming*, which solves a complex problem by solving simpler cases first and storing those solutions to avoid needless recalculations, as well as the *brute-force* (or *exhaustive*) *search* strategy, where you find a solution to a problem by systematically trying all possible solutions. Finally, we'll explore *greedy algorithms* that apply a heuristic of choosing the best local option at each junction of a problem, with the hope that the given methodology will lead to the solution. Unlike the other strategies mentioned in this list, greedy algorithms may not always arrive at the best solution.

The strategies explored here are successfully applied to develop algorithms used along with data structures for the implementation of specific abstract data types (ADTs), so focusing on how to design a new solution for any given problem is worthwhile. The techniques covered in this chapter aren't exhaustive, but they lie below the surface in many of the algorithms that we'll explore later.

Recursion

The simplest definition of recursion goes something like this: "A function calls itself over and over, again and again, until it doesn't." In other words, when facing a problem, if it's small enough, it can be solved without any further recursive calls, but if it isn't, the function calls itself to solve smaller problems, and out of those solutions, it finds the solution for the original, larger problem.

NOTE *For a bit of computer humor, here's a dictionary definition: "recursion: (n) see recursion." A common saying is also "In order to understand recursion, you must first understand recursion."*

As discussed in Chapter 2, recursion is a key technique in functional programming. Some languages, like Haskell, for instance, don't even provide common "loops" and instead work exclusively with recursion. In computer science, recursion is all-sufficient for any algorithm, and anything you can do with loops you can also do with recursion. In fact, using recursion is much easier for many algorithms and definitions.

Recursion appears naturally in several areas:

Mathematics   Definitions such as the factorial of a number or the Fibonacci series are naturally recursive. We'll explore both later in this chapter.

Data structures   Many structures are defined in a recursive fashion. For example, as you'll see in Chapter 10, a *list* may either be empty or consist of a special node, the head of the list, followed by another list; another example from Chapter 13 may be a *tree*, consisting of a parent node called the *root*, connected to any number of trees as its children.

Procedures   Several algorithms can be expressed logically in a recursive fashion. An example from everyday life is searching your house for an object. You first look in one room, and if you find the object, you're done; if you don't find it, you search the rest of the place, applying the same logic. If you have nowhere left to search, you failed.

A recursive function always has two kinds of cases: simple ones that can be solved directly without any recursion and complex ones that need to use the function itself as an aid. The key to solving something recursively is to assume that the problem has already been solved and then code it using the (supposedly) available function. It's a four-step procedure—which may seem to be circular:

1. Assume you already have a function that solves your problem.
2. Find some simple base cases that you can solve directly without any complications.
3. Figure out how you can solve the original problem by first solving one or more smaller versions of it.
4. Apply your assumed function from step 1 to solve the minor problems of step 3, or if they are small enough, solve them as in step 2.

Let's look at a couple of recursion techniques to show how to design clear, simple-to-understand algorithms.

The Divide-and-Conquer Strategy

As mentioned, the basic idea in recursion is to base the solution of complex cases on the solution of simpler ones. You *divide* the problem into smaller versions of itself, and *conquer* it using the solutions to all of them. Often you'll solve a problem by recursively solving an "only one smaller" version, and that strategy gets its own name, *decrease-and-conquer,* but it's still the same idea: reduce the original problem to smaller versions of itself. The only difference is that you solve the big problem by just solving *one* (smaller) version first. We'll start with a look at some simpler decrease-and-conquer examples and then move on to the divide-and-conquer strategy.

Calculating Factorials

The most often quoted example of a recursive calculation is likely the factorial of a number, $n!$, which is an example of the decrease-and-conquer strategy. The factorial of a non-negative integer number n is defined as follows: for $n = 0$, $0! = 1$, and for $n > 0$, $n! = n \times (n - 1)!$, which is the recursive definition.

This formula comes from a recursive problem; namely, how many ways can you order n books in a row on a shelf? The answer is simple: if no books are on the shelf, there's only one way—an empty shelf. However, if you have $n > 0$ books, you can choose any one of them (there are n options), place it at the leftmost empty space on the shelf, and then place the $(n - 1)$ other books in all possible permutations to the right of the book you just placed, which is $n! = n \times (n - 1)!$ as just described: n ways of choosing the first book multiplied by $(n - 1)!$ ways of ordering the rest.

A quick implementation of the factorial (see also question 5.1) is as follows:

```
const factorial = (n) => {
❶ if (n === 0) {
    return 1;
❷ } else {
    return n * factorial(n - 1);
  }
};
```

This code closely follows the definition, with two clear cases: if *n* is 0 ❶, return 1, and for greater values of *n* ❷, use recursion. It's hard to go wrong with a recursive implementation, because the logic matches the definition closely.

Searching and Traversing

Let's look at a few other decrease-and-conquer examples using searching and traversing. In Chapter 4, we mentioned binary search, which is a way of searching an ordered array. If the array is empty, you're out of luck; the value you want isn't there. If the array isn't empty, check its middle element, and if it's what you want, you succeed. If the element doesn't match and is higher than the value you want, search recursively in the left half of the array; otherwise, search its right half.

As another example, think about sorting a deck of cards. If the deck is empty, you're done. Otherwise, you go through the deck looking for the lowest card and remove it from the deck. Then, you sort the rest of the deck and put it on top of the card you put aside.

Finally, let's consider going through a list of pending tasks. (This is called a *traversal* of the list.) If the list is empty, you've got nothing to do; you're finished. Otherwise, you take the top task from the list, do it, and then recursively go through the remaining tasks.

Considering the Fibonacci Series

For a mathematical divide-and-conquer example, consider the Fibonacci series. This series starts with 0 and 1, and after that, each item is the sum of the two previous ones, so the series goes 0, 1, 1, 2, 3, 5, 8, 13, 21, 34, 55, and so on. (You'll meet a Fibonacci-based structure in Chapter 15, and the usage of the series even applies to estimating the complexity of tasks in agile methodologies, so it's certainly pervasive.)

NOTE *For curious readers, this series was named after an Italian mathematician, Leonardo of Pisa, also known as Fibonacci ("filius Bonacci," or "son of Bonacci"), who is also famous for having introduced Arabic numbers to the Western world. Fibonacci posed (and solved) a question involving the growth of an idealized population of rabbits, but the sequence had already appeared in many other contexts, such as counting patterns of verse.*

In order to implement the series recursively, you need to give a proper definition, and that's not hard: you can say that $F_0 = 0$, $F_1 = 1$, and for $n > 1$, $F_n = F_{n-1} + F_{n-2}$. Given this definition, here's the code:

```
const fibo = (n) => {
❶ if (n === 0) {
    return 0;
❷ } else if (n === 1) {
    return 1;
❸ } else {
    return fibo(n - 1) + fibo(n - 2);
  }
};
```

You have two base cases for 0 ❶ and 1 ❷ and a recursive case ❸ for other values. The code is so simple, it can't go wrong, and testing verifies it. However, it does have a performance defect, which we'll consider when we discuss dynamic programming later in this chapter.

Sorting and Puzzles

Many sorting methods that we'll explore later in this book, such as merge sort or quick sort, are expressed succinctly in a recursive fashion, but let's look at another classic example: the Towers of Hanoi puzzle.

This puzzle, invented by French mathematician Édouard Lucas in the 19th century, has three posts: the first has a stack of disks of decreasing size (largest at the bottom, smallest at the top), and the other two posts are empty. To solve the puzzle, you need to move all the disks from the first post to the last one, following two rules: you can move only the top disk from any post at a time (you can't move two or more disks at once, and you can't move any disks from the middle of a post), and you can move a disk to another post only if the top disk on that post is larger than the one you're moving (a larger disk can never go on top of a smaller one). Figure 5-1 shows the initial setup; all disks are on the leftmost post, and the goal is to move all the disks to the rightmost post.

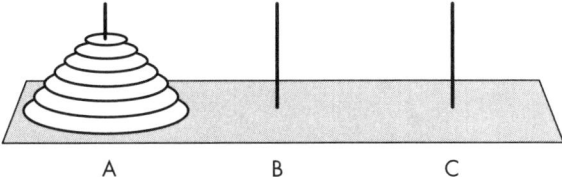

A   B   C

Figure 5-1: Towers of Hanoi

How do you solve this with the divide-and-conquer strategy? You start by thinking you already have the required function, which might be towers(disks, origin, extra, destination), to move a certain number of disks from the origin post to the destination post using the extra post as an auxiliary, and you may use that function to implement the function itself. The base case is simple: if there are no disks to move, nothing needs to be done; otherwise, you make the moves described previously. The code could be as follows:

```
const towers = (disks, origin, extra, destination) => {
❶ if (disks > 0) {
  ❷ towers(disks - 1, origin, destination, extra);
  ❸ console.log(`Move disk ${disks} from ${origin} to ${destination}`);
  ❹ towers(disks - 1, extra, origin, destination);
  }
};
```

You first test for the base case ❶, because if there are no disks to move, you're clearly done. Otherwise, recursively move all disks but the bottom one to the extra pole ❷. Having cleared the large disk, move it to the destination pole ❸, and finish by bringing the other disks on top of it ❹.

A call like towers(4, "A", "B", "C") to move four disks from pole A to pole C will produce the following output:

```
Move disk 1 from A to B
Move disk 2 from A to C
Move disk 1 from B to C
Move disk 3 from A to B
Move disk 1 from C to A
Move disk 2 from C to B
Move disk 1 from A to B
Move disk 4 from A to C
Move disk 1 from B to C
Move disk 2 from B to A
Move disk 1 from C to A
Move disk 3 from B to C
Move disk 1 from A to B
Move disk 2 from A to C
Move disk 1 from B to C
```

Using recursion to solve the simpler steps of a puzzle is a clear example of the divide-and-conquer strategy. (See question 5.2 if you ever need to do this puzzle with no computer.)

NOTE   *There's a coda to this puzzle. In the original version, monks had to move 64 golden disks from one post to another, and the world would end as soon as they accomplished the task. (In the original puzzle, the temple was in India; who knows how it moved abroad and got to Hanoi?) For n disks, $M(n) = 2^n - 1$ moves are required to solve the puzzle, so it's an algorithm of exponential order; the formula can be verified by noting that $M(n) = 2M(n - 1) + 1$ and $M(0) = 0$. At one movement per second, the achievement would require $2^{64} - 1$ seconds, more than 584 billion years, so we're safe!*

The Backtracking Technique

Backtracking is another problem-solving technique that's usually best implemented in a recursive way. When facing multiple options, choose one and try finding a solution with it. If you succeed, you're done. If you fail, backtrack to the point where you made the selection and choose a different option. If at some point you run out of options, there definitely is no solution.

Finding a Path in a Maze

Finding the way out of a maze (such as the one shown in Figure 5-2) is a classic, ancient problem, and you'll see it again in Chapter 17, when working with graphs. It's also the archetypical example for backtracking, so you'll use it here. we'll explore the full algorithms later; this is just the pseudocode.

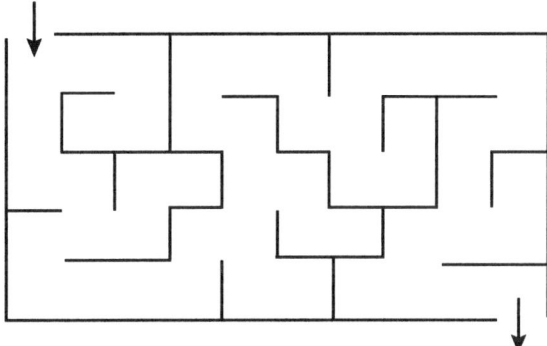

Figure 5-2: A maze to be solved using backtracking

Whenever you get to a junction in the maze with two or more options available, you have to choose one, but obviously, you might end up choosing the wrong way. The idea is to follow the choice: if you get out of the maze, you succeeded, but if not, you backtrack to the last junction and select a different option. If no options are left, you'll need to backtrack again, and again, until you either find a solution or decide there isn't one. Here's the pseudocode for such a recursive algorithm:

```
❶ solveMaze(fromCell, toCell, maze, path=[])
❷ if(fromCell === toCell) {
     return path // success!
   }
❸ mark fromCell as visited
❹ for all nextCell cells adjacent to fromCell {
   ❺ updatedPath = solveMaze(nextCell, toCell, maze, path + fromCell)
     if updatedPath is not null {
       return path
     }
   }
```

```
            // All adjacent cells were tried, and failed...
❻ return null    // failure
}
```

The parameters for this function ❶ are the starting point of the path, the final goal, the maze, and the path you'll with your journey. If you reach the goal ❷, you have succeeded; otherwise ❸, mark the cell as visited so you don't choose it again in the future and start trying all the available options ❹. If a path is shown again ❺, you've succeeded. When all options have been discarded ❻, you know you have to backtrack because you've failed. Figure 5-3 shows an intermediate position in the search.

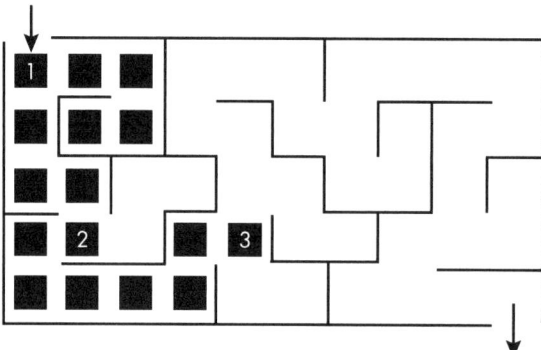

Figure 5-3: An intermediate step when solving the maze

At position 1, the algorithm had two possible options; it chose the left one and failed, and then it backtracked to choose the other one. (All the cells reached from that incorrect option were left marked.) At position 2, another selection was made; in this case, it chose the option on the right, and the left wasn't (yet) considered, so the cells in that path remain unmarked. The algorithm currently stands at 3. If it doesn't find a way out from here, it will backtrack to 2 to try the pending option. Whether it gets from position 3 to the exit quickly depends on the algorithm's "luck" at choosing the correct option at each junction, but in any case, the algorithm is guaranteed to find a path eventually, if there is one, by recursively backtracking.

Solving the Squarest Game on the Beach Puzzle

Let's apply this technique to the Squarest Game on the Beach puzzle developed by American puzzler Sam Loyd, shown in Figure 5-4. In this puzzle, the players throw balls at dolls, and if they manage to knock over dolls whose numbers add up to 50, they win a cigar. (See question 5.3 for a similar puzzle you can also solve with backtracking.)

Figure 5-4: Sam Lloyd's Squarest Game on the Beach puzzle (public domain)

You can implement a recursive backtracking algorithm as follows:

```
❶ const solve = (goal, standing, score = 0, dropped = []) => {
❷  if (score === goal) {
      return dropped;
❸  } else if (score > goal || standing.length === 0) {
      return null;
    } else {
❹    const chosen = standing[0];
❺    const others = standing.slice(1);
❻    return (
        solve(goal, others, score + chosen, [...dropped, chosen]) ||
        solve(goal, others, score, dropped)
      );
    }
  };
❼ console.log(solve(50, [15, 9, 30, 21, 19, 3, 12, 6, 25, 27]));
```

In the function ❶, goal is the number of points you try to make, and standing represents the available options, an array with the still-standing dolls. The points you have gotten so far will be held in score, and the dolls you knock over go in the dropped array. If you reach the goal exactly, you are finished ❷, and dropped has the list of dolls to drop. If you exceed our goal, or if there are no more dolls to drop ❸, you fail. Otherwise, you pick a doll ❹ (taking the first is simplest in terms of coding), remove it from future consideration ❺, and then attempt to solve the puzzle, including the recently chosen doll. If that fails, you backtrack and try again without including that doll ➏. To find the solution to the puzzle ➐, call `solve()` providing the goal (50) and list of doll points.

Dynamic Programming

Dynamic programming (DP) is a technique for solving a problem by first solving other (smaller) problems and storing those results, so they don't need to be recalculated if they're needed again. Dynamic programming comes in two flavors: top down, which solves the problem logically by checking whether it's already been solved before dealing with a subproblem, and *bottom up*, which requires first looking at the smaller subproblems and then solving the original problem. In other words, with top-down DP, you try to solve the original problem directly and then recursively solve smaller problems first, and in bottom-up DP, you start with the simplest problems and move upward, solving harder problems step by step.

This description begs a question: What's the best way to save previous results? We'll look at two methods: *memoization*, based on a higher-order function in functional programming and probably best suited for top-down DP, and *tabulation*, based on arrays or matrices, which is typically best for bottom-up DP. Memoization is usually linked to recursive implementations, while tabulation is more useful for straightforward, nonrecursive solutions. The trade-off is that tabulation is probably quicker (not needing recursion) but may solve subproblems that aren't actually needed, while memoization is slower (because of recursion) but will calculate strictly what's needed.

Calculating Fibonacci Series with Top-Down DP

Let's return to the Fibonacci numbers discussed earlier in the chapter. Here's the code:

```
const fibo = (n) => {
  if (n === 0) {
    return 0;
  } else if (n === 1) {
    return 1;
  } else {
    return fibo(n - 1) + fibo(n - 2);
  }
};
```

This is a divide-and-conquer case, but as noted, that implementation has a problem, and we'll solve it with dynamic programming. The code is clear, simple, and correct, but it can be quite slow. When experimenting with increasingly higher values of n, the required time to calculate the nth Fibonacci number grows exponentially as shown in Figure 5-5. What's happening?

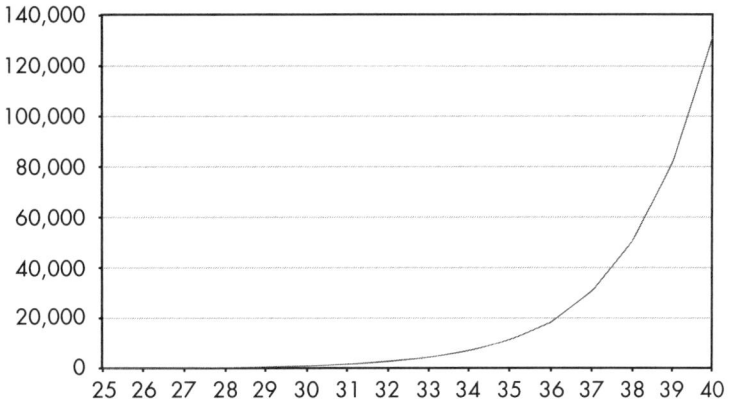

Figure 5-5: The number of additions needed to calculate Fibonacci numbers recursively grows exponentially.

To understand the problem, consider the calculations involved for fibo(7). Figure 5-6 shows all the required calls.

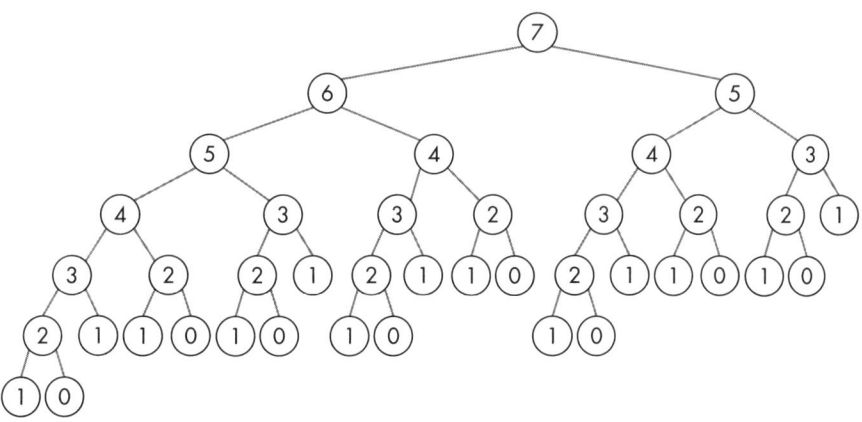

Figure 5-6: Required calls to calculate fibo(7)

Clearly too many calls are repeated. Calculating fibo(7) implies summing fibo(6) and fibo(5), but the former is calculated as fibo(5) plus fibo(4), so you are repeating fibo(5). The diagram shows that the calculation of other Fibonacci numbers implies even more repetitions; how many times are fibo(3) or fibo(2) called? (See also question 5.4.) This implementation is exponential in order, so how do you solve it?

Memoizing is a functional programming technique that may be applied to any pure function (those with no side effects that always return the same results for the same arguments, as discussed in Chapter 2). The idea is that when a memoized function is called, it first checks an internal cache to see whether the calculation was already made. If so, it returns the cached value instead of redoing the calculation. If the requested value hasn't been calculated before, the memoized function does its work, but before returning the result to the caller, it stores it in the internal cache for future reference.

Higher-order functions, such as `fast-memoize` (from *https://www.npmjs.com/package/fast-memoize*), are publicly available, but it's not hard to whip up one yourself:

```
❶ const memoize = (fn) => {
  ❷ const cache = {};
  ❸ return (...args) => {
      ❹ const strX = JSON.stringify(args);
      ❺ return strX in cache
        ❻ ? cache[strX]
        ❼ : (cache[strX] = fn(...args));
    };
};
```

This higher-order function ❶ receives a function as an argument and returns a new one. It uses a closure to maintain a cache of previous calls and calculated values; you use a simple object here ❷, but you could also use a set (see Chapter 11 for other possible structures). The returned function ❸ first creates a string out of the arguments to the original `fn` function ❹. If that string is already used as a key in the cache ❺, directly return the previously calculated value from it ❻; otherwise, call the original function, store the returned value in the cache, and return it ❼.

Given `memoize()`, you can speed up the calculations straightaway with a minor change, wrapping the original function:

```
const fibo = memoize((n) => {
  ...
});
```

If you now try something like `fibo(100)`, the results will be immediate. To understand why, you'll need `fibo(99)` and `fibo(98)`, but after calculating `fibo(99)`, the value of `fibo(98)` will have been calculated, so it won't be evaluated again. Each possible Fibonacci number between 0 and 100 will be calculated, but only once. The algorithm has become linear instead of exponential just by applying the dynamic programming technique of storing previously calculated values.

Line Breaking with Top-Down DP

Let's look at a practical problem you can solve by applying top-down DP: building a nice-looking web form. Say you want a web page to be able to generate multiple forms onscreen, each with different sets of fields. If you had a fixed number of forms with predetermined, fixed sets of fields, it wouldn't be a problem. However, in this case, the number of forms grows unpredictably, and fields need to be added or removed, as well as moved around, so you need a more flexible solution. What you need is a "form creator" that takes in a list of fields in a given order and produces a suitable form as output. For instance, a part of the form to be generated might look like the one shown in Figure 5-7.

2	**Account Registration**								
Account Holder: First Name, Middle Initial		Last Name		Date of Birth (mm/dd/yyyy)	Gender ○ Male ○ Female		Email Address ☐ No Email Address		
Primary Address (cannot be P.O. Box) ○ Home ○ Business				Social Security Number ○ SSN ○ TIN			Home Phone		Mobile Phone
City		State/Province		Zip/Postal Code		Country	Business Phone		
Mailing Address ○ Home ○ Business ○ Other				Marital Status ○ Single ○ Divorced ○ Married ○ Widowed ○ Domestic Partner			Country of Citizenship ○ U.S. ○ Other ___		▼
City		State/Province		Zip/Postal Code		Country	Country of Legal Residence (if different from mailing address) ○ U.S. ○ Other ___		▼
ID#: (Driver's Lic., Passport, Alien, Gov't.) (attach copy if Non-US Citizen)		Expiration Date	State/Country of Issuance		OFAC/FATF Comparison performed	○ Yes ○ No	Photo ID verified? ○ Yes ○ No		

Figure 5-7: An example web form

The problem is you want a justified right margin, but the widths of the fields are inconsistent, so you'll need to break rows and stretch some fields to make everything even. You need to be careful when deciding where to break rows and what fields to put in each row.

NOTE *The TeX typesetting system implements the Knuth-Plass algorithm to determine line breaks for paragraphs so they look nice. The problem here is essentially the same, but we'll use DP to solve it instead.*

Consider five fields of widths 7, 2, 5, 3, and 6 (see Figure 5-8). You need to arrange them in rows of width 10.

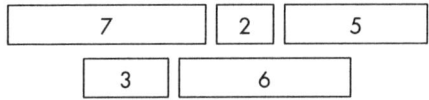

Figure 5-8: Example fields with various widths

You can't manage with fewer than three rows, and having four or more rows results in too much wasted space (although we'll need to quantify this concept later). You won't add white space between fields (as TeX does between words); instead, you'll expand the fields themselves. First decide how much empty space to leave in each row before expanding blocks or separating words. You have three possible layouts of three rows (see Figures 5-9, 5-10, and 5-11; blocks in gray represent extra added white space at the end of rows; you'll have to share that space among all blocks in the same row).

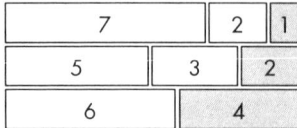

Figure 5-9: Field layout 1

Figure 5-10: Field layout 2

Figure 5-11: Field layout 3

What solution is best? Assuming that adding smaller spaces in many rows is better than adding larger spaces in fewer rows, consider the "row cost" to be the square of the added white space for that row, and the total cost will be the sum of all the row costs. (To better understand why squaring is used, imagine you need to add two spaces; putting all of them on the same line would cost $2^2 = 4$, but placing one space in each of two lines would cost $1^2 + 1^2 = 2$, so squaring the costs before adding them implements a policy that favors smaller spaces.) Given this definition, the cost of the layouts would be $1^2 + 2^2 + 4^2 = 21$, $3^2 + 0^2 + 4^2 = 25$, and $3^2 + 3^2 + 1^2 = 19$, so the third diagram represents the design the algorithm should produce. Let's program it.

Consider a list of block widths (which in this case would be 7, 2, 5, 3, and 6) and a maximum width (MW) to achieve. The following logic would work: calculate the sum s of all widths, and if s is not greater than MW, the cost is $(MW - s)^2$. You can't make it better by splitting the list into two or more rows. Otherwise, if you have more fields than you can fit in a single row, you can try splitting the list into two fragments in all possible ways and then choose the split that produces the lowest cost.

The following logic does that, but it leaves out the code to distribute white space among the fields in a row, since that's only needed later. This code finds the cost of the best set of line breaks and where those breaks should be made:

```
const costOfFragment = (p, q) => {
❶ const s = totalWidth(p, q);
  if (s <= MW) {
  ❷ return [(MW - s) ** 2, [q]];
  }
```

```
❸ let optimum = Infinity;
❹ let split = [];
❺ for (let r = p; r < q; r++) {
  ❻ const left = costOfFragment(p, r);
    const right = costOfFragment(r + 1, q);
  ❼ const newTry = left[0] + right[0];
  ❽ if (newTry < optimum) {
     optimum = newTry;
     split = [r, ...right[1]];
    }
  }
}
❾ return [optimum, split];
};
```

The function finds the best split for a set of blocks from p through q, inclusive, and also returns the list of splits to be made. Assume we have a totalWidth(x,y) function that calculates the width of blocks x through y (you'll see how to best implement it later). First calculate the width of the whole list of blocks ❶; if it's less than the available space, you won't need any splits and you are done. Calculate the cost per the definition and return that a split is done after the q position ❷. If you need a split, set up a search; optimum will be the best possible cost ❸, and split will be the place to split the list ❹. Loop through all possible breaks ❺ and find the costs of fragments p through r and fragments r + 1 through q ❻. The cost of each split is stored ❼, and if it's better than the previous optimum ❽, r is preferred as the new split. The end result ❾ is the best cost found, together with the list of split points.

Figure 5-12 shows how this algorithm would deal with your list of blocks.

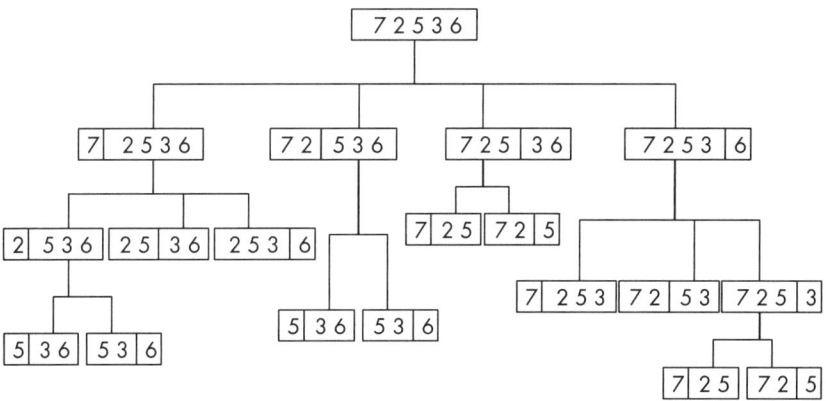

Figure 5-12: All the possible splits evaluated by the algorithm

Calculating costs, Figure 5-13 shows the optimum solution.

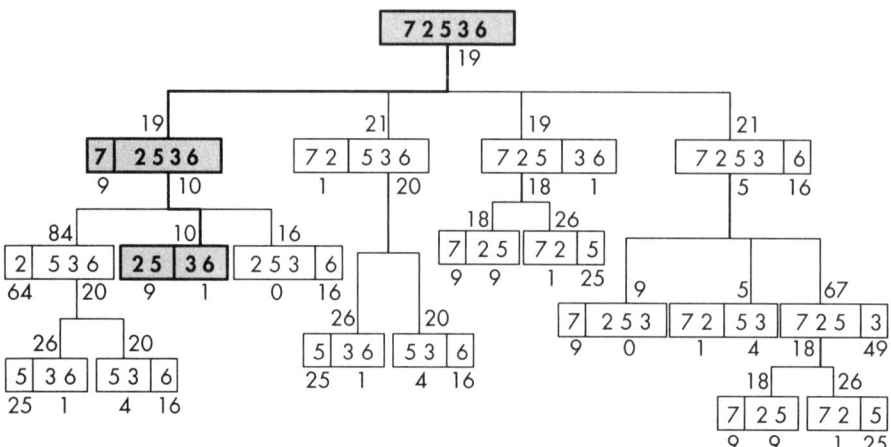

Figure 5-13: The optimum solution

The cost is shown underneath each block. If a block is split into several blocks, its cost is the sum of the costs of its parts. The highlighted path shows how to achieve the optimum solution: leave 7 on its own in the first row, place 2 and 5 in the second row, and place 3 and 6 in the last row, for a total cost of 19. Running the algorithm produces the following result:

```
❶ const blocks = [7, 2, 5, 3, 6];
   const costOfFragment = ...
❷ const result = costOfFragment(0, blocks.length - 1);
❸ console.log(result[0], result[1]);
   // 19 [ 0, 2, 4 ]
```

You can define the list of block widths ❶, and using costOfFragment(...) ❷ produces the result: the best total cost is 19, and you split lines at positions 0 (just the 7), 2 (the 2 and the 5), and 4 (the 3 and the 6) as expected ❸.

You are done, but if you look closely at Figure 5-13, you will notice the same problem as with the Fibonacci calculations: the cost of some blocks is calculated multiple times, for example, (5, 3, 6), (2, 5, 3), and (7, 2). You can apply memoization to avoid this problem, and that produces the needed algorithm:

```
const costOfFragment = memoize((p, q) => {
  ...
});
```

How would the optimized algorithm deal with this example? Figure 5-14 shows how little would actually be calculated.

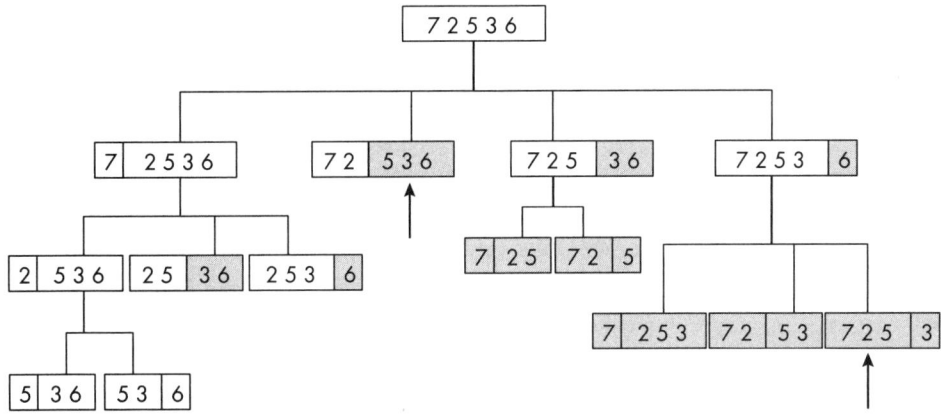

Figure 5-14: The optimized calculation reduces work in a significant way.

The grayed-out blocks don't need recalculation; due to memoization, you just reuse the previously calculated costs. At several places (marked with arrows) no recursion was needed. All in all, the algorithm worked faster, but see question 5.5 for a further optimization.

Calculating Fibonacci Series with Bottom-Up DP

Let's consider DP the other way round, from the bottom up. When working in a top-down fashion, you have to hold off on calculating values until some calculations for lower values are finished. For example, in the section "Calculating Fibonacci Series with Top-Down DP" on page 72, you couldn't calculate fibo(7) until the calculations for fibo(6) and fibo(5) were done. Using the bottom-up method, you start at the lowest cases and work your way up. To find a Fibonacci number from the bottom up, make the calculations the same way that the series is defined, starting with 0 and 1, and always adding the last two numbers to create the next number in the sequence:

```
const fibo = (n) => {
❶ if (n < 2) {
    return n;
❷ } else {
    let a = 0;
    let b = 1;
  ❸ while (n > 1) {
      [a, b] = [b, a + b];
      n--;
    }
  ❹ return b;
  }
};
```

It's simple code: for 0 or 1 ❶ you need no calculations. For other values ❷, set up a loop starting with a = 0 and b = 1 (a and b represent the two latest numbers in the sequence), and loop enough times ❸ until b becomes the number you are seeking ❹.

You might notice that all the previously calculated numbers aren't saved, and it's true, but that's because you don't need them for this particular case. The algorithm works in a bottom-up fashion, calculating later numbers by using previous ones; it so happens that to do this, you always need only the two latest numbers, so there's no need to store all the others.

Summing Ranges Recursively with Bottom-Up DP

In the line-breaking algorithm (see "Line Breaking with Top-Down DP" on page 74), you needed the `totalWidth(x,y)` function that would add together the widths of values in positions x through y (both inclusive) of an array of block widths. This function needs to be as fast as possible so as not to impact an algorithm's performance negatively. The trivial version (looping through the array, summing as it goes) has linear $O(n)$ performance, if n is the number of blocks. However, you can improve upon this, as you'll see in a couple of alternative implementations, which will focus not only on DP but also on other techniques that you've seen in this chapter.

The first algorithm, using a loop to get the sum, is straightforward. Now include another parameter to the function, arr, with the block widths, to make it more general and independent of its caller:

```
const totalWidth1 = (arr, from, to) => {
  let sum = 0;
  for (let i = from; i <= to; i++) {
    sum += arr[i];
  }
  return sum;
};
```

To optimize it using memoization requires only a small change:

```
const totalWidth1 = memoize((arr, from, to) => {
  ...
});
```

This version will be faster if (and this is a big if) you call the function two or more times, with the same arguments. Calling it with different arguments every time will slow it down instead of speeding it up, because of the extra caching work. Suppose you had already calculated the sum of the range from 10 to 20, and now you wanted the sum of the range from 10 to 21. You could add the 21st value to the sum of the range from 10 to 20, with no more work.

This concept is the key to DP: base a problem's solution on the solution of previous, smaller problems. To implement it, you need to define the sum of a range of values in terms of sums of previous ranges. If you want to calculate the sum of a range of values of array arr consisting of a single element (from p to p), the result is just arr[p]. If you want the sum of values from position 0 to position q (greater than zero), first sum the range from

0 to q-1 and then add arr[q] to that result. Finally, to find the sum of values from position p (greater than zero) to position q (greater than p), find the sum of range 0 to q and subtract the sum of range 0 to p-1.

You can also use memoization to keep track of previously calculated values; the logic is as follows:

```
const totalWidth2 = memoize((arr, from, to) => {
  if (from === to) {
    return arr[from];
  } else if (from === 0) {
    return totalWidth2(arr, 0, to - 1) + arr[to];
  } else {
    return totalWidth2(arr, 0, to) - totalWidth2(arr, 0, from - 1);
  }
});
```

This function works better and does less work. For instance, if you had asked for the sum of range 10 through 20, all sums from 0 to 0, 0 to 1, 0 to 2, and so on, up to 0 to 20 would need to be cached. If you then asked for the range 10 through 21, it would try to calculate the sum of range 0 to 21, which would be done immediately (as the sum of range 0 to 20, plus element 21), and subtract the sum of range 0 to 9 (which was already available). You still have an $O(n)$ algorithm, but over time, it becomes an $O(1)$ process; initial delays become amortized. But you can do even better.

Summing Ranges by Precomputing with Bottom-Up DP

Seeing how totalWidth2(...) in the previous section needs the sums of ranges from 0 to all possible other positions, you could use tabulation to precompute all those values, and then all queries would be $O(1)$. You can use an internal cache (partial) for those values:

```
const totalWidth3 = ((tab) => {
❶ const partial = [0];
  tab.forEach((v, i) => {
  ❷ partial[i + 1] = partial[i] + v;
  });
❸ return (from, to) => partial[to + 1] – partial[from];
❹ })(arr);
```

This is a bit trickier, because you are using a closure for the partial array, initialized in an immediately invoked function expression (IIFE). The precalculation sets up partial[k] to be the sum of the first k elements in the original array, which correctly implies that partial[0] should equal 0 ❶. (You are wasting an extra array place, but that's irrelevant in comparison with the speedy algorithm that you'll get.) You also use DP to calculate these partial sums: partial[i+1] is calculated based on the previous calculation of partial[i] ❷. The function you want will calculate the total between two elements ❸ by taking the sum up to the rightmost one (partial[to+1]) and subtracting the sum of elements up to, but not including, the leftmost one (`partial[from]`), which produces the desired $O(1)$ algorithm. The IIFE trickery is done by providing the original array of widths as a parameter ❹. (See question 5.6 for yet another way of doing this work.)

You've seen two different ways to use DP in a bottom-up fashion to optimize an algorithm, eventually reaching $O(1)$ performance. Given that width calculations are frequently used to calculate line breaks, this is a game-changer for your code's performance and usability.

Brute-Force Search

Brute-force algorithms attempt to find a solution to a problem by systematically trying all possible combinations of values. The main issue with this kind of logic is the combinatorial explosion of the number of cases to try. The order of the resulting algorithms usually goes into exponential or factorial classes (as discussed in Chapter 4) that makes them potentially impossible to use for even modest-sized input.

We'll look at a problem in each category, going from worse to worst. Given the resulting order of algorithms in this category, there's no surprise that we'll avoid this kind of code in the rest of the book.

Detecting Tautologies

In terms of logic, a tautology is a boolean expression that is always true. For instance, if X, Y, and Z are boolean variables, two of the following JavaScript expressions are tautologies:

- X OR Y OR (NOT X AND NOT Y)

- X OR (NOT X AND Y) === X OR Y

- (NOT X) OR (X AND Z) OR (NOT Y) OR (Y AND Z) OR Z

Even for readers well versed in logic and expressions, it may not be immediately obvious which of these expressions are always true.

Recognizing whether a function of n boolean parameters is a tautology potentially requires 2^n tests for each possible combination of true/false values, verifying for each one whether the function produces true as its result. Alternatively, you could try to find some combination of arguments that would make it false, and upon finding such a case, you'd know that the function isn't a tautology. That kind of search would require a logic similar to what you used to solve the Squarest Game on the Beach puzzle.

Using recursion comes in handy: if a function of n variables is a tautology, setting the first variable to false should also be a tautology, and the same would happen if the first variable were set to true. To see whether the original function is a tautology, you need to test a couple of functions with one fewer argument, which leads to a simple implementation:

```
❶ const isTautology = (fn, args = []) => {
❷ if (fn.length === args.length) {
  ❸ const result = !!fn(...args);
```

```
❹ if (!result) {
    console.log("Failed at", ...args);
  }
  return result;
} else {
❺ return (
    isTautology(fn, [...args, false]) && isTautology(fn, [...args, true])
  );
  }
};
```

The isTautology() function receives the original function to test, fn, and a list of arguments ❶. The latter will be the combination of values with which you'll test whether the function is true. If you have the right number of arguments ❷, you evaluate the function ❸, and if it produces a false value ❹, you'll log the fact and return false, which will short-circuit all future and pending evaluations. If the function returns true, the search will continue. If not enough arguments were provided ❺, you'll test the function twice: adding a true and adding a false to the list of arguments, so all combinations will be tested eventually.

The following tests the three boolean expressions mentioned earlier:

```
const f = (x, y) => x || y || (!x && !y);
console.log(isTautology(f)); // true

const g = (x, y) => (x || (!x && y)) === (x || y);
console.log(isTautology(g)); // true

const h = (x, y, z) => !x || (x && z) || !y || (y && z) || z;
console.log(isTautology(h)); // false: Failed at true true false
```

The first two functions actually were tautologies, but the last one wasn't. The search listed at least one case where the failed function evaluates to false.

Solving Cryptarithmetic Puzzles

Cryptarithmetic puzzles (also known as *cryptarithms*) are puzzles that provide a mathematical equation where the digits have been replaced by letters of the alphabet. The goal for the solver is to find which letter stands for which digit. Usually no numbers may start with zero, all letters should have different values, and the equation should translate to a phrase that makes sense. Figure 5-15 shows an early example of this, which was invented by British writer, puzzlist and mathematician Henry Ernest Dudeney in 1924.

Figure 5-15: A classic cryptarithmetic puzzle

You could solve this sort of puzzle with careful analysis (see question 5.7 for another example), but here you will write a solver that goes through all possible combinations of digits, checking whether any work. In this example, given that there are 10 digits, you need to check 10! (3,628,800) combinations, but some puzzles have numeric bases other than 10, so in general, this is an $O(n!)$ algorithm. A similar example (in terms of its solution) is the traveling salesman problem that provides a list of n cities and the distances between each pair of cities; you need to find the shortest possible route that visits each city only once, returning to the starting city. The solution to this is likewise $O(n!)$, and the algorithm is similar to the one you'll see next. (You'll also see a different type of solution using a greedy algorithm for this problem, later in this chapter.)

What algorithm do we need? The idea is simple: try all combinations of digits from 0123456789 to 9876543210, and check for each one whether the puzzle is solved. (In this case, you'll use only the first eight digits, but that really doesn't change anything.) You could design the main logic as follows, assuming that puzzle() is a function to test whether a combination is valid:

```
❶ const solve = (puzzle, digits = [0, 1, 2, 3, 4, 5, 6, 7, 8, 9]) => {
❷ const d = [...digits].sort();

❸ for (;;) {
  ❹ if (puzzle(...d)) {
      console.log("SOLUTION: ", ...d);
      return true;
    }
  ❺ // Try generating the next combination of d.
    // If there are no more combinations left, return false.
  }
};
```

The digits parameter ❶ will have the set of digits that you'll use for the problem; although in this case, 0 to 9 are the possible values, you may as well write code that could be used for cryptarithmetic puzzles in other bases. Make a local copy of the set of digits ❷ to avoid modifying the original argument and to avoid a side effect (as discussed in Chapter 2), and sort it to go through the combinations in ascending order. Then set a loop ❸ that will exit when you either find a solution or decide none exists. If the current combination of digits works out ❹, log it and exit; otherwise, generate the next combination of digits ❺ and keep looping, unless you reach the last combination, and then you'll know that the problem has no solution.

Generating the next permutation of a given set is a well-known algorithm, likely discovered by Indian mathematician Narayana Pandita in the 14th century. Assuming that the current permutation is stored in array d, it requires four steps, in order:

1. Find the rightmost index p such that d[p] < d[p + 1]; if no such p exists, you were already at the last permutation, and the algorithm finishes.

2. Find the rightmost index q such that d[p] < d[q]; d[q] is the least value to the right of d[p]greater than it.

3. Swap the values of d[p] and d[q]; now the values from d[p + 1] to the end of d will be in descending order.

4. Reverse the list of values from d[p + 1] to the end of d.

Figure 5-16 shows a working example, starting with permutation 8403976521.

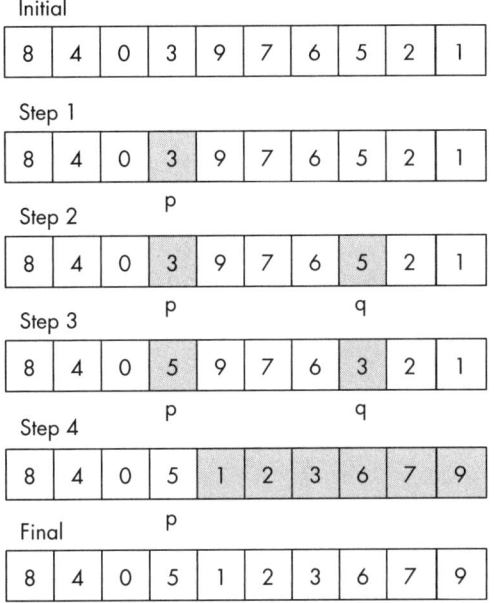

Figure 5-16: Generating the next permutation

Step 1 sets p to point at 3, because 3 < 9; all other elements from the right (976521) are in descending order. Step 2 sets q to point at 5, the least value to the right of 3 that is greater than it. Step 3 swaps the values pointed by p and q; values to the right of p are again in descending order (976321). Step 4 ends by reversing the values to the right of p, which end in ascending order (123679), and you get the next permutation: 8405123679.

With this logic, you can see the complete version of the cryptarithmetic puzzle solver by adding the code to generate permutations:

```
const solve = (puzzle, digits = [0, 1, 2, 3, 4, 5, 6, 7, 8, 9]) => {
  const d = [...digits].sort();

  for (;;) {
    if (puzzle(...d)) {
      console.log("SOLUTION: ", ...d);
      return true;
    }
  }
```

```
    let p = d.length - 2;
    while (p >= 0 && d[p] > d[p + 1]) {
      p--;
    }

    if (p === -1) {
      console.log("No solution found");
      return false;
    }

    let q = d.length - 1;
    while (d[p] > d[q]) {
      q--;
    }

    [d[p], d[q]] = [d[q], d[p]];

    let l = p + 1;
    let r = d.length - 1;
    while (l < r) {
      [d[l], d[r]] = [d[r], d[l]];
      l++;
      r--;
    }
  }
};
```

This code is the same as earlier, with steps 1 through 4 of the permutation algorithm highlighted.

You can now write a function to test whether a given combination of values is actually a solution:

```
❶ const sendMoreMoney = (s, e, n, d, m, o, r, y) => {
❷  if (s === 0 || m === 0) {
     return false;
❸  } else {
     const SEND = Number(`${s}${e}${n}${d}`);
     const MORE = Number(`${m}${o}${r}${e}`);
     const MONEY = Number(`${m}${o}${n}${e}${y}`);
     return SEND + MORE === MONEY;
   }
 };

❹ solve(sendMoreMoney);
 // SOLUTION:  9 5 6 7 1 0 8 2 3 4
 // 9567+1085=10652
```

The function is called with all 10 digits ❶, but you use only the first 8, ignoring the last 2. If a leading digit is 0 ❷, the solution isn't valid, so reject that out of hand. If there are no leading zeros ❸, compute the values of the three words (SEND, MORE, and MONEY) and we check whether they fulfill the original equation. Given this function, all you need to do is pass it to the solve() function ❹ and wait (very little) for the solution to appear.

Greedy Algorithms

Let's conclude with a set of algorithms that have a rather curious characteristic: they may not always work. The basic definition of an algorithm implies that it's a well-defined procedure to solve a problem or accomplish some task. *Greedy algorithms* may (or may not) achieve that.

Sometimes a heuristic is used to describe a way to get a (hopefully not too bad) solution in a faster way by applying some arbitrary choices instead of doing a thorough search. For instance, a chess algorithm could, in principle, always find the best move by considering all possible moves and all possible opponent responses, and all possible responses to those responses, and so on, but that approach grows exponentially and isn't feasible. The alternative is a heuristic. Reconsidering the chess example, instead of going to the maximum depth, you would stop the search short after a few moves, do a ballpark evaluation of the resulting board positions, and choose the move that leads to the best evaluated move. This method doesn't *guarantee* making the best move, but it at least provides some solution.

Greedy algorithms are usually applied to optimization problems. You've seen algorithms that use brute force to try all possibilities; greedy algorithms don't do that. At each step where a decision needs to be made, these algorithms make the best possible choice at the time. On one hand, this approach ensures that the algorithm proceeds quickly without needing to backtrack. On the other hand, the algorithm doesn't necessarily make the best choice, because it doesn't look far enough ahead. However, under certain conditions, which you'll explore in the following sections, these algorithms perform well and are successful.

How to Make Change

How do you make change using the fewest bills and coins? In other words, suppose you have to pay out some amount using today's US currency: $100, $50, $20, $10, $5, and $1 bills and $0.25 (quarter), $0.10 (dime), $0.05 (nickel), and $0.01 (penny) coins. How would you pay $229.60? You could use so many combinations to reach that amount, but with a greedy algorithm, you would follow this simple rule: at each step, choose as many as possible units of the largest possible denomination, and keep going until you're done.

The method starts by using two $100 bills, then one $20, a $5, four $1s, two quarters, and a dime. No other solution involves fewer bills and coins. This greedy algorithm is guaranteed to succeed, but it depends on the available denominations. Paying out $16 (the greedy way) in a country that had $9, $8, and $1 bills would end up with one $9 and seven $1 bills, instead of using just two $8 bills. Greedy algorithms may (or may not) succeed depending on the case.

The Traveling Salesman Problem

Let's consider a problem that requires a brute-force search but that a greedy algorithm usually solves quite well. The traveling salesman problem works like this: imagine a salesman has to do a tour, visiting each city on a list once and then returning to the starting point. (In graph terms, this is called a *Hamiltonian cycle*.) Distances (or costs) for traveling between cities is known. What's the shortest (or cheapest) way to achieve the task?

As is, an algorithm for this problem would require testing all possible permutations of cities (as you did previously for the SEND + MORE = MONEY cryptarithmetic puzzle). If the number of cities grows, the problem becomes intractable because the required time to run the algorithm takes too long.

A greedy algorithm for this problem (which may not find the best solution but performs quickly) would proceed as follows: at each step visit the nearest not-yet-visited city. This method won't necessarily find the best possible path, and several heuristics may discover an even better one, but under some conditions, the algorithm finds the optimal solution.

Minimum Spanning Tree

Let's wrap up our discussion of greedy algorithms by considering a problem you'll explore in Chapter 17. Imagine a cable TV company must provide service to several houses. The company cannot place cables just anywhere and must follow existing roads. Where should it put cables to minimize the total cost?

The solution for this is technically called a *minimum spanning tree*, and Kruskal's algorithm (which you'll implement in Chapter 17) is a greedy algorithm that solves it and is guaranteed to find the optimal solution. Start by choosing the cheapest segment of road until all houses are connected, and always add the cheapest possible segment that won't generate a loop; after all, what good would having a closed cable loop be?

You can solve other graph-related problems using greedy algorithms, so this technique may also be a valid one to consider when trying to write code for a specific problem.

Summary

In this chapter we considered several techniques (recursion, DP, and brute-force and greedy algorithms) that will help you develop algorithms on your own, and they'll appear again several times in the rest of the book.

In the next chapter, we'll explore several common problems, such as sorting, selecting, shuffling, sampling, and searching—that's a lot of alliteration but also a lot of interesting code and plenty of opportunities to study how to write algorithms.

Questions

5.1 Factorial in One

The code for factorial() is totally correct, but it's seven lines long! Not that it matters (having a long correct function is better than a short incorrect one), but can you write it in a more compact way?

5.2 Hanoi by Hand

The recursive algorithm for the Towers of Hanoi is good for computers, but not so much for normal human beings. Can you design a simple algorithm to solve the puzzle that doesn't involve recursion?

5.3 Archery Backtracking

Sam Loyd devised another puzzle (see Figure 5-17) similar to the Squarest Game on the Beach that you solved earlier in this chapter. In this puzzle, you need to get 100 points by aiming arrows at the target. The important difference is that in this puzzle, you can hit a ring two or more times, while in the other problem you could drop a doll only once.

Figure 5-17: Another classic puzzle by Sam Loyd, where players must get exactly 100 points with their arrows (public domain)

Can you modify the backtracking algorithm to deal with this variation? And, even with the differences, can you still use the solve() function that you used for the other puzzle to find the solution?

5.4 Counting Calls

If you call $C(n)$, the total number of calls needed to calculate the nth Fibonacci number with your recursive implementation, you will see, for example, that $C(7) = 41$. Can you give a recurrence for $C(n)$ and find an explicit solution for it? Hint: the answer will again involve Fibonacci numbers.

5.5 Avoid More Work

When considering how to arrange blocks in rows (in "Line Breaking with Top-Down DP" on page 74) and when considering splits, you analyzed them as (7, 2, 5) and (3, 6) or (7, 2, 5, 3) and (6). However, this wasn't really needed because blocks $7 + 2 + 5$ or $7 + 2 + 5 + 3$ couldn't fit in a line. Figure 5-18 shows crossed-out options that an enhanced algorithm wouldn't have considered.

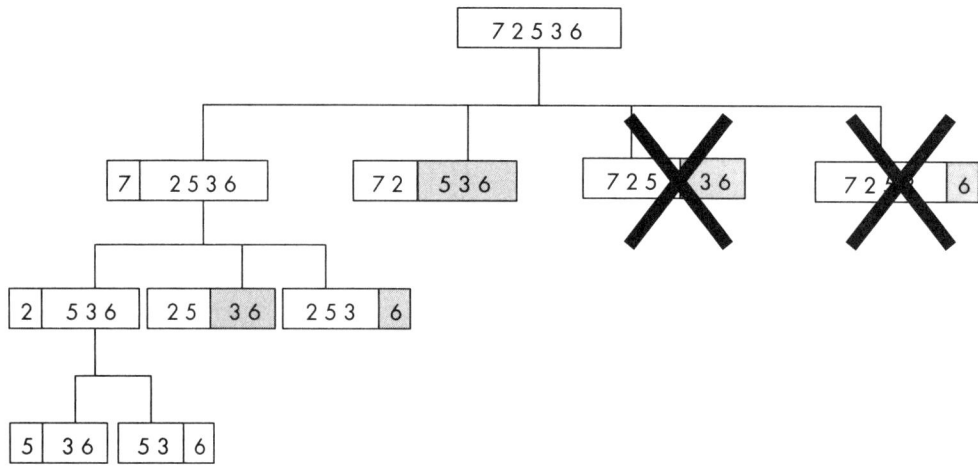

Figure 5-18: An even more efficient way to find row breaks

Can you add this optimization to the code?

5.6 Reduce for Clarity

The totalWidth3(...) function (in the section "Summing Ranges by Precomputing with Bottom-Up DP" on page 81) used a common loop to generate the partial array. Can you do the same using .reduce(...) instead?

5.7 Got GOUT?

Figure 5-19 shows another well-known cryptarithmetic puzzle; find what each letter stands for, and then the value of GOUT is your answer. You can solve this with the techniques shown earlier in the chapter, or you could try your hand at directly working it out.

Figure 5-19: A simple cryptarithmetic puzzle with only four letters to find

6

SORTING

In previous chapters we discussed concepts related to programming and designing algorithms. Now we'll start considering their actual application. The problem we'll explore is how to sort a set of records into order, where each record consists of a key (alphabetical, numerical, or several fields) and data.

The algorithm's output should include the exact same set of records, but shuffled so that the keys are in order. You usually want the keys in ascending order, but descending order requires only a minor change in sorting algorithms—namely, reversing comparisons—so you won't see it here. (See question 6.1 at the end of the chapter.)

We'll first consider general aspects of the sorting problem and then moves on to look at several algorithms based on comparisons of keys (the most common algorithms), followed by a few algorithms based on other principles. We'll consider the performance of all algorithms and even toss in some humorous algorithms for comparison.

The Sorting Problem

A *sorting algorithm* is basically an algorithm that, given a list of records containing a key and some data, reorders the list so that the keys are in nondecreasing order (no key is smaller than its preceding key), and the output list is a permutation of the input list, retaining all original records. Forgetting the second condition is easy, but ignoring it would mean that the following would be a valid sorting function:

```
const wrongWayToSort = (inputData) => [];
```

Sorting is important in and of itself, but it also affects the efficiency of other algorithms. For instance, in Chapter 9 we'll see how working with sorted data allows for more efficient search procedures.

For our examples, we'll usually assume single-field keys that you can directly compare using the < and > operators. For more generic cases, you could modify the algorithms to use the compare(a,b) comparison function, as JavaScript's sorting algorithm does (see the section "JavaScript's Own Sort Method" on page 95). In the code examples in this book, you'll always write tests as a>b, so modifying the code for generic sorting requires only changing that comparison to compare(a,b)>0. (See question 6.2 for a variation.) In Chapter 14, you'll actually use this kind of solution by applying a goesHigher(a,b) function to decide which of a or b should be higher in a heap.

Internal vs. External Sorting

An important consideration when sorting data is whether it can all be stored in memory at the same time, or whether it's so large that it must reside in a storage device. The first case is called *internal sorting*, and the second is called *external sorting*. All the algorithms in this chapter fall into the first category, but what if you need to sort more data than can fit in memory?

External sorting breaks up all the input into runs that are as large as possible to fit in memory, then uses internal sorting to sort the runs, saves them to external storage, and merges the sorted runs into the final output. That said, it's highly likely that for large sorting tasks like this, you'll be better off using a standard system sort utility, which also might be optimized to use parallel threads, multiple central processing units (CPUs), and so on. In any case, should you decide to roll out your own external sort procedure, the algorithms in this section cover the needed internal sorting, and using a heap (as in Chapter 14) would help with writing efficient merge code.

Adaptive Sorting

A sorting algorithm is called *adaptive* if it somehow takes advantage of whatever existing order already exists in its input. Shell sort, which you'll learn about in the section "Making Bigger Jumps with Comb and Shell Sort" on page 103, is such a case: the algorithm performs better when data is partially sorted. On the other hand, quicksort, which you'll learn about in the section "Going for Speed with Quicksort" on page 105, could be considered anti-adaptive. Its worst performance happens when data is already in order (though there are ways around this).

In-Place and Out-of-Place Sorting

Another consideration for sorting algorithms is whether they require extra data structures (and thus extra space). This requirement is often relaxed to allow for constant, less than $O(n)$ extra memory—the key rule is whether extra space proportional to the input size is needed. We don't take into account the $O(n)$ space needed to store the n elements to be sorted. Algorithms that don't require such extra space are called *in-place*, and those that do require more memory are known as *out-of-place* or *not-in-place* algorithms. This doesn't mean that out-of-place algorithms return a new list; they may perfectly well reorder the input list in place, but they require $O(n)$, or more, extra space to do so.

Consider carefully how much memory an algorithm uses: some recursive algorithms like quicksort require internally using a stack that is $O(log\ n)$ but that is also allowed to count as in-place. Merge sort usually requires extra space to merge sequences, so it has $O(n)$ needs and thus falls into the out-of-place category.

Online and Offline Sorting

Another distinction to make when considering algorithms is whether they can process the input data in a serial stream-like fashion or whether all the data needs to be available from the beginning. Algorithms in the first category are called *online algorithms*, and those in the second are *offline algorithms*. This distinction applies not only to sorting but to other problems as well; for example, you'll see it again when discussing sampling in Chapter 8.

In terms of sorting, an online algorithm will always have a sorted list, adding new elements to it as they come in, while an offline algorithm will have to wait until all elements are available. Offline algorithms usually have better performance, though. Online algorithms don't know the whole input, so they have to make decisions that may turn out later to be suboptimal, which is the same kind of situation as with greedy algorithms (see Chapter 5).

As an example of this distinction, consider how you could sort a set of playing cards. If you keep the cards you've sorted so far in your hand and then every time you get a new card you insert it into place among the previous ones, you are implementing an online algorithm—in fact, it's an *insertion sort*, which we'll study in the section "Sorting Strategies for Playing Cards" on page 100. If you wait until you have all the cards and then sort them somehow, that's an offline sort.

Sorting Stability

Sorting data with possibly equal keys raises a question: In what relative order do the elements with equal keys end up? A *stable sorting* algorithm maintains the same order as the input, so if one element preceded another and both had the same key, in the ordered output, the first one will precede the second.

Why does stability matter? Imagine you want to have a drop-down element in an HTML page that shows your contacts but with this rule: starred contacts (favorites) should appear first, in alphabetical order, followed by nonstarred contacts, also in alphabetical order.

To achieve the required ordering, you could first order the whole list by name and then reorder it so starred contacts are first. Figure 6-1 illustrates this method.

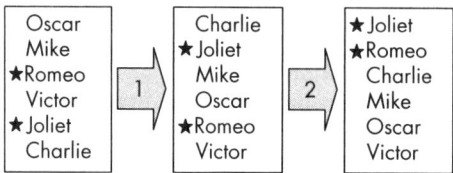

Figure 6-1: Sorting by two fields with a stable sorting algorithm

The first sort reorders the list by name, alphabetically, and the second sort places starred names before the ones without stars. If the second sort is stable, this ordering won't affect the previous alphabetical sorting. With an unstable sort, that might not be true. Stability is the reason Joliet precedes Romeo in the final list. Joliet preceded Romeo when sorting by name, and when sorting by star, they keep the same relative order.

You can modify any sorting algorithm to force it to be stable. No matter what the key for ordering is, consider a new extended key formed by the original key followed by the item's position in the list. Ordering this array by the new extended key, items that shared the same (original) key value will be sorted together, but because of the added position, they will keep the same original relative order, as shown in Figure 6-2.

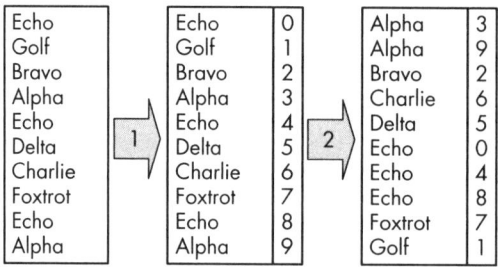

Figure 6-2: Sorting made stable by using an extended key

The first step adds the item's position as an extra key; the second step sorts by name and position. Elements that had the same original key (Alpha and Echo in the example) are kept in their original relative positions to each other. You would finish by dropping the added field.

JavaScript's Own Sort Method

When sorting data in JavaScript, don't forget that the language already provides a .sort(...) method, and despite considering more (and possibly better) sorting algorithms later in this chapter, in many cases using JavaScript's own sort might be most effective. Let's quickly review how this sort works (see *https://developer.mozilla.org/en-US/docs/Web/JavaScript/Reference/Global _Objects/Array/sort* for more information).

Given an array, the .sort(comparisonFunction) method reorders the array in place using an optional comparison function. (The newer .toSorted() method doesn't sort in place, but rather produces a new, sorted version of the array.) If that function is omitted, JavaScript converts elements to strings and then sorts lexicographically, which may not be what you wanted:

```
const a = [22, 9, 60, 12, 4, 56];
a.sort();
console.log(a); // 12 22 4 56 60 9
```

To accommodate other ways of sorting, you need to provide a function that will receive two elements, a and b, and return a negative value if a should precede b, a positive value if a should follow b, and zero if both keys are equal and if a and b could be in any order. You can fix the previous example quite easily:

```
const a = [22, 9, 60, 12, 4, 56];
a.sort((a, b) => a - b);
console.log(a); // 4 9 12 22 56 60
```

You can also implement more complex comparisons; the following example shows how you would sort objects by date and name:

```
❶ const people = [
    { d: 22, m: 9, y: 60, n: "alpha" },
    { d: 12, m: 4, y: 56, n: "bravo" },
    { d: 22, m: 3, y: 56, n: "hotel" },
    { d: 9,  m: 1, y: 60, n: "foxtrot" },
    { d: 22, m: 4, y: 56, n: "echo" },
    { d: 22, m: 3, y: 56, n: "delta" },
    { d: 22, m: 3, y: 56, n: "india" },
    { d: 14, m: 1, y: 34, n: "charlie" },
    { d: 9,  m:12, y: 40, n: "golf" }
  ];

  const dateNameCompare = (a, b) => {
❷ if (a.y !== b.y) {
    return a.y - b.y;
❸ } else if (a.m !== b.m) {
    return a.m - b.m;
❹ } else if (a.d !== b.d) {
    return a.d - b.d;
❺ } else if (a.n < b.n) {
```

```
      return -1;
  } else if (a.n > b.n) {
      return 1;
  } else {
❻ return 0;
  }
};
```

The data to sort ❶ has dates as three separate fields (d, m, and y, for day, month, and year) and name (n). If two persons are from different years ❷, you return the correct negative or positive value by subtracting years. If the years are equal, you can compare months with the same kind of logic ❸, and if the months are also equal ❹, you do the same once more for days. If the dates are equal, you resort to comparing names ❺, and since you cannot use math and just subtract dates, you need to make actual comparisons, date part by date part. The final return 0 is done ❻ only if all fields were compared and found to match.

If you sort the people array with the dateNameCompare(...) function you just wrote, you get the expected result:

```
console.log(people.sort(dateNameCompare));

[
  { d: 14, m: 1, y: 34, n: 'charlie' },
  { d: 9,  m:12, y: 40, n: 'golf' },
  { d: 22, m: 3, y: 56, n: 'delta' },
  { d: 22, m: 3, y: 56, n: 'hotel' },
  { d: 22, m: 3, y: 56, n: 'india' },
  { d: 12, m: 4, y: 56, n: 'bravo' },
  { d: 22, m: 4, y: 56, n: 'echo' },
  { d:  9, m: 1, y: 60, n: 'foxtrot' },
  { d: 22, m: 9, y: 60, n: 'alpha' }
]
```

Finally, consider stability. Originally, the specification for the .sort(...) method didn't require it, but ECMAScript 2019 added the requirement. Be aware, however, that if using an earlier JavaScript engine, you cannot assume stability, so you might have to resort to the solution described in "Sorting Stability" on page 93. Also, keep in mind that any given engine may just not correctly implement the standard.

Sort Performance

If you have to sort n values, your logic has to be able to deal with all possible $n!$ permutations of those values. How many comparisons will be needed for that? Think of the game of 20 questions. In that game, you have to guess a selected object by asking, at most, 20 yes or no questions. If you plan your questions carefully, you should be able to pick any element out of more than a million ($2^{20} = 1,048,576$, actually) possible options. You can apply that logic to sorting n elements.

If you are comparing elements to sort an array, it's indirectly implied that you're deciding which was the original permutation. Well-placed questions divide the range of options in half, so you need to know how many questions are needed for $n!$ possibilities. This is equivalent to asking how many times you should divide $n!$ by 2 until you get down to 1. The answer is log $n!$, in base 2. (Alternatively, you can see it as asking what value of k is such that $2^k > n!$) This section won't go into its derivation, but Stirling's approximation says that $n!$ grows as n^n, so the logarithm of $n!$ is $O(n \log n)$.

This automatically implies that any algorithm based on comparing elements will be $O(n \log n)$ at the very least. No better results are achievable, but worse results are obviously possible. With that in mind, in the next section we'll consider several algorithms, from worst to best performance.

Note, however, the observation about these algorithms being "based on comparing elements." If you manage to sort a list without making actual comparisons, all bets are off. You'll see that some methods allow sorting in $O(n)$ time, without ever comparing keys to each other.

Sorting with Comparisons

As mentioned previously, we'll consider the major sorting algorithms, all of which depend on comparing values to each other. The first algorithms we'll consider are $O(n^2)$, so they're not optimum, but we'll move on to better ones until we reach several that achieve the best $O(n \log n)$ performance.

In all cases you'll write functions that receive an array of values (as stated earlier, you don't have to worry about key + data pairs, as that can easily be accommodated), and you'll also pass parameters to specify which part of the array (from, to) should be sorted. As usual, you'll want to sort the whole array. Those parameters will have default values, so the whole array will be sorted if they're not present.

Bubbling Up and Down

We'll start our review of sorting algorithms with *bubble sort*, which probably has the catchiest name, possibly to compensate for its subpar performance. This algorithm is easy to implement, but you'd use it only for smaller sets of data. It also has generated several variations (you'll look at comb sort in the next section, which actually leads to a better-performing algorithm).

Bubble Sort

The bubble sort algorithm derives its name from the simple idea that larger numbers represent bubbles that bubble up to the top of the list. It starts at the beginning of the array and goes in order through all elements in the array, and if an element is greater than the following element, it swaps them (see Figure 6-3).

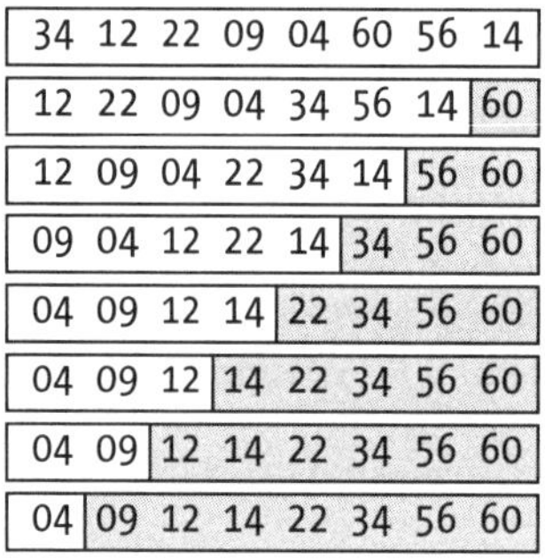

```
34 12 22 09 04 60 56 14
12 22 09 04 34 56 14 60
12 09 04 22 34 14 56 60
09 04 12 22 14 34 56 60
04 09 12 14 22 34 56 60
04 09 12 14 22 34 56 60
04 09 12 14 22 34 56 60
04 09 12 14 22 34 56 60
```

Figure 6-3: With bubble sort, each pass moves another element to its place.

The first pass at the top of Figure 6-3 goes from left to right, comparing adjacent values and swapping if needed so that the higher value is always to the right. After the first pass, 60 goes to the top of the array. You proceed in the same way with the rest of the array, and after the second pass, 56 goes to the next-to-last position, so you have at least two elements in the right place. After the third pass, three elements will be in place, and so on. The last two rows required no swapping, because previous passes had already moved the elements to the correct places, which frequently happens.

Here's the logic for this algorithm:

```
❶ const bubbleSort = (arr, from = 0, to = arr.length - 1) => {
❷   for (let j = to; j > from; j--) {
❸     for (let i = from; i < j; i++) {
❹       if (arr[i] > arr[i + 1]) {
          [arr[i], arr[i + 1]] = [arr[i + 1], arr[i]];
        }
      }
    }
    return arr;
};
```

All sorting functions share the same signature: an array to sort (arr) and the limits for sorting (from, to) that, by default, will be the array's extremes ❶. The outer loop ❷ goes from the right to the left; after each pass, the element in position j of the array will be in the right place. The inner loop ❸ goes from the left extreme to the right up to (but not reaching) the outer loop j; you compare each element with the next ❹, and if the second is smaller, you swap them.

You can improve performance in most sorted arrays (a not uncommon case) by checking whether any swaps occurred on each pass through the array. If none were detected, it means the array is in order (see question 6.7).

The performance of this algorithm is $O(n^2)$, which is easy to calculate. First count comparisons: the first pass does $(n-1)$ comparisons, the second pass does $(n-2)$, the third $(n-3)$, and so on. The total number of comparisons is then the sum of all numbers from $(n-1)$ down to 1, which is $n(n-1)/2$, so $O(n^2)$.

Sinking Sort and Shuttle Sort

Bubble sort quickly moves the greatest values to the end of an array, but the smallest values may take a while to reach their final positions. Similarly, *sinking sort* (see question 6.6) makes the lowest values quickly sink to the beginning of the array, but correspondingly, it takes longer for the greatest values to go to their places. You can alternate a pass of bubbling with a pass of sinking to get an enhanced algorithm, called *shuttle sort* (also known as *cocktail shaker sort* or *bidirectional* bubble sort). In comparison with bubble sort, the first passes of the shuttle sort proceed as shown in Figure 6-4.

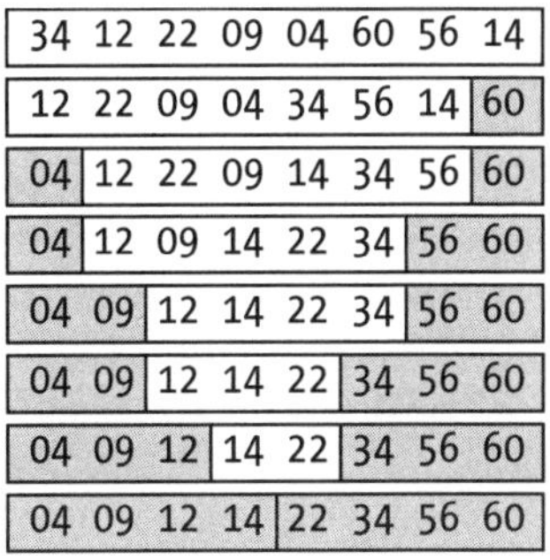

34	12	22	09	04	60	56	14
12	22	09	04	34	56	14	60
04	12	22	09	14	34	56	60
04	12	09	14	22	34	56	60
04	09	12	14	22	34	56	60
04	09	12	14	22	34	56	60
04	09	12	14	22	34	56	60
04	09	12	14	22	34	56	60

Figure 6-4: Shuttle sort alternates left-to-right and right-to-left passes.

Starting with the same elements, the first pass is the same as bubble sort's, moving 60, which is the greatest value in the array, to the rightmost position. The second pass goes right to left and moves 04, the smallest value in the array, to the leftmost position. The third pass again goes left to right and moves 56 to its place; after that, it goes right to left, then left to right, and so on, alternating direction every time.

Here's the corresponding code:

```
❶ const shuttleSort = (arr, from = 0, to = arr.length - 1) => {
❷ let f = from;
  let t = to;

❸ while (f < t) {
❹   for (let i = f; i <= t - 1; i++) {
      if (arr[i] > arr[i + 1]) {
        [arr[i], arr[i + 1]] = [arr[i + 1], arr[i]];
      }
    }
❺   t--;

❻   for (let i = t - 1; i >= f; i--) {
      if (arr[i] > arr[i + 1]) {
        [arr[i], arr[i + 1]] = [arr[i + 1], arr[i]];
      }
    }
```

```
❼ f++;
  }

  return arr;
};
```

As mentioned earlier, the signature for this sort function is always the same: an array to sort and the portion to put in order ❶. You have two variables ❷ that mark how far to the left and right the array is already sorted: f (as in *from*) starts at the left and grows by 1 after each right-to-left pass, and t (as in *to*) starts at the right and decreases by 1 after each left-to-right pass. When these variables meet ❸, the sort is done. You perform a left-to-right pass as shown earlier ❹, and then you decrement t ❺, since you've placed a new value in the right place. After this pass, you do the same ❻, but right to left, and you increment f ❼ to finish.

The algorithm is still $O(n^2)$, but the actual implementation typically is double the speed or even better if you include testing for swaps (see question 6.7). In any case, it's easy to show that it can't do any worse, for in each pass, it places one number at its final position, so after having placed $(n-1)$ numbers at their place, it will be done, the same as bubble sort.

Nevertheless, despite the catchy name, this sort algorithm is not good enough in comparison with those that we'll explore later in the chapter.

Sorting Strategies for Playing Cards

Thinking about how you do simple tasks can provide tips for developing an algorithm. For example, suppose you have a few playing cards in your hand and want to order them from lowest to highest. You could apply a couple of different strategies, which we'll look at next: selection sort or insertion sort.

Selection Sort

A simple solution is to look for the lowest card and place it farthest to the left in your hand. Then look for the next lowest card and place it after the first, and keep doing that, always selecting the lowest remaining card and placing it next to the already sorted cards. This process is the basis for the *selection sort* algorithm, which adds a small detail: when placing a card to the left, you do a swap with the other card (see Figure 6-5).

34	12	22	09	**04**	60	56	14
04	12	22	**09**	34	60	56	14
04	09	22	**12**	34	60	56	14
04	09	12	22	34	60	56	**14**
04	09	12	14	34	60	56	**22**
04	09	12	14	22	60	56	**34**
04	09	12	14	22	34	**56**	60
04	09	12	14	22	34	56	60

Figure 6-5: Selection sort looks for the smallest element and swaps it to get it into place.

In the first pass at the top, you find that the minimum number is 04, and you do a swap to move it to the first place in the array. The second pass finds 09 and swaps it with 12, so you now have two numbers in order. The process continues the same way; an exception is in the next-to-last line, in which no swap is needed because 56 was already in the correct place.

Here's an implementation:

```
const selectionSort = (arr, from = 0, to = arr.length - 1) => {
❶ for (let i = from; i < to; i++) {
  ❷ let m = i;
  ❸ for (let j = i + 1; j <= to; j++) {
    ❹ if (arr[m] > arr[j]) {
        m = j;
      }
    }
  ❺ if (m !== i) {
      [arr[i], arr[m]] = [arr[m], arr[i]];
    }
  }

  return arr;
};
```

Go in order ❶ from the first place in the array to the last. The m variable ❷ keeps track of the position of the minimum value already found. As you loop through the yet unsorted numbers ❸, if you find a new minimum candidate ❹, you update m. After finishing this loop, if the minimum isn't already in place ❺, do a swap.

The order of this algorithm is, again, $O(n^2)$. You have to look at n elements to find what should go in the first place; then look at $n - 1$ for the second place, $n - 2$ for the third, and so on. You already know this sum is $O(n^2)$. The algorithm in the next section is also based on how you'd sort playing cards, but it has slightly better performance.

Insertion Sort

With selection sort, we thought about sorting playing cards, but you could have considered another method. Take the first card; that's clearly already in order by itself. Now look at the second card, and either place it before the first (if it's lower) or leave it where it is (if it's higher). You now have two cards in order. Look at the third card, decide where it should go among the previous two, and place it there. As you go through all the cards in your hand, you'll end up putting them in order, and this is called an *insertion sort*, because of the way you insert new cards among the previously sorted ones (see Figure 6-6).

34	12	22	09	04	60	56	14

12	34	22	09	04	60	56	14

12	22	34	09	04	60	56	14

09	12	22	34	04	60	56	14

04	09	12	22	34	60	56	14

04	09	12	22	34	60	56	14

04	09	12	22	34	56	60	14

04	09	14	12	22	34	56	60

*Figure 6-6: Sorting by insertion works
the way one sorts playing cards.*

Start with a single card in order, in this case, number 34. Then consider the next value, 12, and place it to the left of 34, so the two numbers are in order. Then consider 22, which goes between 12 and 34, and now three values are ordered. Continue working this way, always inserting the next number where it belongs among the previously sorted ones, until you reach the last line. After placing 14 among the already sorted numbers, the whole array becomes ordered.

The following code implements this method:

```
const insertionSort = (arr, from = 0, to = arr.length - 1) => {
❶ for (let i = from + 1; i <= to; i++) {
  ❷ for (let j = i; j > from && arr[j - 1] > arr[j]; j--) {
    ❸ [arr[j - 1], arr[j]] = [arr[j], arr[j - 1]];
    }
  }
  return arr;
};
```

Set up a loop that starts at the second place in the array and goes to the end ❶, and loop back as long as the list isn't in order ❷, swapping to get new numbers in place ❸.

Looking at this carefully, you'll notice it's doing too many swaps to get the new element to its place.

You can quickly optimize the code to avoid that and do just one swap per loop:

```
const insertionSort = (arr, from = 0, to = arr.length - 1) => {
❶ for (let i = from + 1; i <= to; i++) {
  ❷ const temp = arr[i];
  ❸ let j;
    for (j = i; j > from && arr[j - 1] > temp; j--) {
      arr[j] = arr[j - 1];
    }
  ❹ arr[j] = temp;
  }
  return arr;
};
```

The first loop ❶ is exactly the same as earlier, but the difference lies within. You set the number to be inserted among the previously sorted aside ❷, and you loop to find where it should go ❸, pushing values that are greater to the right. At the end ❹, you place the new value in its final position.

Insertion sort is a simple algorithm, which makes it a good choice for smaller arrays. Later in the chapter we'll look at how it's sometimes used in hybrid sorting algorithms as a replacement for theoretically more convenient, but practically slower, alternative methods.

Making Bigger Jumps with Comb and Shell Sort

Bubble sort and its variants are not the best-performing sorting algorithms. However, the idea of swapping elements to make them bubble up or sink down isn't bad, and applying the idea of making larger jumps (for example, swapping elements that are farther apart) eventually leads to a better algorithm, *Shell sort*. You'll explore this idea with a bubble sort variant called *comb sort* first.

Comb Sort

Let's go back to bubble sort and consider how keys move in an array like rabbits and turtles. Rabbits represent the large values near the beginning of the list, which quickly move to their places at the end of the array, swap after swap. On the other hand, turtles represent the small values near the end of the list, which slowly move to their places in a single swap per pass. You want both turtles and rabbits to move quickly to their respective sides of the array.

The idea is to perform some passes with swaps, but instead of comparing one element with the next one, you'll consider larger gaps. Thus, rabbits will jump further distances toward the right, but turtles will correspondingly jump further distances toward the left. You'll do passes with successively smaller gaps, and when the gap becomes 1, you'll apply the common bubble sort to finish.

The logic is as follows:

```
const combSort = (arr, from = 0, to = arr.length - 1) => {
❶ const SHRINK_FACTOR = 1.3;

  let gap = to - from + 1;
  for (;;) {
❷   gap = Math.floor(gap / SHRINK_FACTOR);
❸   if (gap === 1) {
      return bubbleSort(arr, from, to);
    }
❹   for (let i = from; i <= to - gap; i++) {
      if (arr[i] > arr[i + gap]) {
        [arr[i], arr[i + 1]] = [arr[i + 1], arr[i]];
      }
    }
  }
};
```

It has been determined empirically that the first gap should equal the array's length divided by 1.3, the "shrink factor" ❶, and successive gaps will always be 1.3 times smaller ❷. When the gap becomes 1 ❸, just apply bubble sort, and you're done. While the gap is greater than 1 ❹, you do what's essentially the central logic of bubble sort, but instead of comparing elements one place apart, you compare elements gap places apart.

Comb sort usually performs better than bubble sort, but it's still $O(n^2)$ in the worst case and becomes $O(n \log n)$ in the best case. However, that's not why we're considering this idea; rather, the concept of sorting elements that are far apart provides real benefits, and you'll see that Shell sort that does exactly that in a way similar to comb sort.

Shell Sort

To understand how Shell sort works, assume you want to order the array shown in Figure 6-7.

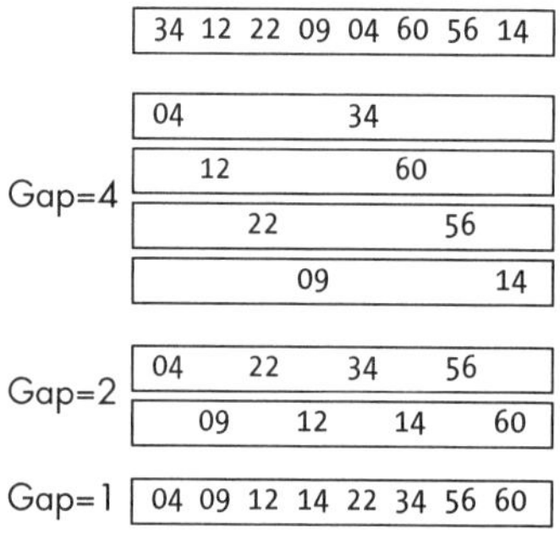

```
34 12 22 09 04 60 56 14
```
```
        04              34
        12              60
Gap=4      22              56
              09              14
```
```
        04   22   34   56
Gap=2   09   12   14   60
```
```
Gap=1   04 09 12 14 22 34 56 60
```

Figure 6-7: Shell sort works similarly to insertion sort, but with larger gaps.

In the first pass, do an insertion sort, but for elements set four places apart, which leads to an array consisting of four short-ordered sequences. Then lower the gap size to 2 and repeat the sort. The array now consists of two ordered sequences. Eventually, you reach a gap size of 1, and in that case, you're just doing an insertion sort, but because of the previous partial sorts, it doesn't do as many comparisons or swaps as with the normal algorithm, which is the advantage of Shell sort.

Here's the Shell sort implementation:

```
const shellSort = (arr, from = 0, to = arr.length - 1) => {
❶ const gaps = [1]; // Knuth, 1973
  while (gaps[0] < (to - from) / 3) {
    gaps.unshift(gaps[0] * 3 + 1);
  }

❷ gaps.forEach((gap) => {
  ❸ for (let i = from + gap; i <= to; i++) {
      const temp = arr[i];
      let j;
```

```
❹     for (j = i; j >= from + gap && arr[j - gap] > temp; j -= gap) {
         arr[j] = arr[j - gap];
      }
      arr[j] = temp;
   }
  });

  return arr;
};
```

First select what gaps to use ❶, keeping in mind that the last one to be applied must be 1. You'll find many suggestions online as to which sequence to use, but this example will use Knuth's proposal (1, 4, 13, 40, 121, . . . , with each term being triple the previous one, plus 1), which leads to an $O(n^{1.5})$ algorithm. Then, you take gaps in decreasing order ❷ and essentially do an insertion sort ❸ but for elements gap spaces apart ❹. With larger gaps, you're ordering sequences of fewer elements, but as you decrease the gap size, you deal with longer sequences that tend to be almost in order, so insertion sort behaves well.

Going for Speed with Quicksort

Let's move on to the quicker algorithms that achieve the $O(n \log n)$ theoretical speed limit—albeit with a problematic worst-case quadratic performance! *Quicksort* (also known as *partition-exchange sort*) was created by Tony Hoare in the 1960s and is a divide-and-conquer algorithm with high speed. We'll consider the standard version first and then discuss some possible enhancements.

Standard Version

How does quicksort work? The idea is first to select a "pivot" element from the array to be sorted and redistribute all the other elements in two subarrays, according to whether they are smaller or larger than the pivot. The array ends with lower values first, then the pivot, and then higher values. Then, each subarray is sorted recursively, and when that's done, the whole array is sorted (see Figure 6-8).

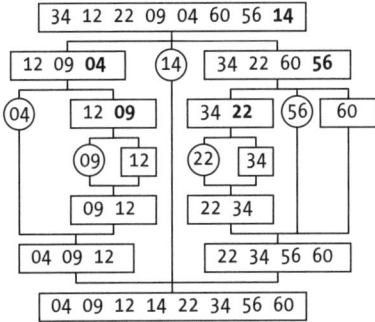

Figure 6-8: Quicksort works by partitioning arrays and recursively sorting the parts.

Let's say we always choose the rightmost element of the array as the pivot. (It won't prove to be a very wise option, as you'll see.) In this case, the first choice is 14, and you rearrange the array so all values less than 14 come first, then 14 itself, and finally all values greater than 14. The same procedure (select the pivot, rearrange, and sort recursively) is applied to each subarray until the whole array is sorted.

Here's a direct implementation of the procedure:

```
const quickSort = (arr, left = 0, right = arr.length - 1) => {
❶ if (left < right) {
  ❷ const pivot = arr[right];

  ❸ let p = left;
    for (let j = left; j < right; j++) {
      if (pivot > arr[j]) {
        [arr[p], arr[j]] = [arr[j], arr[p]];
        p++;
      }
    }
  ❹ [arr[p], arr[right]] = [arr[right], arr[p]];

    // Recursively sort the two partitions
  ❺ quickSort(arr, left, p - 1);
    quickSort(arr, p + 1, right);
  }

  return arr;
};
```

First, check whether there's actually anything to sort; if the left pointer is equal to or greater than the right one, you're done ❶. The rightmost element will be the pivot ❷. Next, go through the array from left to right ❸ in a fashion reminiscent of the insertion sort, exchanging elements if needed so smaller elements move to the left, greater ones to the right, and the pivot ends at position p ❹. It would be a good idea to simulate a run of the pivoting code by hand. Despite its short length, it's a bit tricky to get right. (What happens if the pivot value appears several times in the array? See question 6.10.) Finally, apply recursion to sort the two partitions ❺.

Analysis shows that *on average*, quicksort works in $O(n \log n)$ time. However, the worst case is easy to find. Consider sorting an already sorted (in ascending or descending order) array. Examining the code shows that partitioning will always end with just one subarray, and you'll have the equivalent of a selection sort or bubble sort, which means performance goes down to $O(n^2)$. But you can fix that.

Pivot Selection Techniques

How you choose the pivot can have a serious impact on quicksort's performance. In particular, if you always choose the largest (or smallest) element in the array, you'll get a negative hit in speed, so consider some alternative pivot-selecting techniques.

The first solution to avoid problems with sorted arrays is to choose the pivot randomly. Select a random position between left and right inclusive and, if needed, swap the selected element to move it to the rightmost position, so you can go on with the rest of the algorithm without any further changes:

```
❶ const iPivot = Math.floor(left + (right + 1 - left) * Math.random());
❷ if (iPivot !== right) {
    [arr[iPivot], arr[right]] = [arr[right], arr[iPivot]];
  }
```

We'll look at random selection in more detail in Chapter 8, but the way you calculate iPivot (the position of the pivot) ❶ selects a value from left to right inclusive with equal odds. The rest of the sorting algorithm assumes that the chosen pivot was at the right of the array, so if the chosen pivot is elsewhere ❷, just do a swap.

This random selection solves the worst-case behavior for almost-sorted arrays, but there's still the (assuredly low) probability that you'll always just happen to pick the highest or lowest value in the array to be sorted, and in that case, performance will suffer.

What's the ideal pivot? Choosing the array's median (the value that splits the array in two) would be optimum. A rule that comes close is called the *median of three*: choose the median of the left, middle, and right elements of the array:

```
const middle = Math.floor((left + right) / 2);
if (arr[left] > arr[middle]) {
  [arr[left], arr[middle]] = [arr[middle], arr[left]];
}
if (arr[left] > arr[right]) {
  [arr[left], arr[right]] = [arr[right], arr[left]];
}
if (arr[right] > arr[middle]) {
  [arr[right], arr[middle]] = [arr[middle], arr[right]];
}
```

Testing this code with all possible permutations of three values shows that arr[right] always ends with the middle value. Even better, you might pick the "ninther," defined as a "median of medians": divide the array in three parts, apply the median of three to each third, and then take the median of those three values.

You can help quicksort become faster by selecting pivots carefully, but you can enhance it even further.

Hybrid Version

Quicksort is fast, but all the pivots and recursion have an impact on running times, so for small arrays, a combination of simpler algorithms may actually perform faster. You can apply a *hybrid algorithm* that uses two distinct methods together. For instance, you may find that for arrays under a certain cutoff limit, an insertion sort performs better, so whenever you want to sort an array smaller than the limit, switch to that algorithm:

```
const CUTOFF = 7;

const quickSort = (arr, left = 0, right = arr.length - 1) => {
  if (left < right) {
    if (right - left < CUTOFF) {
      insertionSort(arr, left, right);
    } else {
      //
      // quicksort as before
      //
    }
  }

  return arr;
};
```

The lines in bold are all you need to change. Define the cutoff limit, and when sorting, if the array is small enough, apply the alternative sort.

Dual-Pivot Version

You can extend the idea of splitting an array to be sorted in two parts, separated by a pivot, to splitting the array in three parts, separated by two pivots. This dual-pivot version is usually faster. (Java uses it as its default sorting algorithm for primitive types.) Choose the leftmost and rightmost elements as pivots, as shown in Figure 6-9.

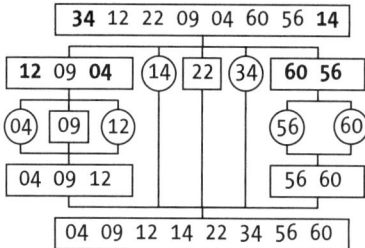

Figure 6-9: Dual-pivot sort is like quicksort, but it splits the array in three parts instead of two.

Start by choosing 34 and 14 as pivots, and rearrange the array so that all values less than 14 (12, 9, 4) come first, then 14 itself, then values between 14 and 34 (just 22), then 34, and finally values greater than 34 (60, 56). Each subarray is then sorted again with the same method.

The algorithm is similar to a basic quicksort; the main differences are in the selection of pivots and partitioning. For performance reasons, you'll use the hybrid approach and turn to an insertion sort if the array to be sorted is small enough; for example:

```
const dualPivot = (arr, left = 0, right = arr.length - 1) => {
  if (left < right) {
    if (right - left < CUTOFF) {
      insertionSort(arr, left, right);
    } else {
      // Choose outermost elements as pivots.
❶    if (arr[left] > arr[right]) {
        [arr[left], arr[right]] = [arr[right], arr[left]];
      }
      const pivotLeft = arr[left];
      const pivotRight = arr[right];

      let ll = left + 1;
      let rr = right - 1;
❷    for (let mm = ll; mm <= rr; mm++) {
❸      if (pivotLeft > arr[mm]) {
          [arr[mm], arr[ll]] = [arr[ll], arr[mm]];
          ll++;
❹      } else if (arr[mm] > pivotRight) {
          while (arr[rr] > pivotRight && mm < rr) {
            rr--;
          }
          [arr[mm], arr[rr]] = [arr[rr], arr[mm]];
          rr--;

          if (pivotLeft > arr[mm]) {
            [arr[mm], arr[ll]] = [arr[ll], arr[mm]];
            ll++;
          }
        }
      }
❺    ll--;
      rr++;
      [arr[left], arr[ll]] = [arr[ll], arr[left]];
      [arr[right], arr[rr]] = [arr[rr], arr[right]];

❻    dualPivot(arr, left, ll - 1);
      dualPivot(arr, ll + 1, rr - 1);
      dualPivot(arr, rr + 1, right);
    }
  }

  return arr;
};
```

You're choosing the leftmost and rightmost elements as pivots, but, of course, you could take any two values and swap them so they end up in the extremes of the array, with the smaller on the left ❶. (Actually, when dealing with arrays nearly in order, choosing two middle elements is better.) Next, you start swapping elements, maintaining these invariants:

- pivotLeft is at the left of the array.
- From positions left + 1 to ll - 1, all values are less than pivotLeft.

- From positions ll to mm - 1, all values are strictly between `pivotLeft` and `pivotRight`.

- From positions mm to rr, the status of values is yet unknown.

- From positions rr + 1 to right - 1, the values are greater than `pivotRight`.

- `pivotRight` is at the right of the array.

You can establish this invariant from the beginning by setting `mm` to `left + 1` and making it go up until it reaches the end of the array ❷. If the element at `mm` is less than `pivotLeft` ❸, a mere swap maintains the invariant. If the element at `mm` is greater than `pivotRight` ❹, you have to do a bit more work to maintain the invariant, moving `rr` to the left. (Remember, the idea is to keep the invariants; this loop ensures the next-to-last one.) After the loop is done ❺, swap the pivots to their final places and apply recursion to sort the three partitions ❻.

Quicksort is a great algorithm with several variants, but it always comes with the possibility (albeit remote) of bad performance.

Merging for Performance with Merge Sort

We'll wrap up our study of comparison-based sorts with the *merge sort* algorithm that guarantees a constant performance, but with the cost of a higher need for memory. Merge sort basically does all sorting by merging. If you have two ordered sequences of values, n in total, merging them into a single-order sequence can be done in an $O(n)$ process. The key idea of a merge sort is to apply recursion. First, split the array to be sorted into two halves, then recursively sort each half, and finally merge both ordered halves into a single sequence (see Figure 6-10).

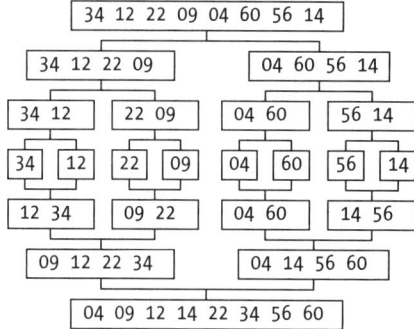

Figure 6-10: Merge sort splits the array into two parts, sorts them, and ends by merging them.

Each array to be sorted is split into two parts, which are sorted and then joined back together. To sort an 8-element array, you need to sort two 4-element arrays, which means you have to sort four 2-element arrays, and that requires sorting eight 1-element arrays. Sorting the latter is trivial (nothing to do), and doing the merge reconstructs the original array.

Here's a recursive implementation:

```
const mergesort = (arr, left = 0, right = arr.length - 1) => {
❶ if (right > left) {
  ❷ const split = Math.floor((left + right) / 2);

  ❸ const arrL = mergesort(arr.slice(left, split + 1));
     const arrR = mergesort(arr.slice(split + 1, right + 1));

  ❹ let ll = 0;
     let rr = 0;
     for (let i = left; i <= right; i++) {
       if (
         ll !== arrL.length &&
         (rr === arrR.length || !arrR[ll] > arrL[rr])
       ) {
         arr[i] = arrL[ll];
         ll++;
       } else {
         arr[i] = arrR[rr];
         rr++;
       }
     }
  }

  return arr;
};
```

First, check whether you even need to sort ❶, which could include a hybrid approach, and if the array is small enough, you'd apply some other method, not merge sort. Then split the array in half ❷ and recursively sort each half ❸. Next, merge both sorted arrays ❹: ll and rr will traverse each array, and the output will go into the original array. Finally, return the sorted array.

Merge sort has very good performance (despite the extra space needed to perform the merge), and it's actually the basis of *Tim sort*, a stable adaptive method that's widely used. Java utilizes it, JavaScript also applies it in the V8 engine, and other languages use it as well. We won't delve into the actual implementation, as the algorithm is quite longer than the ones we've been considering (a couple of implementations in GitHub run to almost 1,000 lines each). Tim sort takes advantage of runs of elements that are already in order, merging shorter runs to create longer ones, and applying an insertion sort to make sure runs are long enough. You've already studied all the pieces that make up the complete Tim sort algorithm.

NOTE *There's more to learn about comparison-based sorting methods, but we'll postpone considering more algorithms until we've seen some data structures. In Chapter 14 we'll explore priority queues and heaps. Likewise, in Chapter 12 we'll study binary trees and, in particular, binary search trees. By adding all elements to be sorted into such a structure, you can traverse it in order later, thus producing another sort, although the performance and relative complexity of that solution don't make it very attractive. Binary search trees are more oriented toward searching; sorting is just a* *by-product. In the same way, other structures such as skip lists (which we'll analyze in Chapter 11) could also provide a sorting method, but as with binary search trees, sorting isn't the intended goal.*

Sorting Without Comparisons

In the previous section, all of the sorting algorithms depended on comparing keys and using that information to move, swap, or partition values. But there are other ways to sort. As an example, imagine you're in charge of customer assistance and receive email for many different reasons. How can you simplify the classification task? You could use a different email address for each category so that messages are automatically sorted into the correct bins for processing.

This simple solution provides a glimmer of what we're going to do. Basically, you won't compare keys; instead, you'll use their values to figure out where they should go in the final, ordered list. It's not always possible, but if you can apply the methods here, performance becomes $O(n)$, which is impossible to beat. After all, no algorithm can sort n values without at least looking at them once, and that's already an $O(n)$ process. In this section, we'll consider a couple of methods, *bitmap sort* and *counting sort*, and we'll also look at a very old sorting method, *radix sort*, whose origins are on par with tabulating machines that used punched cards to do census work.

Bitmap Sort

Let's start with a sorting method that has excellent performance but some limitations, if you can live with them. We have to make three suppositions. First, you're going to sort only numbers on their own (no key + data). Second, you know the possible range of numbers, and it's not very big. (For instance, if all you knew was that they were 64-bit numbers, the range from lowest to highest numbers would make you forget about attempting this algorithm.) And, third, the numbers are never going to be duplicated; all numbers to sort will be different.

With these (too many) restrictions in mind, you can easily use a bitmap. Assume you are starting with all bits turned off, and whenever you read a number, set that bit to on. After you're done, go through the bits in order, and whenever a bit is set, output the corresponding number, and you're done (see Figure 6-11).

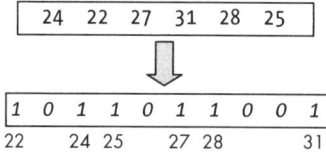

Figure 6-11: Bitmap sort takes advantage of knowing the range of values to be sorted.

You must go through all the numbers to find the minimum and maximum values to define the size of the bitmap. After that, go through the numbers again, setting bits whenever a number appears. In Figure 6-11, bits corresponding to numbers 22, 24, 25, 27, 28, and 31 are set. (JavaScript mandates that all arrays start at position 0, so you have to remember that position 0 actually corresponds to key 22, position 1 to key 23, and so on.) Finally, go through the bitmap, outputting the numbers whose bits are set; it's simple.

This algorithm is limited, but it's the basis for a different, enhanced algorithm. For simplicity, this example will use an array of booleans instead of a bitmap and write the following code:

```
const bitmapSort = (arr, from = 0, to = arr.length - 1) => {
❶ const copy = arr.slice(from, to + 1);
❷ const minKey = Math.min(...copy);
  const maxKey = Math.max(...copy);

❸ const bitmap = new Array(maxKey - minKey + 1).fill(false);
❹ copy.forEach((v) => {
  ❺ if (bitmap[v - minKey]) {
      throw new Error("Cannot sort... duplicate values");
  ❻ } else {
      bitmap[v - minKey] = true;
    }
  });

❼ let k = from;
  bitmap.forEach((v, i) => {
  ❽ if (v) {
      arr[k] = i + minKey;
      k++;
    }
  });

  return arr;
};
```

First make a copy of the input array ❶ to simplify the next step, which is determining the minimum and maximum keys ❷. (This could be done in a single loop a tad more efficiently.) Then create a bitmap array of the right length ❸, but in reality you'll be using common booleans, not bits. You need to be careful with indices, because JavaScript's arrays always start at zero; a bit of index math will be needed to relate keys to array positions. Then go through the input array ❹ and check whether the key already appeared. If so ❺, there's a problem. If not ❻, just mark that the number did appear. Finally, go through the bitmap ❼, and whenever you find a set flag ❽, output the corresponding number.

Not being able to allow for duplicate keys is a serious limitation, and dealing with numbers only is another; you need to be able to sort elements consisting of a key + data, as in all the other algorithms you've explored so far.

Counting Sort

The previous sort is quite effective but applies in only limited cases. You can make improvements by calculating where each sorted element should go. To do that, you need to count how many times each key appears and then use that information to decide where to place sorted elements in the output array (see Figure 6-12).

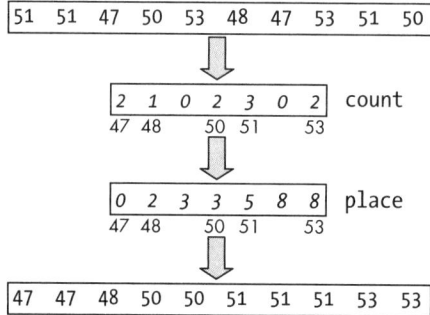

Figure 6-12: Count sort is somewhat similar to bitmap sort, but it can deal with repeated keys.

In the same way as with bitmap sort, you need to find the minimum and maximum values in the array to be sorted and set up an appropriate array with counters, all initialized to zero. (Again, remember that position 0 corresponds to the minimum key, which is 47 in this case; position 1 corresponds to 48, and so on.) Then go through the array again, incrementing the corresponding counters. After you have all the counts, you can follow an easy procedure to determine where each key goes. For instance, elements with the minimum key (47) start at position 0 of the output array; elements with the next key (48) follow two places later (because there were two 47s) at place 2. Each new key is placed to the right of the previous key, leaving as many empty spaces as needed to place all the previous elements.

The implementation for this algorithm follows:

```
const countingSort = (arr, from = 0, to = arr.length - 1) => {
❶ const copy = arr.slice(from, to + 1);
  const minKey = Math.min(...copy);
  const maxKey = Math.max(...copy);

❷ const count = new Array(maxKey - minKey + 1).fill(0);
❸ copy.forEach((v) => count[v - minKey]++);

❹ const place = new Array(maxKey - minKey + 1).fill(0);
❺ place.forEach((v, i) => {
    place[i] = i === 0 ? from : place[i - 1] + count[i - 1];
  });

❻ copy.forEach((v) => {
    arr[place[v - minKey]] = v;
```

```
❼ place[v - minKey]++;
  });

  return arr;
};
```

The first three lines of this algorithm are the same as the bitmap sort ❶, and you create a copy of the input array and determine the minimum and maximum keys. You then create an array with the counts for all keys (initialized to zero and needing the same kind of index math as in bitmap sort ❷). Then go through the input data ❸ and increment counts for each key value. Now generate a new array ❹ to calculate the starting place for elements with each key. The minimum key starts at position 0, and each key is a few spaces away from the previous one, according to the count of the previous key ❺. (For example, if the previous count was 5, you'll have the new key 5 places away from the first occurrence of the previous key.) Finally, use the place array to start positioning sorted elements in their right places ❻; each time an element goes into the output array, the corresponding place is incremented by 1 ❼ for the next element with the same key.

Radix Sort

The last sorting algorithm in this chapter is probably the oldest. It was used with Hollerith punch cards (see Figure 6-13) when tabulating census data, back in the days when IBM was founded.

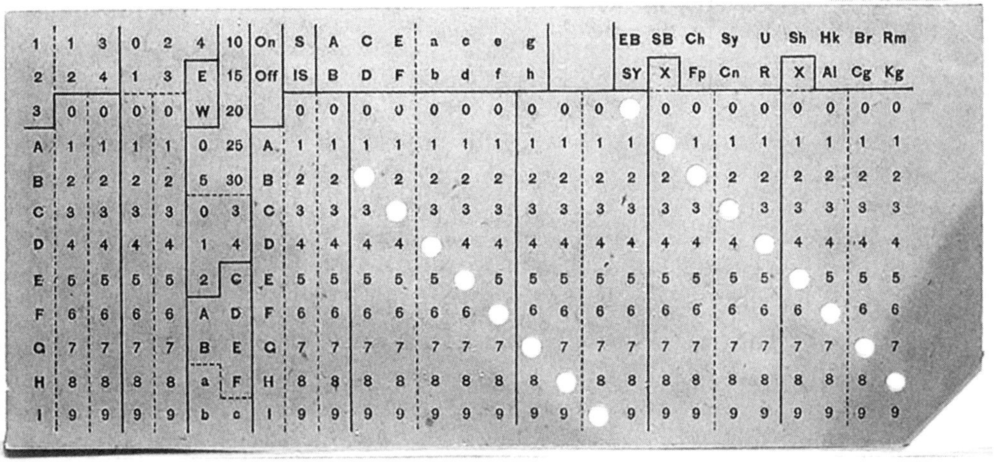

Figure 6-13: An original Hollerith card (public domain)

Suppose you have a disordered set of punch cards, numbered in columns 1 to 6, and you want to sort them. Using a *classifier*, a machine that processes cards and separates them into bins according to the value on a specific column, you would follow these steps:

1. Separate cards into bins according to column 6 and choose cards with a 0, then cards with a 1, and so on, finishing with cards with a 9. You have sorted the cards by the sixth column, but you have to keep working.

2. Redo the same process, but use column 5. When you pick the cards up, you'll find that they are sorted by two columns (refer back to the "Sorting Stability" section on page 93 to understand why).

3. Do the process again for columns 4, 3, 2, and 1, in that order, and you'll end up with a totally sorted deck of cards.

You'll explore this algorithm in more detail in Chapter 10 when looking at lists, which will be the way you'll emulate the bins.

Inefficient Sorting Algorithms

We'll finish on a not-too-serious note by considering some algorithms that are really inefficient, going from bad to worse. These algorithms are not intended for actual use!

Stooge Sort

The name of this algorithm comes from the Three Stooges comedy group, and if you're familiar with them, its inefficiency will remind you of their antics. The process to sort a list starts by comparing its first and last elements and swapping them (if needed) to ensure the greater one is at the end. Next, it recursively applies Stooge sort to the initial two-thirds of the list, then it sorts the last two-thirds of the list (which ensures that the last third will have the greatest values, in order), and finally, it sorts the first two-thirds of the list again. The number of comparisons needed for n elements satisfies $C(n) = 3C(2n / 3) + 1$, so the algorithm has a complexity of $O(n^{2.71})$, which makes it perform worse than bubble sort, but there's even worse.

Slow Sort

This algorithm was designed as a joke. Rather than divide and conquer, it's based on "multiply and surrender." The authors were proud to have found an algorithm worse than any that were previously created. To sort an array with two or more elements, the algorithm first splits it in half, and then it uses recursion to sort each half. Finally, it compares the last element of each half and places it (swapping if needed) at the end of the original array. After doing that, the algorithm proceeds to sort the list with the maximum extracted. The number of comparisons for this algorithm satisfies $C(n) = 2C(n / 2) + C(n - 1) + 1$, and its time is $O(n^{\log n})$. It's not even polynomial!

Permutation Sort

In Chapter 5, you saw how to go forward from one permutation of values to the following one, which suggests an even worse algorithm for sorting a sequence: repeatedly attempting to produce the next permutation of the elements until the algorithm fails because the last permutation was reached and then reversing the sequence. For a random order, this algorithm requires testing on average $n!\ /\ 2$ permutations, which means its time is at least factorial. For almost any size, the algorithm becomes impossible to run because of its running time.

Bogosort

The last algorithm derives its name from a portmanteau of the words *bogus* and *sort*, and it's a probabilistic algorithm that sorts its input with probability 1, but without any certainty as to its running time. The idea also has to do with permutations: if the list to be sorted isn't in order, it shuffles its elements randomly (we'll look at such algorithms in Chapter 8) and tests again. If you were to apply this method to sorting a deck of cards, the logic would be as follows: if the cards are not in order, throw them into the air, pick them up, and check again—the odds of getting it right are $1/52!$, so roughly around one in a hundred million million million million million million million million million million million. Not good!

Sleep Sort

The last sort is specifically meant for JavaScript, and its running time depends on the maximum key to be sorted. It works with numeric keys, and the idea is that if an input key is K, wait K seconds and output its value. After enough time has passed, all values will be output in order:

```
const sleepSort = (arr) =>
  arr.forEach((v) => setTimeout(() => console.log(v), v * 1000));
```

Even if this algorithm seems to work, with a sufficiently large dataset, it may crash (too many timeouts waiting) or fail. The algorithm goes through the list and starts to output numbers—think of processing a list such as 1, 2, 2, 2, . . . , 2, 2, 0, and with enough 2s, the initial 1 may be output before the last 0 is processed.

Summary

In this chapter we've explored several sorting algorithms with different performance levels. In the next chapter, we'll touch on a similar subject, the selection problem, which is akin to sorting only part of an array, because instead of getting all elements in order in their proper place, you care only about placing a single element in its final place, not necessarily sorting the whole list.

Questions

6.1 Forced Reversal

Suppose you want to order a set of numbers in descending order, but you have a sorting function that sorts only in ascending order with no options whatsoever to change how it works. How can you manage to sort your data as you wish?

6.2 Only Lower

Suppose you had a boolean function `lower(a,b)` that returns true if a is lower in sorting order than b and false otherwise. How can you use it to decide whether a is higher in sorting order than b? And how can you use it to see whether both keys are equal in order?

6.3 Testing a Sort Algorithm

Imagine you're trying out a new sorting algorithm of your own. How would you test that it actually sorted correctly?

6.4 Missing ID

Imagine you got a set of six-digit IDs, but the count is under 1,000,000, so at least one ID is missing. How can you find it?

6.5 Unmatched One

Say you have an array with transaction numbers, and each number should appear twice somewhere in the array, but you know there was a mistake, and there's a single transaction that appears only once. How do you detect it?

6.6 Sinking Sort

This is a variant of bubble sort. Instead of starting at the bottom of the array and making higher values bubble to the top, sinking sort starts at the top of the array and makes smaller values sink to the bottom. In terms of performance, it's the same as bubble sort, but it may be used if you want to find only the k lowest elements of the array, as you'll see in Chapter 7. Can you implement sinking sort?

6.7 Bubble Swap Checking

Add a test to bubble sort after each pass through the array to exit earlier if no swaps were detected. This test will speed things up if you deal with arrays that were practically in order and just a few swap passes get everything in its place.

6.8 Inserting Recursively

Can you implement insertion sort in a recursive way?

6.9 Stable Shell?

Is Shell sort stable?

6.10 A Dutch Enhancement

The Dutch National Flag Problem requires you to arrange an array with elements that are either red, white, or blue, so all red elements come first, followed by all white ones, and finishing with all blue ones, as in the Dutch national flag. Show how you may similarly enhance quicksort's performance with repeated elements by rearranging the array to be sorted into three parts: all elements less than the pivot, all elements equal to the pivot, and all elements greater than the pivot. The middle part won't need any further sorting.

6.11 Simpler Merging?

When merging halves in merge sort, you wrote the following (look specifically at the text in bold):

```
for (let i = left; i <= right; i++) {
  if (ll !== arrL.length && (rr === arrR.length || !arrR[ll] > arrL[rr]))
{
    ...
  } else {
    ...
  }
}
```

Why is it written that way? You always want to compare using the greater-than operator to be able to easily substitute a function for more complex comparisons, but why not write arr[rr]>arr[ll] instead?

6.12 Try Not to Be Negative

What happens with radix sort if some numbers are negative? Also, what happens if you have noninteger values? Can you do something about this?

6.13 Fill It Up!

In radix sort, imagine you wanted to initialize the buckets array with 10 empty arrays, and you did it as follows:

```
const buckets = Array(10).fill(0).map(() => []);
```

Why wouldn't the following alternative work?

```
const buckets = Array(10).fill([])
```

And what about this other possibility?

```
const buckets = Array(10).map(() => [])
```

6.14 What About Letters?

How would you modify radix sort to work with alphabetical strings?

7

SELECTING

In the previous chapter, we looked at the sorting problem, and here we'll consider a related problem with many similar algorithms: *selection*. The basic situation is given a number k and an array with n items, we want to find the value at the array's kth place if we ordered the array. But we don't actually need the array to be sorted; we just need to know its kth element. Unlike the sorting problem, JavaScript doesn't provide a "ready-made" solution for selection, so if you're in need of this kind of function, you'll have to use some of the algorithms in this chapter.

The way this problem relates to sorting is simple: if you just sort the list of values (using any of the algorithms in the previous chapter), you can quickly produce the kth value of the sorted list for all possible values of k; you just look at the kth place in the sorted array. That would be a good solution if you actually needed to make many selections from the same array; an $O(n \log n)$ sort followed by many $O(1)$ selections. However, there's no requirement to actually sort the list, and we'll try to avoid doing that. The selection algorithms that we'll explore in this chapter perform better than sorting algorithms because they don't need to sort everything.

In the selection problem, if you ask for $k = 1$, you're asking for the minimum of the list; $k = n$ asks for the maximum, and $k = n/2$ asks for the median. Keep in mind that in "real life" k goes from 1 to n, but because of JavaScript's 0-based arrays, k goes from 0 to one less than the array's length.

NOTE   *Formally, if the list of values is of even length, the definition of median would ask for the average of the two center values of the sorted array, but we're not doing that. In order for your selection code to produce the median of arrays with even length, you'd need to call the selection algorithm twice to get the two center values and only then calculate their mean. We'll just deal with the problem of finding the value at any given position.*

Selection Without Comparisons

In the same way you could implement sorting without comparisons (meaning you never have to compare one key with another), you can use variations of the bitmap and counting sorting methods to find the kth value of a list quickly, without even attempting a partial sort of the data. Remember that these algorithms are limited; they work only for numbers (not key + data of any kind) and preferably numbers in a not very extensive range.

Bitmap Selection

The bitmap sort worked by reading all data and setting bits on in a bitmap; after that, outputting the sorted numbers just required walking through the bitmap. You'll do the same here, except you won't output all numbers; you need only the kth value in the bitmap. Figure 7-1 shows the method; assume you want to find the 4th element in the same array used as an example in Chapter 6.

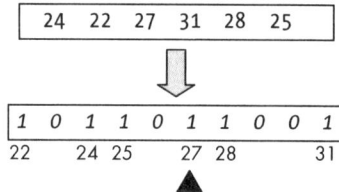

Figure 7-1: A variation of bitmap sort allows a fast selection algorithm.

First generate the bitmap and then traverse it looking for the 4th set bit, which in this case corresponds to 27.

The code is as follows:

```
❶ const bitmapSelect = (arr, k, from = 0, to = arr.length - 1) => {
❷   const copy = arr.slice(from, to + 1);
    const minKey = Math.min(...copy);
    const maxKey = Math.max(...copy);

    const bitmap = new Array(maxKey - minKey + 1).fill(false);
    copy.forEach((v) => {
      if (bitmap[v - minKey]) {
        throw new Error("Cannot select... duplicate values");
      } else {
        bitmap[v - minKey] = true;
      }
    });

❸   for (let i = minKey, j = from; i <= maxKey; i++) {
❹     if (bitmap[i - minKey]) {
❺       if (j === k) {
❻         return i;
        }
❼       j++;
      }
    }
};
```

The parameters for this algorithm are the same as when sorting ❶ with the addition of k, the place of interest. The logic to create the bitmap ❷ is exactly the same as for sorting; the only difference comes in the final output ❸. Set a counter j to the first position in the array, and every time you find a set bit ❹, test whether j reached the desired place at k ❺; if so, you're done ❻. Otherwise, keep looping, after counting one more found number ❼.

This algorithm is obviously $O(n)$, and if it weren't for the limitations mentioned earlier, it would be one of the best for solving the selection problem.

Counting Selection

Under the same circumstances as for bitmap sort, in Chapter 6 we considered the counting sort, which didn't have issues if numbers were repeated in the input. This situation, however, was a problem when using a bitmap.

You can apply the same kind of solution here: go through the array, generate the list of counts, and finish by going through the counts from left to right until you find what value is at the kth place.

Consider an example using the same numbers from Chapter 6 (see Figure 7-2); you want to find the value at the 4th place in the array.

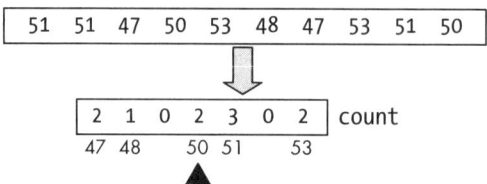

Figure 7-2: Counting sort also provides a simple selection algorithm.

First find all the counts and then sum through them, from left to right, until the sum equals or exceeds 4; in this case, that happens at value 50 when the accumulated sum goes from 3 to 5. (Keep in mind the case where the sum exceeds k because of repeated values in the input array.)

Here's the logic:

```
❶ const countingSelect = (arr, k, from = 0, to = arr.length - 1) => {
❷ const copy = arr.slice(from, to + 1);
  const minKey = Math.min(...copy);
  const maxKey = Math.max(...copy);

  const count = new Array(maxKey - minKey + 1).fill(0);
  copy.forEach((v) => count[v - minKey]++);

❸ for (let i = minKey, j = from; i <= maxKey; i++) {
  ❹ if (count[i - minKey]) {
    ❺ j += count[i - minKey];
    ❻ if (j > k) {
        return i;
      }
    }
  }
};
```

The parameters are the same as earlier ❶, and all the logic to generate the counts ❷ is the same as in Chapter 6. The changes appear when preparing output. First initialize a counter j at the first position of the input array ❸, and every time you find a nonzero count ❹, update the counter ❺ and see whether you reached or passed k with that sum. If so, return the corresponding value ❻; otherwise, just keep looping.

Again, we have an $O(n)$ algorithm, but we want to be able to handle more general conditions, so let's move on to selection algorithms based on key-to-key comparisons that will work in every case.

Selecting with Comparisons

Most algorithms for the selection problem are based on sorting algorithms. The first one we'll explore is based on selection sort, but we won't sort the whole array—just its first k values. Selection sort works by finding the minimum of the array and exchanging it with the value at the first place; then it looks for the minimum of the remaining values and exchanges it with the value at the second place, and so on, until the whole array is sorted. We'll do the same, but stop after finding the *k*th minimum:

```
❶ const sortingSelect = (arr, k, from = 0, to = arr.length - 1) => {
❷   for (let i = from; i <= k; i++) {
      let m = i;
      for (let j = i + 1; j <= to; j++) {
        if (arr[m] > arr[j]) {
          m = j;
        }
      }
      if (m !== i) {
        [arr[i], arr[m]] = [arr[m], arr[i]];
      }
    }

❸   return arr[k];
  };
```

The parameters for this algorithm are the same as before ❶. We made a small change in the loop. When sorting, you went through the entire array, but now you'll stop after having reached the *k*th place ❷. The rest of the logic is exactly the same as for the sorting algorithm, except that you return the desired value instead of the sorted array ❸.

The performance of this algorithm is $O(kn)$, which is an efficient result for low values of k and an asymptotically bad one if k grows and is proportional to n. (See question 7.3 for a unique case.) In particular, if you want to find the middle element of the array, then $k = n / 2$ and performance becomes $O(n^2)$; you'll do better with different algorithms.

The Quickselect Family

Many selection algorithms are derived from the quicksort code, in particular, the way it partitions an array in relation to a pivot, moving values around so that the array ends up consisting of values lower than the pivot on one side, then the pivot itself, and values greater than the pivot on the other side. In the case of quicksort, after partitioning the array this way, the algorithm continues recursively by sorting each of the two parts; in this case you'll continue the search in only one of the parts. See Figure 7-3 for an example where you want to find the 6th element of the array.

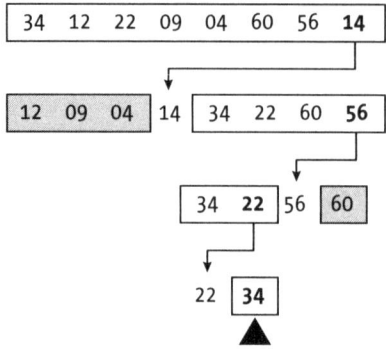

Figure 7-3: The pivot technique used in quicksort provides a selection algorithm.

You'll use the same first pivot scheme as for quicksort and choose the rightmost value (14) as the pivot. After repartitioning the array around 14, the pivot ends at the 4th place in the array. You want the 6th element, so continue searching on the right side of the pivot. There you choose 56 as the pivot, and after repartitioning, 56 ends at the 7th place in the array. That comes after the place you want, so continue searching the left part. You then choose 22 as the pivot. It ends in the 5th place, and you continue searching the right side, which now consists of a single element, so you know for sure 34 is the 6th value in the array. To the left of 34 there are lower values (but not necessarily ordered), and to the right there are greater values.

As mentioned in Chapter 6, the performance of quicksort on average is $O(n \log n)$, but in the worst case, it becomes $O(n^2)$. Quickselect's average performance has been proven to be $O(n)$, but it could become $O(n^2)$ if you generally make unlucky pivot selections, so rather than study a single algorithm, we'll consider a whole family of them by varying how we choose the pivot.

Quickselect

Let's start with the basic logic. As in Chapter 6, assume single-field keys that can be compared with the < and > operators. Always write tests as a > b, so adapting the code for a more generic comparison would just require writing compare(a,b) > 0, assuming a user-provided compare(x,y) function that returns a positive value if x is greater than y.

The following code implements the basic structure of the quickselect family; the pivot selection part is in bold, and we'll make changes to that section to get other, enhanced versions of the selection function:

```
❶ const quickSelect = (arr, k, left = 0, right = arr.length - 1) => {
    if (left < right) {
      const pick = left + Math.floor((right + 1 - left) * Math.random());
      if (pick !== right) {
        [arr[pick], arr[right]] = [arr[right], arr[pick]];
      }
```

```
    const pivot = arr[right];

    let p = left;
    for (let j = left; j < right; j++) {
      if (pivot > arr[j]) {
        [arr[p], arr[j]] = [arr[j], arr[p]];
        p++;
      }
    }
❷ [arr[p], arr[right]] = [arr[right], arr[p]];

❸ if (p === k) {
    return;
❹ } else if (p > k) {
    return quickSelect(arr, k, left, p - 1);
❺ } else {
    return quickSelect(arr, k, p + 1, right);
  }
  }
};
```

The parameters for quickselect ❶ are the same as for selection sort and all the algorithms in this chapter. The start of this algorithm is exactly like quicksort's, with the option of using a random choice for the pivot, up to and including how you split the array, having the chosen pivot end at position p ❷. The only difference is how to proceed after that. If the pivot ends in the kth position ❸, you're done, because that's the value you want. Otherwise, use recursion to examine the left ❹ or right ❺ partition, whichever includes the kth position. (Actually, you don't need to use recursion; see question 7.4.)

As is, quickselect reorders (partitions) the input array to ensure the element in the kth place isn't lower than any element before it or greater than any element after it. You can easily get the value itself by writing an auxiliary function:

```
const qSelect = (arr, k, left = 0, right = arr.length - 1) => {
❶ quickSelect(arr, k, left, right);
❷ return arr[k];
};
```

Use quickselect to repartition the array ❶, and then return the value at the desired position ❷. (See question 7.5 for a simple modification.) On average, this algorithm can be shown to be linear, but if it happens to choose the worst pivot every time, it becomes quadratic instead. Now consider some alternative pivot-choosing strategies.

Median of Medians

The previous version of quickselect could become slow, but you can split the array better. For example, you don't want either of the two possible partitions to be small in case you have to recurse on the large one.

One strategy you can apply is called *median of medians*, and the idea is as follows:

1. Divide the array in groups of up to five elements.
2. Find the median of each group.
3. Find the median of the medians found in the previous step.
4. Use that value to split the array.

Figure 7-4 illustrates this concept; each rectangle is a set of five values in order from low to high from bottom to top (as the vertical arrow shows) with the median in the middle. The medians themselves grow from left (lowest median) to right (highest median) according to the horizontal arrow. The pivot you'll choose is the median of the set of medians—the center value in the diagram.

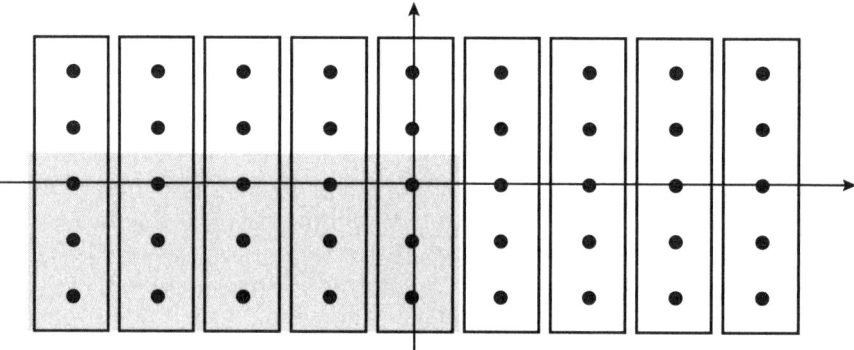

Figure 7-4: The middle element in each column is its median; medians are sorted from left to right, and the center value is not less than the shaded values, a third of the array.

In Figure 7-4, all the gray values (15 out of 45, a third of the complete set) are *guaranteed* not to be greater than the chosen pivot. Similarly, the chosen pivot is also guaranteed not to be greater than the other third of array values (see Figure 7-5).

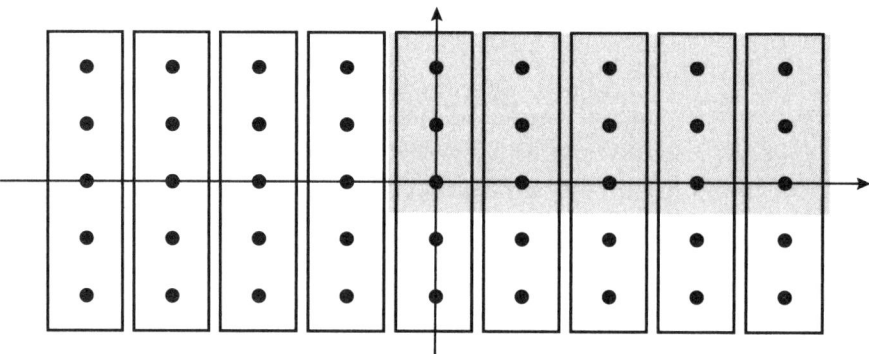

Figure 7-5: In the same situation as Figure 7-4, the center value is also not greater than the shaded values, a third of the array.

This means the chosen pivot will be such that it splits the array somehow between 33/66 percent and 50/50 percent. At worst, you'll have to apply recursion in a new array that is two-thirds the size of the original array (and, at best, one that's only one-third the size) and that can be shown to produce $O(n)$ performance.

The following code implements this method (the bold indicates the parts that changed):

```
const quickSelect = (arr, k, left = 0, right = arr.length - 1) => {
  if (left < right) {
    let mom;
❶  if (right - left < 5) {
      mom = simpleMedian(arr, left, right);
    } else {
❷    let j = left - 1;
      for (let i = left; i <= right; i += 5) {
❸      const med = simpleMedian(arr, i, Math.min(i + 4, right));
        j++;
❹      [arr[j], arr[med]] = [arr[med], arr[j]];
      }
❺    mom = Math.floor((left + j) / 2);
❻    quickSelect(arr, mom, left, j);
    }
❼  [arr[right], arr[mom]] = [arr[mom], arr[right]];

    const pivot = arr[right];

    let p = left;
    for (let j = left; j < right; j++) {
      if (pivot > arr[j]) {
        [arr[p], arr[j]] = [arr[j], arr[p]];
        p++;
      }
    }
    [arr[p], arr[right]] = [arr[right], arr[p]];

    if (p === k) {
      return;
    } else if (p > k) {
      return quickSelect(arr, k, left, p - 1);
    } else {
      return quickSelect(arr, k, p + 1, right);
    }
  }
};
```

If the array is short enough (five elements or fewer) ❶, you can use another algorithm to find the median of medians (mom). If the array has more than five elements ❷, consider sets of five elements at a time. You find the median of the set ❸ and move it to the left of the original array ❹ by swapping, so all medians end up together starting at position left of the array. You now want the median of this (smaller) set, so you calculate its position ❺ and use recursion ❻ to find the desired pivot. Once you've found it, swap it with the value at the right of the array ❼, and from that point onward, it's the same pivoting logic as shown earlier.

Now complete the code. You need a fast `simpleMedian(...)` algorithm to find the median of an array of up to five elements, and an insertion sort does the job (you also could use the `sortingSelect(...)` code from the section "Selecting with Comparisons" on page 124):

```
const simpleMedian = (arr, left, right) => {
❶ insertionSort(arr, left, right);
❷ return Math.floor((left + right) / 2);
};
```

Sort the whole array ❶, which isn't very slow because an insertion sort is quite speedy for such a small set of values, and then choose the middle element of the sorted array ❷.

This logic works well and has guaranteed results, unlike the original quickselect that had a worst case different from the average case.

Repeated Step

Another variation on how to select the pivot is called *repeated step*. This algorithm seemingly does a worse job of partitioning an array, but it has advantages in terms of speed. Choosing the median of three elements is quite quick using the "ninther" technique (as described in Chapter 6): first go through the array, generating a set by choosing the median out of every trio of values; then, go through that set of medians to create a second set by choosing the median out of every trio of medians. Figure 7-6 shows how this would work for an array with 18 elements. The idea is exactly the same for larger arrays, but there's not enough space to show it here.

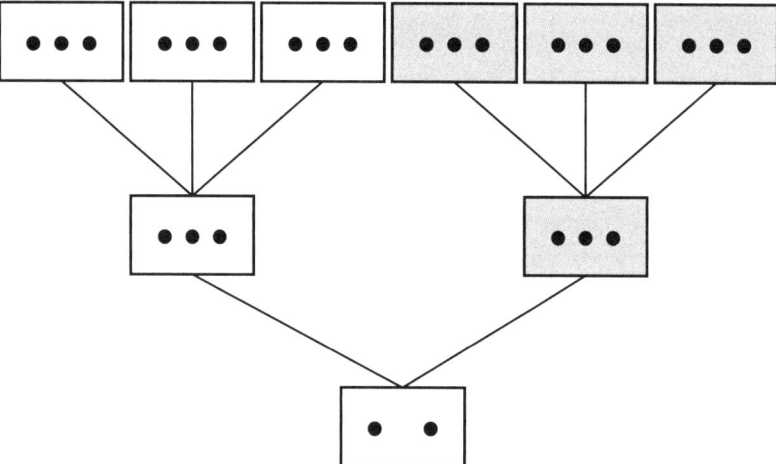

Figure 7-6: Repeatedly applying the "median of three" process reduces the original array to one-ninth of its size.

The repeated step of choosing the median of three reduces the original array to one-ninth of its size and makes recursion go very quickly. (In a sense, you are choosing the median of medians of medians.) The implementation is as follows:

```
const simpleMedian = (arr, left, right) => {
  insertionSort(arr, left, right);
  return Math.floor((left + right) / 2);
};

const quickSelect = (arr, k, left = 0, right = arr.length - 1) => {
  if (left < right) {
  ❶ let mom;
  ❷ if (right - left < 9) {
      mom = simpleMedian(arr, left, right);
    } else {
    ❸ let j1 = left - 1;
    ❹ for (let i = left; i <= right; i += 3) {
        const med = simpleMedian(arr, i, Math.min(i + 2, right));
        j1++;
        [arr[j1], arr[med]] = [arr[med], arr[j1]];
      }

    ❺ let j2 = left - 1;
    ❻ for (let i = left; i <= j1; i += 3) {
        const med = simpleMedian(arr, i, Math.min(i + 2, j1));
        j2++;
        [arr[j2], arr[med]] = [arr[med], arr[j2]];
      }

    ❼ mom = Math.floor((left + j2) / 2);
      quickSelect(arr, mom, left, j2);
    }
  ❽ [arr[right], arr[mom]] = [arr[mom], arr[right]];

    const pivot = arr[right];

    let p = left;
    for (let j = left; j < right; j++) {
      if (pivot > arr[j]) {
        [arr[p], arr[j]] = [arr[j], arr[p]];
        p++;
      }
    }
    [arr[p], arr[right]] = [arr[right], arr[p]];

    if (p === k) {
      return;
    } else if (p > k) {
      quickSelect(arr, k, left, p - 1);
    } else {
      quickSelect(arr, k, p + 1, right);
    }
  }
};
```

The mom variable ends up with the median of medians position in the array ❶. If the array is less than nine elements long ❷, you don't need to do any fancy work; just use a sort-based algorithm to find the desired median. Variable j1 keeps track of the medians you've swapped to the left of the array ❸. A simple loop goes through the array's elements, three at a time, finding the median of that trio and swapping it to the left ❹. You then perform the same logic again, using a new j2 variable ❺ and another loop ❻. After these loops, positions from left to j2 have the medians of medians ❼, and you apply the algorithm recursively to find its median, which you swap with the element at right ❽, so you can proceed with the rest of the otherwise unchanged quickselect algorithm.

This algorithm can also be proved to have $O(n)$ performance, so it's a good option. Why use recursion after two rounds of finding the medians of medians and not go on? (See question 7.6.)

So far you've explored algorithms that find the kth element for any value of k; this chapter finishes by explicitly considering the problem of finding the center element of an array.

Finding the Median with Lazy Select

If you want to find the median (remember the working definition isn't the one used in statistics; you just choose the element closest to the center of the array without any particular considerations for arrays of even length), you could obviously use any of the algorithms in this chapter, letting k be half the length of the input array. However, there are some other ways to find the center value, and in this section, we'll consider an interesting one that's based on random sampling (you'll study sampling algorithms in Chapter 8) and probability calculations. The lazy select algorithm uses sampling and may find the right value with a single pass with an $O(n^{-1/4})$ probability of failure, looping again and again as needed until it succeeds.

The algorithm to find the median of set S of size n works as follows:

1. Choose a random sample R of $n^{3/4}$ values from S.
2. Sort R using any algorithm.
3. Choose two values, d and u, in R that will satisfy $d < median < u$ *with high probability* (you'll see how to do this shortly).
4. Let $dSize$ be how many values of R are $< d$; if $dSize > n/2$, you failed and must try again.
5. Let $uSize$ be how many values of R are $> u$; if $uSize > n/2$, you must try again.
6. Let m be the set of values x of S that are $d < x < u$; if the count exceeds $4n^{3/4}$ you must try again.
7. Sort m and return the value at its $n/2 - dSize$ position.

The proof of performance for this algorithm depends highly on probabilistic arguments, and you won't see those here. The key concept is that a random choice of values *R*—but not too many, so sorting *R* is $O(n)$—should usually be good enough to find lower and upper limits to the median (*d* and *u* in the previous list) and that the set of values between *d* and *u* should be small enough so that, again, sorting doesn't go above $O(n)$ in performance. This algorithm may fail, but the probability is low $O(n^{-1/4})$, meaning that in the worst case, a few new attempts should succeed. As an example, if the odds of failure were 10 percent (which means the algorithm may succeed at first 90 percent of the times), the odds of two failures in a row would be 1 percent (10 percent of 10 percent, resulting in 99 percent odds of success), and three successive failures would happen once every 1,000 times, and so on.

The implementation is straightforward, but with lots of math:

```
❶ const sort = require("../sorting/mergesort");

  const lazySelectMedian = (arr, left = 0, right = arr.length - 1) => {
❷ const len = right - left + 1;
❸ const sR = Math.floor(len ** 0.75);
❹ const dIndex = Math.max(0, Math.floor(sR / 2 - Math.sqrt(len)));
  const uIndex = Math.min(sR - 1, Math.ceil(sR / 2 + Math.sqrt(len)));
❺ let dSize, uSize, m;
  do {
❻   const r = [];
    for (let i = 0; i < sR; i++) {
      r.push(arr[left + Math.floor((right - left) * Math.random())]);
    }
❼   sort(r);

    dSize = uSize = 0;
    m = [];
    for (let i = left; i <= right; i++) {
      if (r[dIndex] > arr[i]) {
        dSize++;
      } else if (arr[i] > r[uIndex]) {
        uSize++;
      } else {
        m.push(arr[i]);
      }
    }
❽ } while (dSize > len / 2 || uSize > len / 2 || m.length > 4 * sR);

❾ sort(m);
  return m[Math.floor(len / 2) - dSize];
};
```

You use merge sort ❶ to sort arrays when needed; it's important to choose an $O(n \log n)$ algorithm, because you'll use it with arrays that are at most $4n^{3/4}$ size, so the performance becomes $O(4n^{3/4} \log 4n^{3/4}) < O(n)$. Then you define several variables for the rest of the code: len is the size of the input array ❷, sR is the size of the sample ❸, dIndex and uIndex are the positions of d and u in the sorted r array ❹, and dSize, uSize, and m ❺ correspond with the description listed earlier in this section.

Use a "sampling with repetition" algorithm ❻ to choose sR random values from the input array into the r array; making sure no repeated values are sampled would work as well, but the logic would be more complex, as you'll see in Chapter 8. After having chosen and sorted r ❼, calculate dSize and uSize (how many values in the input array are smaller than d or greater than u; note that you never actually define d and u; you just refer to them by their indices) and m (with values between d and u).

Finally, you want to know whether the results are as expected ❽. If dSize or uSize includes more than half the input array, the median isn't in m, as was expected; you failed. Likewise, if m is too large, you also failed. If all tests pass, m has a proper size that allows you to sort it and choose the median from it ❾. Note that you account for the dSize values lower than d, which precede the array m.

This algorithm is quite different from most of what you've considered in this book, because it depends on probabilistic properties to work, but performance is usually quite good, and it finds the median with few iterations, if any.

Summary

In this chapter you studied several algorithms for selection, most of which are closely related to the sorting algorithms examined in Chapter 6. The selection problem isn't as common as sorting, so it's no surprise that JavaScript doesn't provide a ready-made method for it, so the implementations in this chapter cannot be avoided if you need this functionality. Most of the algorithms covered here have $O(n)$ performance, which is optimum, but the proofs of their behaviors are often complex, so they were omitted.

Questions

7.1 Tennis Sudden Death

Suppose 111 tennis players enter a knockout tournament to find the champion. In each round, random pairs of players play each other, and the loser is out of the tournament, while the winner passes to the next round. If there's an odd number of players, one player gets a free pass to the next round. How many matches will be necessary to find the champion? How many extra matches will you need to find the second-best player? (And no, whoever lost to the champion in the last game isn't necessarily the second-best player.) Can you generalize your answer for n players?

7.2 Take Five

"Take Five" is the name of a jazz piece that Dave Brubeck made famous, but in this case you want to take the median of five elements. What's the absolute minimum number of comparisons that guarantees finding that median? Can you provide an appropriate medianOf5(a,b,c,d,e) function that will return the median of its five arguments? You could be achieving a better `simpleMedian()` function with this!

7.3 Top to Bottom

If k is close to n, the length of the input array, your selection sort–based algorithm would have a bad quadratic performance, but you can make it quite better with a simple trick; can you see how?

7.4 Just Iterate

Quickselect does a single tail recursive call and may be rewritten to avoid all recursion; can you do it?

7.5 Select Without Changing

As is, `qSelect` returns the desired kth value, but it has a side effect: the input array will be changed. Can you modify `qSelect` to avoid this secondary effect?

7.6 The Sicilian Way

The repeated step selection algorithm does two rounds of choosing medians of three, and finally, it uses recursion to find the median of the resulting array of medians of medians. Implement the following variation: instead of recursion, keep applying the same method (grouping by three, choosing the median, and so on) until the resulting array is less than 3 in length, and then choose the pivot from that small array without any recursion.

8

SHUFFLING AND SAMPLING

Consider this chapter to be a complement of the two previous chapters, but instead of sorting values into some kind of order, you want to shuffle them into a random, disordered sequence (as for a card game). And rather than select a value at a given position, you want to choose a set of values randomly (as for statistical sampling algorithms). Chapters 6 and 7 revolved around order and consistency, but this chapter works with disorder and randomness instead.

Choosing Numbers Randomly

First, consider a basic function that you'll need to use throughout this chapter: generating a random number in a given interval. JavaScript already provides `Math.random()`, which produces a pseudorandom number r such that $0 \leq r < 1$. (For more information, see *https://developer.mozilla.org/en-US/docs/Web/JavaScript/Reference/Global_Objects/Math/random*.) The distribution of the numbers this function produces is *uniform*, which means that each value is equally possible, and no value is more likely than another.

NOTE *Why is it pseudorandom? The random numbers are actually produced by an algorithm in such a way that the properties of the generated sequence are approximately that of a sequence of truly random numbers. However, the fact that the numbers are generated by a procedure automatically means that they aren't truly random; they just look like they are. For the sake of simplicity, though, in this chapter we'll consider the produced numbers to be random.*

Using this function, you can scale its results to produce numbers in any given range. The following functions will come in handy for the rest of the chapter:

```
const randomBit = () => Math.random() >= 0.5;
const randomNum = (a, b) => a + (b - a) * Math.random();
const randomInt = (a, b) => Math.floor(randomNum(a, b));
```

The first function is useful when you want to decide between two alternatives randomly, as if simulating a coin toss. If a random number is less than 0.5 (0 to 0.4999 . . .), you return `false` (heads), and you return `true` (tails) otherwise (0.5 to 0.9999 . . .). Given a range of values from a to b (not necessarily integers, but $a < b$), the second function produces a random floating-point number r such that $a < r < b$. This is easily verified by noting that `(b - a)*Math.random()` is greater than or equal to zero but strictly less than `(b - a)`. The last function is meant to be called with integer arguments, and it produces a random integer r such that $a < r < b$. You can also write it as follows:

```
const randomInt = (a, b) => a + Math.floor((b - a) * Math.random());
```

Some people have difficulty with `randomInt(...)`. For instance, to simulate rolling a die, they might write `randomInt(1,6)`, but that won't work: `randomInt(1,7)` does the job. (See question 8.2 for another take on this.) You could obviously rewrite `randomInt(...)` to do it another way, but you're following JavaScript's lead as in the `array.slice(start,end)` method, whose parameters work exactly as these do, taking elements from `start` up to (but not including) `end`.

With these basic tools, let's turn to the problems of shuffling and sampling, both of which will be based on random numbers in one way or another.

Shuffling

The first problem we'll consider is *shuffling* an array of values in order to produce a random sequence of values—or to use a mathematical term, a permutation. This is equivalent to shuffling a deck of cards before playing a game to start anew with a different sequence of cards every time.

An important requisite is that every possible permutation should be equally likely, which presents a thorny problem: How can you make sure that the shuffling code ran correctly? For instance, when sorting an array, you can check that the sorted array is actually in order and that its elements are the same before and after sorting. Similarly, for selection algorithms, you can check that it worked by sorting the array separately and then checking whether the selected value is correct. Shuffling is harder to check.

First, you should prove (somehow) that the logic is correct so that all results are equally probable. However, what if you implement the algorithm badly, with some bug? (Don't ask me how I know.) An empirical suggestion is to run the algorithm many times with a known input sequence and test statistically whether the observed outcomes suggest a uniform distribution; we'll leave the mathematical aspects of this solution to the textbooks and instead try an easier way (see question 8.1).

Shuffling by Sorting

We'll start with a sorting-based algorithm. It doesn't have the best performance, but it's the simplest implementation. In order to shuffle a set of values, you associate a random number with each value, sort the set on that random value, and the result will be a totally random shuffle (see Figure 8-1).

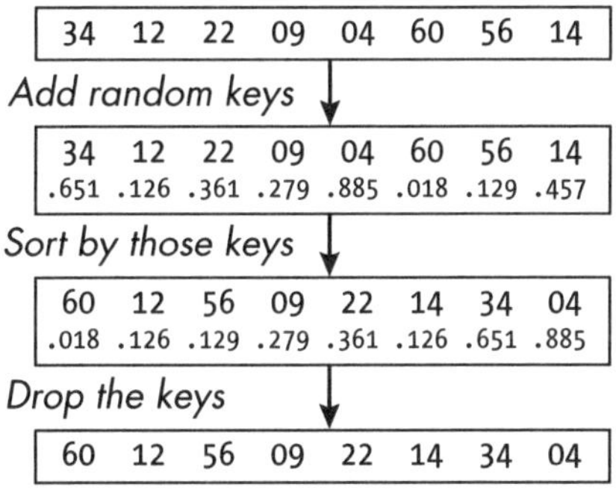

Figure 8-1: Sorting an array by a randomly assigned key produces a totally random shuffle.

You can implement this solution with any of the algorithms discussed in Chapter 6. Let's go the simplest possible direction and use JavaScript's own `.sort(...)` method. The shuffle code ends up being a single line, even though more lines are used here to show it clearly:

```
const sortingShuffle = (arr) =>
  arr
```

```
❶ .map((v) => ({ val: v, key: Math.random() }))
❷ .sort((a, b) => a.key - b.key)
❸ .map((o) => o.val);
```

The code directly matches the steps in Figure 8-1. Given the array of values, create a new array where objects have the original value in val and a random value in key ❶. Then sort it by this random key ❷ and produce a new array with only the values ❸.

This algorithm is probably the shortest one in the book, and it produces a shuffled list of values easily. However, it's easy to make a mistake when implementing random sorting (see question 8.3 for an example).

The performance of this code is $O(n \log n)$, but you can do better. First, however, we'll consider something you could have designed based on an interesting mix of concepts from Chapters 5 and 6.

Shuffling by Coin Tossing

Let's explore other ways to shuffle a set of values. Imagine a divide-and-conquer procedure where you split a set in two (using a simulated coin toss to decide what goes where), recursively shuffle each part, and join them back together. Empty sets or sets with only one element would need no shuffling. You can shuffle a set with exactly two elements by randomly deciding (again, tossing a coin) which elements will be first and last. For sets with more than two elements, apply the recursive procedure illustrated in Figure 8-2.

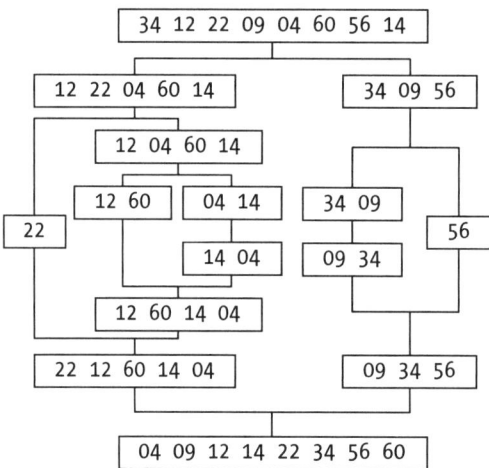

Figure 8-2: Randomly splitting an array, shuffling each part, and joining the results produces a shuffle that's reminiscent of merge sort.

The first split at the top separates the array into two parts with five and three elements. The subsequent steps follow moving downward. The five-element array splits into arrays with one and four elements. The single element doesn't need any further shuffling, and the array with elements splits into two parts. Of those two parts, one is left as is (12, 60), and the other is swapped. The joined pairs create a random shuffle of the original four-element array, which then joins with the single element (22) to produce a random shuffle of the initial five-element array. A similar process occurs on the right side of the array, and the final result is at the bottom.

Here's the implementation:

```
❶ const coinTossingShuffle = (arr, from = 0, to = arr.length - 1) => {
    const len = to - from + 1;
❷ if (len < 2) {
    // nothing to do
❸ } else if (len === 2) {
    if (randomBit()) {
      [arr[from], arr[to]] = [arr[to], arr[from]];
    }
❹ } else /* len > 2 */ {
    let ind0 = from - 1;
    let ind1 = to + 1;
    let i = from;
    while (i < ind1) {
      if (randomBit()) {
        ind1--;
        [arr[i], arr[ind1]] = [arr[ind1], arr[i]];
      } else {
        ind0++;
        i++;
      }
    }
  ❺ coinTossingShuffle(arr, from, ind0);
  ❻ coinTossingShuffle(arr, ind1, to);
  }
❼ return arr;
};
```

The parameters for shuffling functions will be an array arr and the portion of it (from, to) you are shuffling ❶. If the length of the array is less than 2 ❷, nothing needs to be done. If the array has exactly two elements ❸, flip a coin to decide whether to leave it be or swap the two elements. If the array has more than two elements ❹, apply a logic reminiscent of partitioning in quicksort: flip coins to decide where each value goes. If the coin flip is true, the value goes into the ind1 to to section, and if the coin flip is false, the value goes in the from to ind0 section. After moving every element to its place (at which time ind0 and ind1 will point to positions next to each other), use recursion to shuffle the elements that received a false bit ❺ and those that got a true bit ❻. Finally, return the shuffled-in-place array ❼.

This algorithm can be proven to have an average $O(n \log n)$ performance with a worst case of $O(n^2)$. Figure 8-2 should remind you of merge sort and quicksort, algorithms with similar workings, so you haven't really done better than with sorting.

Shuffling in Linear Time

How fast can we shuffle? The best possible result with shuffling is $O(n)$, where you access each element in the array once. All the methods in the previous section had worse performance (although for small values of n they may be quite suitable), so now you're going to consider linear time shuffling algorithms. And to better match what we did in Chapter 6, we'll shuffle just a portion of an array.

Floyd's Shuffle

Robert Floyd's linear time shuffling algorithm has some interesting ideas. The process has two steps: first, it generates a random permutation of numbers 0 to $n - 1$, and then it uses that generated permutation to shuffle the original array. (You'll also see this technique in Floyd's sampling algorithm, later in this chapter.) Start by generating the permutation, which is similar to an insertion sort (see Figure 8-3).

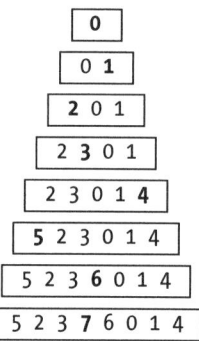

Figure 8-3: Floyd's algorithm produces a shuffle by randomly inserting new values into the previously shuffled ones.

The algorithm works the same way as arranging playing cards by hand. You pick the first card, and that's it. Then you pick the second card and place it to the left or to the right of the previous one. Then pick the third card and place it to the left, in the middle, or to the right of the previous two cards. Each new card goes somewhere among the previous cards, in a random place.

Here's the code:

```
const floydShuffleN = (n) => {
❶ const result = [];
❷ for (let i = 0; i < n; i++) {
   ❸ const j = randomInt(0, i + 1);
   ❹ result.splice(j, 0, i);
   }
   return result;
};
```

For a simple implementation ❶, you can use an array for the generated shuffle. First loop n times starting at 0 ❷, and each time you generate a random position ❸ where you insert the new number among the previous ones ❹ by using the very handy .splice(...) method.

But how do you get from this permutation to a shuffle of the original array? Figure 8-4 shows how to use the previous result to finish the task.

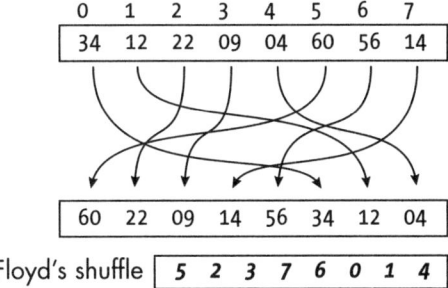

Floyd's shuffle

Figure 8-4: A random permutation of numbers is used to shuffle an array.

Each element of the original array gets moved to a different place, according to the corresponding value that floydShuffleN(...) produces. Implementing the moves requires an extra array. Having generated a shuffled list of numbers from 0 to $n - 1$, here's the code to finish shuffling:

```
const floydShuffle = (arr, from = 0, to = arr.length - 1) => {
❶ const sample = floydShuffleN(to - from + 1);
❷ const original = arr.slice(from, to + 1);
❸ sample.forEach((v, i) => (arr[from + i] = original[v]));
  return arr;
};
```

First generate a sample of numbers the same size as the portion of the array that you want to shuffle ❶, and then take the values in the input array ❷ and replace them according to the method ❸ shown in Figure 8-3.

How the code performs depends on what data structure you choose for the sample. Using an array as shown here means that insertion .splice(...) is $O(n)$, so the whole algorithm becomes $O(n^2)$. You'll see appropriate data structures in future chapters, but Floyd suggests using a hash table of size $2n$, with entries forming a linked list, for an expected average $O(n)$ performance, or a balanced ordered tree with linked nodes, for $O(n \log n)$ assured performance.

Robson's Algorithm

Here's a different take on how to generate a permutation. With an array of n elements, there are $n!$ possible shuffling outcomes. The idea in Robson's algorithm is to randomly select a number between 0 and $n! - 1$ inclusive and use that number to generate a permutation, so each different number produces a different shuffle.

NOTE *This method is related to a mathematical concept called the Lehmer code, which is a way to encode each possible permutation of* n *numbers, but we won't go into that here.*

If you want to shuffle an array with four elements to produce a random permutation out of the 24 (= 4!) possible ones, you'd start with a random number between 0 and 23 inclusive. Then divide that number by 4. The quotient will be a number between 0 and 5, and the remainder will be between 0 and 3. (An important detail is that all possible combinations of quotient and remainder are equally probable. Can you verify that?)

Use the remainder to choose one of the four elements in the array, set it aside, and keep working with the other three. Consider the quotient: it's a random value between 0 and 5.

This time, divide by 3. The new quotient will be 0 or 1, and the remainder will be 0, 1, or 2, which you can use to choose one of the three remaining numbers. Consider the quotient, which is either 0 or 1. If you divide the quotient you had by 2, you'll get a quotient of 0 (no more work to be done). You can use the remainder (0 or 1) to choose one of the two remaining numbers, and you'll have your desired shuffle. (After you've chosen 3 out of 4, the complete shuffle is implied.) Figure 8-5 shows the algorithm if you had drawn 14 as the random number.

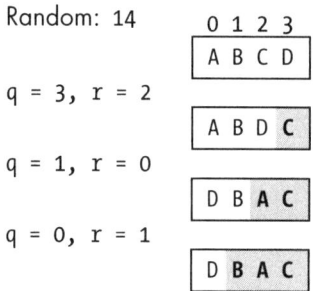

Figure 8-5: Robson's shuffling algorithm is also based on transforming the original array depending on a randomly chosen permutation.

Start with values ABCD in the four positions of the array, from 0 to 3. In the first step, divide 14 by 4, getting a quotient of 3 and a remainder of 2. Then swap the element in position 2 of the array with the last, getting ABDC. The second step divides 3 (the previous quotient) by 3, producing quotient 1 and remainder 0. Then swap the element in position 0 with the next-to-last element, resulting in DBAC. Then, divide 1 (the latest quotient) by 2, which gives quotient 0 and remainder 1. You don't need to swap, because you'd swap the element at position 1 with itself, and you'd still have DBAC. After having shuffled three of the elements of the array, the fourth is also in place, and you're done.

Here's the logic:

```
const robsonShuffle = (arr, from = 0, to = arr.length - 1) => {
❶ const n = to - from + 1;
❷ let r = randomInt(0, fact(n));
❸ for (let i = n; i > 1; i--) {
    ❹ const q = r % i;
    ❺ [arr[from + i - 1], arr[from + q]] = [arr[from + q], arr[from + i - 1]];
    ❻ r = Math.floor(r / i);
  }
  return arr;
};
```

The number of elements to shuffle is n ❶. You generate a random number ❷ between 0 and $n! - 1$ using the factorial function developed in Chapter 5. You then loop through the array from right to left ❸: calculate q ❹, use it to swap elements ❺, and find r to loop again ❻.

The algorithm is obviously $O(n)$, as it follows from its single loop. However, as is, the algorithm has a problem. You wouldn't be able to use it for large arrays, because calculating a factorial may exceed the available precision of JavaScript (see question 8.5). Fortunately, there's a way out.

The Fisher-Yates Algorithm

The problem with Robson's algorithm is the need to calculate $n!$ to get a random number with which to proceed. But if you consider it carefully, you don't really need the factorial. The key to that algorithm was the series of remainders, and you can generate those by using a random function. The first remainder was in the range 0 to $n - 1$, and it was used to choose the initial value of the permutation; the second remainder was in the range 0 to $n - 2$, and it was used to choose the second value of the permutation, and so on. It follows that you just need to generate random values at proper times in Robson's algorithm, and that's the Fisher-Yates algorithm.

Thus, you can write this alternative to Robson's code:

```
const fisherYatesShuffle = (arr, from = 0, to = arr.length - 1) => {
❶ for (let i = to + 1; i > from + 1; i--) {
  ❷ const j = randomInt(from, i);
  ❸ [arr[i - 1], arr[j]] = [arr[j], arr[i - 1]];
  }
  return arr;
};
```

As in Robson's shuffle, loop from right to left ❶, and at each pass calculate a random number ❷ that you use to decide what elements to swap ❸. Basically, any element in positions from to i could be chosen for the swap.

The Fisher-Yates algorithm is frequently written to shuffle from left to right, which is basically the same idea:

```
const fisherYatesShuffle2 = (arr, from = 0, to = arr.length - 1) => {
❶ for (let i = from; i < to; i++) {
  ❷ const j = randomInt(i, to + 1);
  ❸ [arr[i], arr[j]] = [arr[j], arr[i]];
  }
  return arr;
};
```

The code is the same ❶ except that the generated permutation starts from left to right, and any element in positions i to to ❷ may be chosen for swapping ❸. This algorithm is quite efficient, and it's often used for shuffling. Be careful, though, because it's easy to mess up; see question 8.4.

Sampling

Sampling is a technique frequently used in statistics. Basically, out of a set of values (an array), you want to pick a random, smaller set, which is called a *sample*. There are two kinds of sampling procedures: sampling with repetition, in which elements may be chosen more than once, and sampling without repetition, in which no element may be chosen two times or more. In mathematical terms, the latter procedure is called selecting a *combination* of elements. (In the first case, sampling with repetition, the number of chosen elements can be anything. In the second case, the number is limited by the number of elements in the original set.) Don't worry about the order in which the elements are selected.

We'll first consider sampling with repetition, for which we just need a couple of short, optimally efficient algorithms, and then we'll dedicate most of the rest of the chapter to sampling without repetition, which requires more complex logic.

Sampling with Repetition

Sampling with repetition is a simple algorithm, and you'll start by selecting just a single value. Choosing a larger sample will simply be a matter of choosing a value over and over again.

Choosing Only One Value

Choosing a single value is the simplest kind of sampling, and all you need is a random number in the appropriate range. You can use the randomInt(...) function, and for an element of an array, the following works:

```
const singlePick = (arr, from = 0, to = arr.length - 1) =>
  arr[randomInt(from, to + 1)];
```

To select an element between the `from` and `to` positions (both included), you produce a random number in that range and return the corresponding element.

Of course, if you always want to choose values from the whole array (as you'll do in the rest of the chapter), simpler code does the job:

```
const singlePickAll = (arr) => arr[randomInt(0, arr.length)];
```

This is equivalent to setting `from = 0` and `to = arr.length - 1`, so this new function works in the same way.

Choosing Several Values with Repetition

As mentioned, to make multiple selections from a set (maybe to simulate a series of roulette wheel turns, or create a strategy for a game of rock/paper/scissors, or implement the lazy select median-finding algorithm from Chapter 7), doing several single selections is enough:

```
const repeatedPick = (arr, k, from = 0, to = arr.length - 1) => {
  const sample = Array(k);
❶ for (let i = 0; i < k; i++) {
  ❷ sample[i] = arr[randomInt(from, to + 1)];
  }
  return sample;
};
```

The logic is simple: first loop k times ❶, randomly choosing elements one by one ❷. Again, as with choosing a single value, to make selections from an entire array, the code is simpler, and you can reuse the `singlePickAll` code from the previous section as well:

```
const repeatedPickAll = (arr, k) => {
  const sample = Array(k);
  for (let i = 0; i < k; i++) {
    sample[i] = singlePickAll(arr);
  }
  return sample;
};
```

For a related coding challenge, see question 8.7. Next, take a look at sampling without repetition, which has the restrictions of not allowing you to choose any element more than once and doing so in an efficient way.

Sampling Without Repetition

This process is equivalent to what's used in Powerball-style lottery drawings: numbers are removed from (but not returned to) an urn, guaranteeing that all selected numbers are different.

For the algorithms in this section, assume you have an array with n elements, from which you want to pick a combination of k elements. It must be

$k < n$—if k were equal to n, no algorithm would be needed, and k cannot be greater than n if no repetitions are allowed. Algorithms will be faster the fewer elements that you want, so for a cheap optimization, you can assume that $k \geq n/2$; indeed, if $k \geq n/2$, instead of selecting k elements, you could select $n - k$ ones and discard them.

Sampling by Sorting or Shuffling

The first idea is inspired by the "Shuffling by Sorting" section on page 139, plus the selection algorithm explored in Chapter 7. You can assign random keys to all elements, sort them, and then get the elements with the lowest k keys. You've already considered all the necessary code to implement this method, so leave actual development to question 8.8.

A second idea you could try is based on the fact that you already know how to generate a random permutation of a set. Given this, an obvious way to generate a sample could easily be to shuffle the set and then take its first k elements. That works, but you can get the desired sample with more efficient logic without having to shuffle (similar to what you found with selection algorithms, when you saw better ways of selecting that didn't require a previous sort). You won't see the code for this procedure either, as it's derived from what you've already done. Let's move on to new algorithms instead.

Floyd's Algorithm

Often you just need a sample of k integers between 0 (included) and n (excluded). Robert Floyd's `floydSampleKofN(...)` algorithm produces an array with a combination of k such numbers, which is interesting in itself and will help you write a more general sampling algorithm. If you need a sample from the original array, you can use the selected numbers produced by `floydSampleKofN()` for that task, as in Figure 8-6.

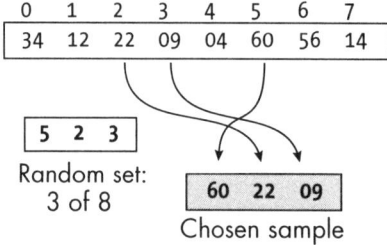

Figure 8-6: A random selection of numbers from 0 to n – 1 produces a random sample.

You can choose values from the original input array using the sample's values as indices, as shown in Figure 8-6. The input array is at the top, the sample produced by Floyd's code is [5, 2, 3], and the final result is shaded in gray.

Here's the code that uses the (yet unseen) floydSampleKofN() function:

```
const floydSample = (arr, k) =>
❶ floydSampleKofN(k, arr.length).map((v) => arr[v]);
```

You generate a random combination of k values out of n ❶ and use those numbers as indices to get values from the original input array.

Let's return to generating the combination and finally see floyd SampleKofN(). The recursive version is the following, and recursion helps you understand how and why the algorithm works:

```
❶ const floydSampleKofN = (k, n) => {
❷ if (k === 0) {
    return [];
  } else {
    ❸ const sample = floydSampleKofN(k - 1, n - 1);
    ❹ const j = randomInt(0, n);
      sample.push(sample.includes(j) ? n - 1 : j);
    ❺ return sample;
  }
};
```

You want a combination of k distinct values between 0 and n - 1 inclusive ❶. If k is 0 ❷, you return an empty sample. Otherwise, you use recursion first to choose a combination of k - 1 values up to n - 2 ❸. Then decide what value to add to that sample ❹. At the end ❺, you return the created sample.

Now examine how to add a new value to the sample, working with getting a sample of three values out of eight, as in Figure 8-6. Suppose you already have a sample of two values out of the set 0 to 6: should you add a 7 value to produce the sample of three values? A possibility (1/8) is that the random number j is exactly 7. It can't be in the previous sample, so it will be added.

The other way to add 7 is if j was one of the two numbers already in the sample (2/8). Thus, the probability that 7 will end up included in the sample of 3 out of 8 is $1/8 + 2/8$, which is exactly 3/8 as you needed. You can apply this argument systematically and find that each of the n values has a probability of k/n of being in the final sample, so the algorithm really produces a correct sample.

Since recursion always happens at the beginning of each pass, you can turn the code into an iterative equivalent version (see question 8.9):

```
const floydSampleKofN = (k, n) => {
❶ const sample = [];
❷ for (let i = n - k; i <= n - 1; i++) {
  ❸ const j = randomInt(0, i + 1);
  ❹ sample.push(sample.includes(j) ? i : j);
  }
❺ return sample;
};
```

First, create an array to return the chosen sample ❶, and you'll return this at the end ❺. A loop executes k times ❷. In each pass choose a random number ❸ and use the same logic (checking whether the randomly selected number was already selected) to decide what to add ❹. The argument to prove that this algorithm works correctly is along the same lines as for the recursive version, so it won't be repeated here.

The key to performance for Floyd's algorithm is how it adds a value to the sample and checks whether a given value is already in the sample. In other words, it needs an efficient implementation of a set. You also could use a bitmap as in Chapters 6 and 7 (we'll leave this for now and consider such options in Chapter 13).

Lottery Drawing

Another method to consider implies actually replicating a lottery drawing. You choose a random element of the set, place it somewhere else, and do it again and again until you get the complete sample. Figure 8-7 shows the process. The set of values is on the left, the selected sample is on the right, and the triangle marks the randomly chosen element at each stage.

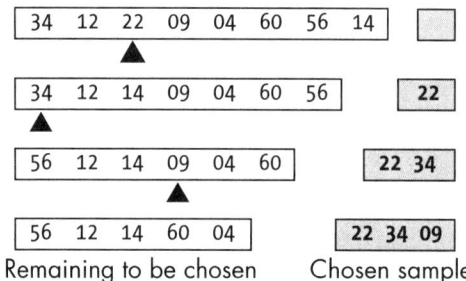

Figure 8-7: A simulated lottery drawing produces a random sample.

When you remove an element, you swap it with the one in the last place of the array to avoid having to shift the whole array, which would slow the code's performance.

Here's a simple implementation:

```
const lotterySample = (arr, k) => {
❶ const n = arr.length;
  const sample = Array(k);

❷ for (let i = 0; i < k; i++) {
  ❸ const j = randomInt(0, n - i);
  ❹ sample[i] = arr[j];
  ❺ arr[j] = arr[n - i - 1];
  }

❻ return sample;
};
```

Start by creating the array that will get the sample ❶. Loop k times ❷, generating a random position ❸ among the first n - i elements of the array, because the already sampled elements will go to the end of the array ❹. The chosen element is added to the sample array, and it's swapped so it won't be considered again in other selections ❺. Finally, return the produced sample ❻.

This algorithm is simple enough, and it has $O(k)$ performance that cannot be improved upon. After all, you want a sample with k elements. However, you can get a bit more speed if you notice that there's no actual need for a separate sample array.

Fisher-Yates Sampling

In the previous lottery sampling algorithm, at any time each element of the array is either chosen or not, so you don't need two arrays. The original one will do. Figure 8-8 illustrates this idea; the shaded numbers are the chosen ones.

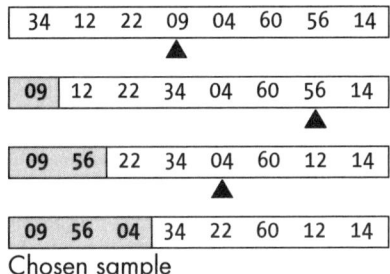

Chosen sample

Figure 8-8: The lottery sampling algorithm can work in place, without extra memory.

Every time an element is chosen, move it to the front of the array, so all of its first elements are in the sample and the rest are the nonchosen ones. This algorithm is a variation of the Fisher-Yates shuffling method (the same logic but applied fewer times, because you don't want to randomize the whole array; only k elements), and you can code it as follows:

```
const fisherYatesSample = (arr, k) => {
  const n = arr.length;
❶ for (let i = 0; i < k; i++) {
  ❷ const j = randomInt(i, n);
  ❸ [arr[i], arr[j]] = [arr[j], arr[i]];
  }
❹ return arr.slice(0, k);
};
```

In this algorithm, the i variable points to the corresponding sampled element. You loop k times ❶, choosing a random position among the yet-unchosen elements ❷, and you do a swap to change the selected element from the unchosen part to the chosen one ❸. After completing the loop ❹, return the initial slice (k elements long) of the original array.

The Fisher-Yates sampling algorithm is also $O(k)$; the only difference is where the sample is stored.

Knuth's Algorithm

Donald Knuth's algorithm has the interesting characteristic that the values in the sample keep their relative order as in the original array. The algorithm is based on probabilities, which directly proves its correctness.

To understand how it works, suppose you want to choose three elements out of eight. The probability that the first element will be included is 3/8. The probability that the second element will be chosen depends on whether the first element was chosen. If it was, the probability of choosing the second is 2/7 (because having selected one of the eight, now you choose two out of the remaining seven), but if skipped, the probability of choosing the second is 3/7 (because now you have to choose three elements out of the remaining seven).

The algorithm chooses or skips elements based on random numbers and probabilities, as follows:

```
const orderedSample = (arr, k) => {
❶ if (k === 0) {
    return [];
❷ } else if (Math.random() < k / arr.length) {
    return [arr[0], ...orderedSample(arr.slice(1), k - 1)];
❸ } else {
    return [...orderedSample(arr.slice(1), k)];
  }
};
```

If you want to choose an empty sample ❶, an empty array is returned; this is the base case for the recursion. Otherwise, you get a random number and compare it with the probability: if it's smaller ❷, you include the first element of the array followed by a k − 1 sized sample of the rest. If it's greater ❸, you select all k elements out of the rest of the array.

A better implementation avoids recursion and all the destructuring and slicing of arrays as follows:

```
const orderedSample = (arr, k) => {
❶ const sample = [];
❷ let toSelect = k;
❸ let toConsider = arr.length;
❹ for (let i = 0; toSelect > 0; i++) {
    if (Math.random() < toSelect / toConsider) {
    ❺ sample.push(arr[i]);
      toSelect--;
    }
  ❻ toConsider--;
  }
  return sample;
};
```

As in other algorithms, sample is the array that will be produced ❶. The variables toSelect ❷ and toConsider ❸ will keep count of how many values you still have to select out of the values not yet considered. You loop until there are no more values to choose ❹. Each time, you decide whether to choose or ignore a value, according to the probabilistic method described. If the test comes out true ❺, add the value to the sample array and decrease the count of pending values to select by 1. Every pass through the loop, we decrease the number of values to yet consider ❻.

You can also write it in a more compact way:

```
const orderedSample2 = (arr, k) => {
  const n = arr.length;
  const sample = [];
❶ for (let i = 0; k > 0; i++) {
  ❷ if (Math.random() < k / (n - i)) {
      sample.push(arr[i]);
    ❸ k--;
    }
  }
  return sample;
};
```

The difference is that you'll use k ❶ instead of a toSelect variable ❸, and you'll calculate how many yet-unseen values there are ❷; otherwise, the algorithm is the same.

Reservoir Sampling

The final algorithm we'll consider was created by Alan Waterman, and it's interesting because it can work in an online mode, without needing the whole array of elements beforehand. All the other algorithms in this chapter work in offline mode. The code goes through the input data, maintaining a suitable random sample at all times; it can be stopped at any moment and would have a proper random sample of the elements seen so far. This algorithm is quite suitable for large streams, where it might be impossible to store all values in memory and then apply one of the previous algorithms considered in this chapter.

Consider a simple case first: choosing just one element out of a sequence of undetermined length. (If you knew the length of the sequence, using randomInt(...) would be the quickest way to pick an element.) The solution to this problem works as follows:

- Choose the first element of the sequence and place it in a reservoir.
- For the ith element in the sequence after the first, use it to replace the reservoir value with probability of $1/i$.

Suppose the sequence had 1,000 elements. What's the probability of choosing the very last element? Obviously, it's 1/1,000. If you didn't choose it, when you have 999 elements, what's the probability of choosing the 999th? It would be 1/999. As more and more elements are processed, the probability of choosing the ith one is always $1/i$.

You can expand this example to choose a sample of *k* elements; the process is quite similar:

- Choose the first *k* elements of the sequence and place them in a reservoir.
- For the *i*th element in the sequence after those *k*, add it to the reservoir value with a probability of k/i by replacing a randomly selected value of the reservoir.

You can code this as follows:

```
const reservoirSample = (arr, k) => {
❶ const n = arr.length;
❷ const sample = arr.slice(0, k);

❸ for (let i = k; i < n; i++) {
  ❹ const j = randomInt(0, i + 1);
    if (j < k) {
    ❺ sample[j] = arr[i];
    }
  }
  return sample;
};
```

We won't work with a stream, but the changes for that are straightforward. Here, to know when the sequence is ended you'll use variable n ❶, and the sample reservoir is initialized with the first k elements of the sequence ❷. You loop through the data ❸ and do a random test ❹ to see whether the number should go into the array; the j variable is used both for the test and to randomly decide what reservoir element to replace ❺.

If you modify the input array (using its first *k* positions for the reservoir), the algorithm looks like this:

```
const reservoirSample2 = (arr, k) => {
  const n = arr.length;
  for (let i = k; i < n; i++) {
    const j = randomInt(0, i + 1);
    if (j < k) {
    ❶ [arr[i], arr[j]] = [arr[j], arr[i]];
    }
  }
❷ return arr.slice(0, k);
};
```

The differences are how you swap a chosen value into the reservoir ❶ and how you return the chosen sample ❷; otherwise, it functions exactly the same way.

Summary

In this chapter we've considered algorithms for generating randomized permutations and combinations of an array, methods that are quite useful for several areas like gaming or statistics, among others. In the next chapter, we'll turn to another common and important task: searching efficiently for a value.

Questions

8.1 Good Enough Shuffling

Implement a logging function that takes a shuffling function as input and runs many tests, counting how often each possible permutation is produced, and then draw a histogram to visualize its results.

Figure 8-9 shows the output from my own tests with good results for a shuffle of an array of four elements.

```
A-B-C-D: 2049 ###############################################
A-B-D-C: 1953 ############################################
A-C-B-D: 2022 #############################################
A-C-D-B: 2028 #############################################
A-D-B-C: 1904 ###########################################
A-D-C-B: 2012 #############################################
B-A-C-D: 1949 ############################################
B-A-D-C: 2013 #############################################
B-C-A-D: 2020 #############################################
B-C-D-A: 2091 ###############################################
B-D-A-C: 1990 ############################################
B-D-C-A: 2002 #############################################
C-A-B-D: 2015 #############################################
C-A-D-B: 1948 ############################################
C-B-A-D: 1938 ############################################
C-B-D-A: 2028 #############################################
C-D-A-B: 2045 #############################################
C-D-B-A: 2041 #############################################
D-A-B-C: 2024 #############################################
D-A-C-B: 2032 #############################################
D-B-A-C: 2033 #############################################
D-B-C-A: 1925 ###########################################
D-C-A-B: 1960 ############################################
D-C-B-A: 1978 ############################################
COUNT= 24
```

Figure 8-9: A histogram showing that a certain shuffling algorithm produces all possible outcomes with similar frequencies

After 48,000 random tries, all permutations (24 = 4!) were generated, and the results seem similar enough. Although this assertion isn't really valid in a statistical way; a χ^2 (that's the Greek letter chi) goodness-of-fit test would be required for that.

8.2 Random Roll

Suppose you have to generate a uniform random triple option: instead of true/false, say high/medium/low. Using Math.random(), it's easy to do, as seen in the randomNum(...) function, but can you do this using only randomBit()? Along the same lines, how can you generate a uniform die roll (1–6) using randomBit()? Or a 1 to 20 roll for a *Dungeons & Dragons* type of game? (This last question is trickier.)

8.3 Not-So-Random Shuffling

After reading the description for random shuffling, a programmer decides to make it simpler: instead of bothering to assign random keys and sorting by them, the programmer took a sorting algorithm (bubble sort, in this case) and changed the comparisons among keys to use a random bit:

```
const naiveSortShuffle = (arr) => {
  for (let j = arr.length - 1; j > 0; j--) {
    for (let i = 0; i < j; i++) {
❶    if (randomBit()) {
        [arr[i], arr[i + 1]] = [arr[i + 1], arr[i]];
      }
    }
  }
  return arr;
};
```

The logic is that of bubble sort (see Chapter 6) but with a single change ❶. Why is this a bad shuffle generator? Where did the programmer go wrong?

8.4 Bad Swapping Shuffle

A developer messed up when implementing the Fisher-Yates shuffling code and wrote the following, which seems good enough at first:

```
const naiveSwappingShuffle = (arr) => {
  const n = arr.length;
  for (let i = 0; i < n; i++) {
    const j = randomInt(0, n);
    [arr[i], arr[j]] = [arr[j], arr[i]];
  }
  return arr;
};
```

The difference is in the line in bold. You always choose a random place from the complete array. What's wrong with this code?

8.5 Robson's Top?

What's the maximum length of array that you can shuffle using Robson's algorithm? Be careful; the answer is tricky.

8.6 Sampling Testing

Can you develop something to visually validate sampling functions, along the lines of what was required in question 8.1?

8.7 Single-Line Repeater

A reviewer of the draft for this chapter mentioned that repeatedPick(...), as shown in the "Choosing Several Values with Repetition" section on page 147, could be written as a single line, in just one statement. What would it be?

8.8 Sort to Sample

Implement the algorithm described in the section "Sampling by Sorting or Shuffling" on page 148.

8.9 Iterate, Don't Recurse

A recursive function along the lines of

```
const something = (p) => (p === 0 ? BASE : other(something(p - 1), p));
```

can be written equivalently in an iterative fashion as the following:

```
const something = (p) => {
  let result = BASE;
  for (let i = 1; i <= p; i++) {
    result = other(result, i);
  }
  return result;
};
```

Explain why this works. Also, try this conversion for the factorial(...) function from Chapter 5 and adapt it for Floyd's sampleKofN(...) algorithm (which will be trickier) to verify what was shown in the text.

8.10 No Limits?

In Knuth's sample code, there's no check to see whether i goes out of bounds; why isn't it needed?

```
const orderedSample = (arr, k) => {
  const n = arr.length;
  const sample = [];
  let toSelect = k;
  let toConsider = n;
  for (let i = 0; toSelect > 0; i++) {
    if (Math.random() < toSelect / toConsider) {
      sample.push(arr[i]);
      toSelect--;
    }
    toConsider--;
  }
  return sample;
};
```

9

SEARCHING

This chapter deals with a common problem: given a set of values, find whether a certain key is in the set. This definition has similar aspects to logic that we'll explore in future chapters when you implement a dictionary abstract data type (ADT), but we'll be concerned only with the search part here. We won't look at adding or deleting keys. In addition, we'll deal only with arrays and explore other data structures in future chapters.

Search Definition

In all cases in this chapter, the problem to solve is that, given an array (ordered or unordered, possibly with repeated values) and a key, you want to learn at which position of the array you can find the key. You'll return -1 to match several of JavaScript's own methods if the key isn't in the array.

As a further optional requirement, you may sometimes want to find the first (or last) occurrence of a key in an array (in the case where the array has repeated keys), or if an array doesn't include a key, you might want to know in which position it should have been.

It's important to keep in mind whether you're doing a single search or many searches. If the latter, you may amortize over time the cost of, say, sorting the data or building some other data structure. If the former, you just want the speediest possible search. (In later chapters you'll explore examples of data structures that help make searches faster.)

There are more efficient algorithms for sorted arrays than for unsorted ones; we'll start with the latter and then move on to better-performing ones.

Searching Unsorted Arrays

The first set of algorithms you'll consider performs a linear search in a disordered array and is the most basic (these are also known as sequential or serial searches). If the array is not in any order, there's no other way to search than to start at the beginning and go through the whole array. This kind of search is obviously slow and $O(n)$, but for small arrays, it's quite reasonable. Furthermore, JavaScript has its own methods for this kind of search, so if the conditions of your problem allow it, the functions in the next section are probably the best bet.

JavaScript's Methods

To find whether a given key is in an array along with its position, JavaScript provides several interesting functions. If you just want to know whether the key is there, you can use the `array.includes(key)` method, which returns true or false depending on whether the key was found. If you want the position of the key in the array (which you'll want throughout this chapter), then `array.indexOf(key)` does the job. It returns the first position at which the key is found or -1 if it wasn't found.

These methods all perform in $O(n)$ time, and they go through the array from the beginning to the end. This performance matches that of the linear search you'll consider next.

Linear Search

The linear search algorithm is basically an implementation of JavaScript's own `.indexOf(...)` method. It searches simply by looping through the whole array checking whether it finds what you want.

Figure 9-1 shows two searches: a successful one for 60 and an unsuccessful one for 50.

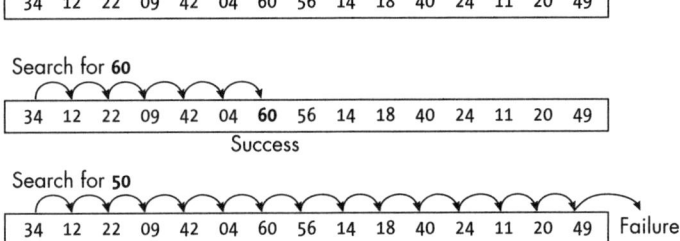

Figure 9-1: A linear search starts at the beginning and advances until it finds the desired key or gets to the end of the array.

The process starts at the beginning and continues until it reaches the desired key or the end of the array. This kind of algorithm is often taught very early to future developers as a basic example of looping. Here's an implementation:

```
const linearSearch = (arr, key) => {
  const n = arr.length;
❶ for (let i = 0; i < n; i++) {
  ❷ if (arr[i] === key) {
      return i;
    }
  }
❸ return -1;
};
```

Loop through the array ❶, and if you find the key you want ❷, return its position. If the loop ends without success ❸, return -1 as defined.

The performance of this algorithm is $O(n)$ in the worst case, and for successful searches, you do an average of $n/2$ probes, so the result is still $O(n)$. There's no way to seriously speed up a search in a disordered array, but a small trick may help a little bit: using a sentinel.

Linear Search with Sentinels

Before searching for a key, append that value at the end of the array, so the search is guaranteed to succeed. Figure 9-2 shows the same two searches from the previous section: the first is successful because it found 60 before the end of the array, but the second is a failure because it found only the 50 that was added, as that value wasn't originally in the array.

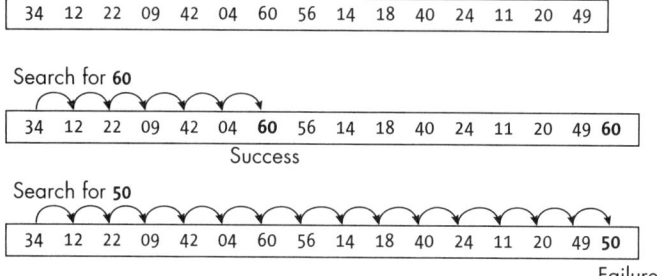

Figure 9-2: Adding a sentinel lets you advance without passing the end of the array.

You can loop through the array without checking for its end, because you know for sure that you'll eventually find the key. The only consideration now is where you find it: if it's at the end, you found only the added sentinel, so the search was unsuccessful. Here's the logic:

```
const linearSearch = (arr, key) => {
  const n = arr.length;
❶ arr[n] = key;
❷ let i = 0;
❸ while (arr[i] !== key) {
    i++;
  }
❹ arr.length = n;
❺ return i === n ? -1 : i;
};
```

Start by adding the key to search for at the end of the array ❶. Then start searching at the first position as before ❷ until you find the key ❸. You don't have to check for the end of the array because you know you'll find the key you want. After finding the key, restore the array ❹ (assigning a new length is enough for this, but the .pop(...) method is probably more common) and then decide what to return depending on where you find the key ❺. If it's at the end, it's a failure, and if it's earlier, it's a success.

The performance of this algorithm is still $O(n)$; the only (slight) advantage may derive from the simpler check in the loop, so the iteration goes faster, but don't expect a big impact from that. In addition, if adding the sentinel causes JavaScript to create a new array and copy the old one, the algorithm could become even slower.

When all is said and done, using linear search can't be enhanced by much, and you've gone as far as possible with unordered arrays. You'll now move on to doing searches in ordered (sorted) arrays, for which far better algorithms are possible.

Searching Ordered Arrays

If the array to be searched is ordered, you can apply better techniques. For instance, if you learn that the value at a certain position of the array is greater than the key you're searching for, you can instantly discard all values after that position, because the key can't be there. All the nonlinear search algorithms (also known as *interval searches*) in this section take advantage of order either to advance more quickly through the array or to discard large portions of it, reducing the area to search.

Jump Search

The basic linear search described earlier potentially goes through the complete array, which makes for $O(n)$ performance with no possibility for enhancement. However, if the array is ordered, you don't have to go through it one by one. Just as when someone is looking for a certain page in a book, they won't turn each page one by one; they'll skip through the book several pages at a time, and then go one by one when closer to the goal.

The idea behind the jump search algorithm is similar to that of Shell sort (see Chapter 6). You start with big jumps to get quickly to the vicinity of the key that you want and then do smaller jumps. Figure 9-3 shows how we'd search for 42, assuming an initial jump size of 4. (We'll get to what the jump size should be later.)

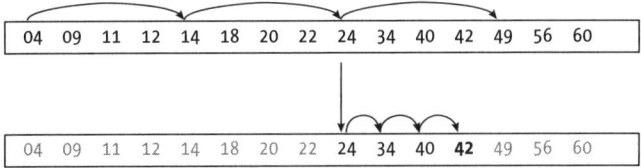

Figure 9-3: A jump search tries to advance faster by making big jumps whenever possible.

At the top you do a linear search but with big jumps, skipping four values at a time. The first value you find is 04, which is too low, so you jump again. Next you find 14 and then 24, but both are still too low. The next jump gets you to 49, so now you know that 42 is (if present) after the 24 and before the 49. You then start a regular linear search, with short jumps advancing one position at a time. You check 34 and 40 and succeed at 42. If you had been searching for 41 instead, when you reached 42, you'd decide that 41 wasn't present and return −1.

What's the expected number of tests for this algorithm? If the step size is s and the array size is n, you can have up to n/s long jumps, followed by s short jumps, which is $n/s + s$ in total. Some calculus proves that this is optimum when s is $\sqrt{n}$, for a maximum of $2\sqrt{n}$ tests, so we'll use that value, as shown in the following implementation:

```
const jumpSearch = (arr, key) => {
  const n = arr.length;
❶ const s = Math.max(2, Math.floor(Math.sqrt(n)));
❷ let i = 0;
❸ while (i + s < n && key >= arr[i + s]) {
    i += s;
  }
❹ while (i + 1 < n && key >= arr[i + 1]) {
    i++;
  }
❺ return i < n && key === arr[i] ? i : -1;
};
```

Start by determining the size of the long jump ❶, making sure that it's at least 2. (The jump size could be 1 only for very short arrays.) The i variable goes through the array ❷. Start by jumping s places every time ❸. If you don't go past the end of the array and the array value you test is not greater than the key you are looking for, you can do a jump by updating i. After this series of jumps, i points to a value not greater than the key you want, and you do a new loop, advancing by 1 ❹. After this loop ends ❺, if you find the key, return its position; otherwise, return -1.

The code has two similar while loops. In one case you jump by s, and in the other you jump by 1.

Consider another implementation that suggests a more enhanced solution. First, here's the code:

```
const jumpSearch = (arr, key) => {
  const n = arr.length;
❶ let s = Math.max(2, Math.floor(Math.sqrt(n)));
  let i = 0;
❷ while (s > 0) {
  ❸ while (i + s < n && key >= arr[i + s]) {
      i += s;
    }
  ❹ s = s > 1 ? 1 : 0;
  }
❺ return i < n && key === arr[i] ? i : -1;
};
```

You're now not using a const for s ❶ because you'll change its value for the second loop. Set up an external loop ❷ depending on s. When it reaches 0, you're done. The internal loop is the same as before, where you always jump s places at a time ❸, but s later becomes 1 to make short jumps. After finishing a loop ❹, reduce s. If it was greater than 1, jump by 1, and if it already was 1, end the loop, setting it to 0. Deciding what value to return ❺ is exactly the same as in the previous version of the algorithm.

You did the search in two stages: long jumps first, followed by shorter jumps, and you got the searches down to $O(\sqrt{n})$. What if you had three stages, with very long jumps first, followed by not-as-long jumps, and shorter jumps to finish? Jumps at one level are proportionally greater to jumps in the next level. (We'll revisit this concept of jumps that decrease in size when looking at skip lists in Chapter 11.) Figure 9-4 shows how it works with three jump levels, searching for a value in an array of 27.

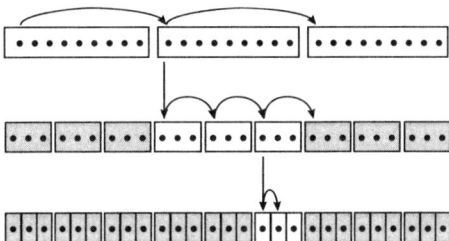

Figure 9-4: A search in three levels

The first jumps are nine elements apart. After finding in which block of nine the key should be, you start with jumps three elements apart; grayed elements are places where the key you want can't be. After you learn in which block of three the key is (allowing you to gray out more elements), you finish with single-element jumps. In this case you'd have, on average, 4.5 tests to find a key, or 9 test tops.

You can code it like this:

```
const jumpSearch = (arr, key, levels = 3) => {
  const n = arr.length;
❶ const b = Math.max(2, Math.floor(arr.length ** (1 / levels)));
❷ let s = Math.floor(n / b);
❸ let i = 0;
❹ while (s > 0) {
    while (i + s < n && key >= arr[i + s]) {
      i += s;
    }
❺ s = Math.floor(s / b);
  }
❻ return i < n && key === arr[i] ? i : -1;
};
```

Start by defining the number of blocks b at each level ❶, and again (as with the jump size in previous algorithms) you want to have at least two blocks at each level. Then set the initial (longest) jump size ❷. And then proceed to search by levels in the same way as before: starting at the beginning ❸ and continuing the search until the number of jumps becomes zero ❹. The difference is how you reduce the jump size ❺, making it b times smaller every time. (Because b > 1, s is guaranteed to eventually reach 0, making the outer loop end.) The final return ❻ is the same as with other versions of jump searching.

It can be shown that this scheme leads to $O(^3\sqrt{n})$ test if you choose jumps that are each $^3\sqrt{n}$ greater than the next, so the algorithm is even better. You could keep adding more and more levels (though, of course, that would be meaningful only with a seriously large array) and get the order of the algorithm to $O(\sqrt[4]{n})$, then $O(\sqrt[5]{n})$, and so on. (Question 9.3 shows just how far you can go.)

We've managed to speed up the search algorithms from $O(n)$ to $O(\sqrt[p]{n})$, if we search in p levels. Let's try a different approach and see if we can do even better.

Binary Search

Now try applying the divide-and-conquer idea to create a search algorithm: given an array to search, check its center value. If it's the value you want, you're done. If the center value is greater than the value you want, you can discard the right side of the array and search the left portion recursively. Similarly, if the center value is greater than the value, discard the left side and search the right. If at some point you have to search an empty array, you know the value wasn't present. Figure 9-5 illustrates the procedure where the search value is 18.

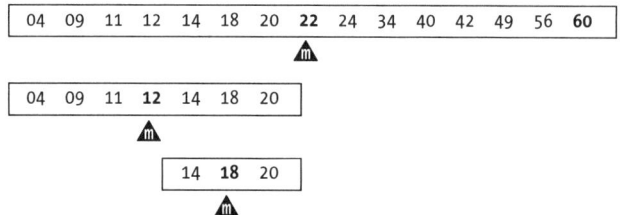

Figure 9-5: A binary search recursively splits the array to be searched in half at every pass.

In Figure 9-5, the triangle with an m points to the middle element of the array. Initially, the middle element is 22, so 18 (if present) had to be on the left side; you can discard the rest. The second line shows how you'd continue: the middle element is 12, so you search on the right. On the third line you succeed, as the middle element is what you wanted. You can code the method as follows:

```
const binarySearch = (arr, key, l = 0, r = arr.length - 1) => {
❶ if (l > r) {
    return -1;
  } else {
  ❷ const m = (l + r) >> 1;
  ❸ if (arr[m] === key) {
      return m;
  ❹ } else if (arr[m] > key) {
      return binarySearch(arr, key, l, m - 1);
  ❺ } else {
      return binarySearch(arr, key, m + 1, r);
    }
  }
};
```

If at any time the interval to search is empty ❶, the search was a failure. If not ❷, compute the middle of the interval. Using the right shift >> operator is an elegant and concise way of doing this rather than the more pedestrian Math.floor((l+r)/2). If the value in the middle is the key that you're looking for ❸, you're done. If the middle value is greater than the key you want ❹, search the left portion of the array; otherwise, search the right ❺.

Since all of the recursion here is tail recursion, you can easily convert this method to an equivalent iterative one. In Figure 9-6, the l and r (left and right) triangles show the portion of the array that you are searching, and the m (middle) triangle indicates the middle point of that portion. Once more, the search value is 18.

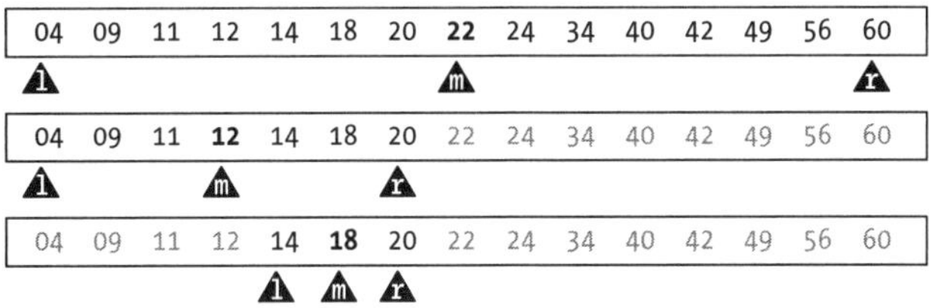

Figure 9-6: An iterative version of the algorithm uses two pointers (l and r) to keep track of the portion of the array that you're searching.

Grayed-out values won't be considered further in the algorithm. Depending on the result of the comparison of the middle value with the key you want, you update l or r and loop again until you succeed or fail.

How do you recognize a failed search? If you were searching for 17 instead, the search procedure would have continued as shown in Figure 9-7.

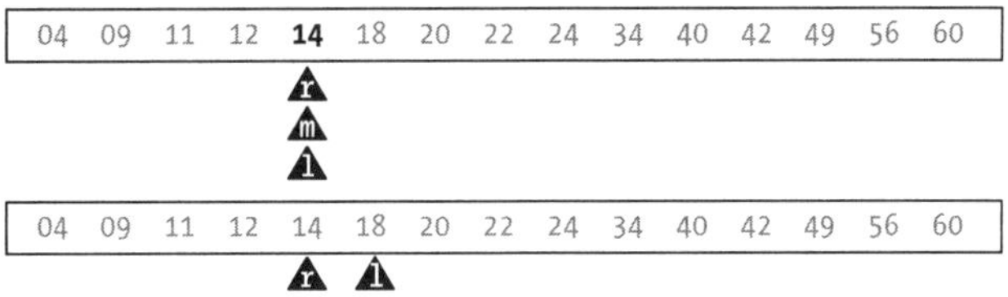

Figure 9-7: If the l and r pointers get "crossed," you can conclude that the search was unsuccessful.

When the search fails, the l and r pointers become "crossed," which means the value can't be in the array. (The missing value, 17, should have been to the left of l and to the right of r.)

You can implement the algorithm as follows:

```
const binarySearch = (arr, key, l = 0, r = arr.length - 1) => {
❶ while (l <= r) {
  ❷ const m = (l + r) >> 1;
  ❸ if (arr[m] === key) {
     return m;
  ❹ } else if (arr[m] > key) {
     r = m - 1;
```

```
❺ } else {
      l = m + 1;
    }
  }
❻ return -1;
};
```

Keep searching ❶ as long as the l and r pointers don't cross. You calculate the middle m the same way ❷ as for the recursive binary search. If the middle value equals the key you want ❸, you're done. If the middle key is greater than the value you want ❹, update the right pointer r to keep looking in the left portion; otherwise, change the l pointer to search on the right ❺. If the loop ends without finding the key you want ❻, we return -1 to show the failure.

What's the performance of this method? We have seen a similar analysis before, and it should remind you of quicksort, for example. Each step halves the size of the array to search, so the order of the binary search is $O(\log n)$, which is a very good improvement on all the previous algorithms you've seen. (For math-oriented readers, question 9.4 calculates the actual average number of tests.) Let's consider another algorithm that shows a similar performance, which actually uses binary search.

Exponential Search

Exponential search (also known as *doubling* or *galloping search*) is a combination of two methods: first you determine in which range of the array you should find the desired key, and then you apply binary search to finish the job. For the first step, you want to find a value in the array that's greater than the key you want to find, and you test the value at position 1; then the value at position 2; then at positions 4, 8, 16; and so on, always doubling, until you decide where to continue searching. Figure 9-8 shows how the algorithm would search for 42.

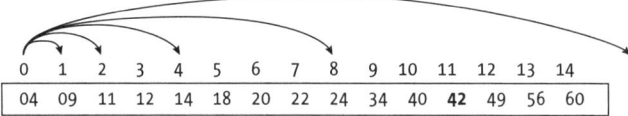

0	1	2	3	4	5	6	7	8	9	10	11	12	13	14
04	09	11	12	14	18	20	22	24	34	40	**42**	49	56	60

Figure 9-8: An exponential search combines ever-longer jumps with binary search.

First, the doubling jumps are taken (size 1, 2, 4, and so on) until finding the section in which 42, if present, should be. (If you were looking for 22, the jumps would have ended after looking at the element in position 8, for instance.) After finding the portion of the array to search, binary search completes the process.

Here's the code:

```
const exponentialSearch = (arr, key) => {
  const n = arr.length;
❶ let i = 1;
❷ while (i < n && arr[i] < key) {
  ❸ i <<= 1;
  }
❹ return binarySearch(arr, key, (i >> 1), Math.min(i, n - 1));
};
```

First initialize the series of jumps at 1 ❶, and although you have neither reached the end nor found a value greater than the key you want ❷, double the jump size ❸ and loop again. If this line looks weird, you could also write it as i = i << 1, and the left shift << operator (you've already used the right shift operator in binary search) makes it equivalent to i = i * 2. After completing all the necessary jumps ❹, do a binary search, again using the shift operator to divide i by 2.

What's the order of this algorithm? Let's start with the worst case, when you look for the last value in an array with up to 2^p elements. The first loop will be executed p times, and that will be followed by a binary search in an array with a size of less than 2^{p-1}: that also is $O(p)$. As p is approximately log n, the total performance is, at worst, $O(\log n)$, but it'll be better the closer the element is to the beginning. In fact, if the key is found at position k, the search will be $O(\log k)$.

Interpolation Search

When searching for a word in the dictionary, no matter how well versed you are in binary search, if you are looking for a word starting with the letter S, you'd open the dictionary near the end, but if you're looking for a word starting with B, you'd open the book nearer the beginning. You can apply this idea to searching in an ordered array, if you can interpolate and estimate the position at which a given value should be, assuming that values in the array are somewhat uniformly distributed.

But first some math. If the value at position l (for left) is L and the value at position r (for right) is R (with $R > L$), the position that would correspond to value V can be calculated as $l + (r - l)(V - L)/(R - L)$. We can check this out. If V equals L, the formula produces l, which is correct. Similarly, if V equals R, the formula produces r, which is correct again. If you are searching an array with values that can be converted to numbers, you can apply this interpolation to more quickly find the value you want.

Consider this in practice. Figure 9-9 shows a search for the value 34.

0	1	2	3	4	5	6	7	8	9	10	11	12	13	14
04	09	11	12	14	18	20	**22**	24	34	40	42	49	56	60

l (0)   m (7)   r (14)

0	1	2	3	4	5	6	7	8	9	10	11	12	13	14
04	09	11	12	14	18	20	22	24	34	**40**	42	49	56	60

l (8)   m (10)   r (14)

0	1	2	3	4	5	6	7	8	9	10	11	12	13	14
04	09	11	12	14	18	20	22	24	**34**	40	42	49	56	60

l (8)   r, m (9)

Figure 9-9: An interpolation search tries to estimate the position of the searched-for value to find it faster.

To start, the left value (at position 0) is 4 and the right value (at position 14) is 60, so the estimate is that 34 should be around position 7. (See Figure 9-10; the dotted line joins the extreme values, and its intersection with a horizontal line at height 34 is between 7 and 8.) Since the value at that position is smaller than 34, move the left pointer to position 8. Then redo the estimation with 24 at 8 and 60 at 14, so 34 should be at 10. The value there is higher than 34, so now move the right pointer to 9. The third iteration is successful, as 34 is estimated to be (and found at) position 9.

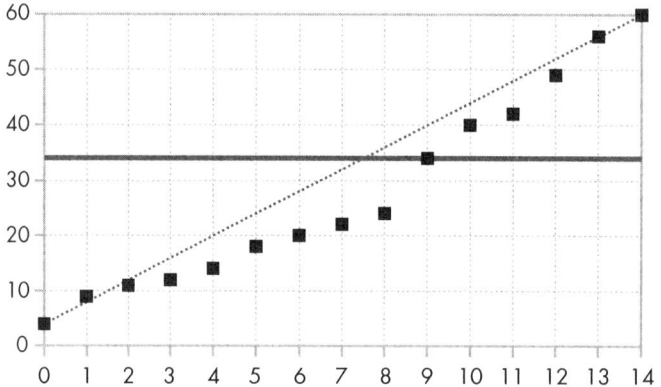

Figure 9-10: Estimating the position for 42, if 4 is the start of the array and 60 is the end

You can directly implement the method:

```
const interpolationSearch = (arr, key) => {
❶ let l = 0;
  let r = arr.length - 1;
❷ while (l <= r) {
  ❸ const m =
      arr[l] === arr[r]
        ? l
        : Math.round(l + ((r - 1) * (key - arr[l])) / (arr[r] - arr[l]));
```

```
❹ if (m < l || m > r) {
      return -1;
❺ } else if (arr[m] === key) {
      return m;
❻ } else if (arr[m] > key) {
      r = m - 1;
❼ } else /* arr[m] < key */ {
      l = m + 1;
   }
}

❽ return -1;
};
```

Start by setting variables l and r to point to the extremes of the search range ❶ as in binary search, and the loop ❷ is the same as in that algorithm. Instead of calculating m as the middle point of l and r ❸, use the interpolation formula, but check whether the values at the extremes are equal because then you'd be dividing by zero. If m ends outside the interval from l to r, the value you want isn't in the array (because the key you want must be smaller than the value at l or greater than the value at r), and you return -1 ❹. On the other hand, if m lies between l and r inclusive, compare the value at its position to the desired key. If the value is equal, you succeeded ❺. If the value is greater, search to the left ❻, and if the value is smaller, search to the right ❼. If the key wasn't found, return -1 ❽.

This method performs well, but it has a couple of drawbacks. First, you must be working with numeric keys (as in the example) or with keys that can be transformed to numbers in order to perform interpolation (one possibility could be transforming characters to their ASCII or Unicode equivalents). This challenge needs to be solved in order to use interpolation search with more general keys.

The second possible drawback has to do with the algorithm's performance. Unlike binary search that always halves the search range at each stage, interpolation might do not so well (quicksort behaved similarly), and the performance would be $O(n)$. (A possible case for bad performance is if values in the array are in a geometric progression, which means linear interpolation won't produce good estimates; however, such a distribution isn't really very likely.) On the other hand, if values are uniformly distributed, it can be shown (you won't see it here) that performance will be $O(\log \log n)$, which is a great improvement.

Summary

In this chapter we've considered algorithms for searching ordered or unordered arrays of values, which is a common function. The methods covered have different rates of performance, and several of them are based on previous methods in order to illustrate interesting algorithm development techniques.

This chapter ends the second part of the book. In Part III, we'll start our exploration of data structures, beginning with lists, an important dynamic structure with plenty of uses.

Questions

9.1 Searching Right?

Implement a framework to test a given search function and see whether it works on every element in the array and on missing elements as well. I used such a test on all of my code for this chapter, and I found some bugs!

9.2 JavaScript's Own

Can you implement an alternative to JavaScript's own `array.indexOf(...)` using some other available array methods?

9.3 Infinite Jump Levels?

In the generalized jump search algorithm, what happens if you want to do the search on an infinite number of levels? (Hint: imagine that `levels` is infinitely large, and see how the algorithm behaves.) What's the resulting algorithm like?

9.4 Exactly How Much?

This question is for the mathematically inclined. Calculate the actual average of number of tests for a successful search. It may help to assume that the array is of length $2^n - 1$. In this case 1 element is found in just one question, 2 elements in two questions, 4 found in three, 8 in four, and so on, up to 2^{n-1} found in n questions.

9.5 Three Tops Two?

Inspired by binary search, which uses a comparison to split the array to search in two, you could think of a ternary search, using comparisons to split the array in three, so you'll have a smaller subarray to work with. How does this compare to binary search? Is ternary search really an improvement?

9.6 Binary First

Assuming that the sorted input array may have repeated values, modify binary search to return the first position of the searched key in the array. If you want to find the *last* position instead, what changes would you need?

9.7 Count Faster

Given an input sorted array and a key, you could find how many times the key appears by writing something like `count = arr.filter(x => x === key).length`, but that would run in $O(n)$ time. Can you find that count in $O(\log n)$ time?

9.8 Rotation Finding

Assume you have an array that was originally sorted, but later it possibly got rotated: for instance [4, 9, 12, 22, 34, 56, 60] might have become [34, 56, 60, 4, 9, 12, 22]. Write a function that determines the position of the lowest value in the rotated array. For instance, in this example, the function should return 3. Make sure your function also works for an array that *isn't* rotated.

9.9 Special First?

I found several implementations of exponential search that specifically tested, before any looping, whether the first position of the array had the key. Does the version in this book need this?

PART III

DATA STRUCTURES

In the third part of the book we'll look at a variety of data structures, the problems they can be applied to, and the corresponding algorithms we'll need.

10

LISTS

In the previous chapters we explored algorithms that perform several generic tasks, and in this chapter we'll study data structures for specific objectives, beginning with the most basic one: a list of elements. Lists are quite simple, but the concepts behind lists appear in many other structures, as you'll learn in the rest of the book. In fact, lists are at the center of the most antiquated language still widely in use: the acronym for LISP, created in 1959, stands for "list processing."

What's a *list*? A simple definition is that a list is a sequence of elements (or values, or nodes), which implies that there's a first element and that every element (except the last) is followed by another element. Another definition, recursive in nature, is that a list is either empty (no elements) or formed by a specific element, called the head of the list, which is followed by the tail—which is another list.

We'll start by defining the basic abstract data type (ADT) for lists and how to implement it in a couple of ways. (See Table 10-1 for all operations.) It happens, however, that the ADT has some more important variants, so we'll also consider those, which will lead to implementing other structures like stacks, queues, deques, and more.

Table 10-1: Basic Operations on Lists

Operation	Signature	Description
Create	→ L	Create a new list.
Empty?	L → boolean	Determine whether the list is empty.
Size	L → number	Count how many elements are in the list.
Add	L × position × value → L	Add a value to the list at a certain position.
Remove	L × position → L	Remove a value from the list at a certain position.
At	L × position → value | undefined	Given a position, return the value at that position.
Find	L × value → boolean	Given a value, find whether it exists in the list.

For some types of lists, such as stacks, queues, or deques, we'll substitute some of the functions in Table 10-2 (possibly with different names) for the add, remove, and at operations. We may also drop some other operations, but we'll consider them case by case. For instance, instead of adding an element at any place in the list, we may want to restrict ourselves to adding new elements only at the front or at the back of the list.

Table 10-2: Extra Operations on Lists

Operation	Signature	Description
Add at front	L × value → L	Add a new value at the front of the list.
Add at back	L × value → L	Add a new value at the back of the list.
Remove from front	L → value | undefined	Remove a value from the front of the list.
Remove from back	L → value | undefined	Remove a value from the back of the list.
At front	L → value | undefined	Get the value at the front of the list.
At back	L → value | undefined	Get the value at the back of the list.

Finally, we'll also be able to use lists to represent other ADTs, such as sets or maps (see Chapter 11).

Basic Lists

Let's start with the most basic implementation of a list, which may be good enough for many applications, and then move on to a dynamic memory version, which is able to deal with more complex situations and structures.

Implementing Lists with Arrays

Given that JavaScript implements *dynamic arrays*, which can grow larger or become smaller as needed, using arrays for lists seems logical, and for most applications that's the case. However, expanding an array often requires moving the whole array to a new, larger space in memory, so operations may not be as instant. (The inner details of how JavaScript allocates space for arrays isn't clear, but if you keep adding elements, at some point, JavaScript will run out of space and have to allocate more space somewhere else and move the array there.) Obviously, with small, short lists, you won't be able to perceive the impact, but for large structures, it could become noticeable.

You can implement all the operations for the ADT in a minimum number of lines, taking advantage of available JavaScript methods as follows. create was renamed newList to make its function clearer, and Empty? was renamed isEmpty because of JavaScript naming rules.

```
❶ const newList = () => [];

❷ const size = (list) => list.length;

❸ const isEmpty = (list) => size(list) === 0;

❹ const add = (list, position, value) => {
     list.splice(list, position, value);
     return list;
   };

❺ const remove = (list, position) => {
     list.splice(list, position);
     return list;
   };

❻ const at = (list, position) => list[position];

❼ const find = (list, value) => list.includes(value);
```

Creating a new list ❶ is just a matter of producing an empty array. The list size is the array's length ❷, and to check whether a list is empty, test whether its size is 0 ❸. Adding an element at a given position ❹ is tailor-made for the splice(...) standard method, which is also used to remove an element ❺. Finally, accessing the element at a given position ❻ is trivial. (The latest version of JavaScript provides an .at(...) method, which is somewhat different from what is defined here because of the possibility of using negative indices; see *https://developer.mozilla.org/en-US/docs/Web/JavaScript/Reference/Global_Objects/Array/at.*) Finally, use the .includes(...) method to see whether a list includes the value ❼.

Table 10-3 shows the performance of these operations.

Table 10-3: Performance of Operations for Array-Based Lists

Operation	Performance
Create	$O(1)$
Empty?	$O(1)$
Size	$O(1)$
Add	$O(n)$
Remove	$O(n)$
At	$O(1)$
Find	$O(n)$

Creating a new list, checking whether it's empty, getting its size, and accessing the element at a given position are all $O(1)$ operations. As expected, finding a value is $O(n)$, because the operation needs to go through the whole list. On the other hand, adding and removing elements are $O(n)$ operations, because they basically move the whole array to a different place in memory. If you implement lists dynamically, these results will change.

Implementing Lists with Dynamic Memory

Languages that support dynamic memory provide a different way to deal with varying-length lists: through pointers. You can include a reference to an object in another object along the lines of the following code:

```
const first = {
  name: "George",
  next: null,
};

const second = {
  name: "John",
  next: null,
};

const third = {
  name: "Thomas",
  next: null,
};

first.next = second;
second.next = third;
```

Given only the pointer to the first object, you can list the next object's name with first.next.name, for example; then first.next.next.name would list the third object's name. All of this is standard JavaScript notation. The last object has its next attribute with a null value, meaning there's no next object in the list.

Figure 10-1 represents pointers with arrows and a null pointer with a line ending in a circle. Of course, you're not limited to having a single pointer in a node; you can have as many as you want. Let's start with a simple case: an example of a list with six elements is shown in Figure 10-1, where first points to the head.

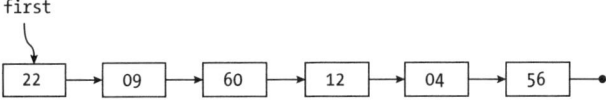

Figure 10-1: A simple list

Adding a new element requires changing a pointer. For instance, Figure 10-2 shows adding an 80 after the 60.

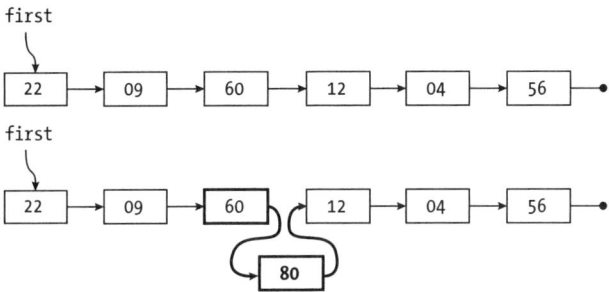

Figure 10-2: Adding a new element to the list requires changing only a single pointer in a node.

The same result occurs when removing an element; you need to change only a single pointer—usually the one from the previous element, or first itself if removing the head of the list. In the next example, let's remove the 60 (see Figure 10-3).

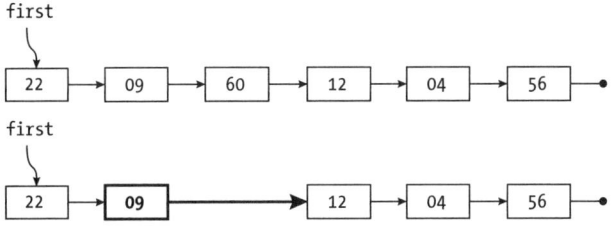

Figure 10-3: Removing an element from the list also requires just changing a single pointer.

Adding and removing elements by themselves are $O(1)$ operations. (Of course, this assumes you already know where to effect the changes and also what other element points to the one you wanted to remove.) Let's consider functioning code for all the possible operations.

Creating a List

A list is just an object, which may have a link to another object, and so on. An empty list is a null pointer. With that in mind, creating a new, empty list is simple, and so is checking whether you have an empty list or calculating a list's size:

```
❶ const newList = () => null;

❷ const isEmpty = (list) => list === null;

❸ const size = (list) => (isEmpty(list) ? 0 : 1 + size(list.next));
```

Creating a list produces a null pointer that eventually points to the list's head ❶. Checking whether a list is empty ❷ means seeing whether the pointer is null. Finally, calculating the list's size is simple with recursion: an empty list has a size of 0, and a nonempty list has a size of 1 (for the list's head) plus whatever the list's tail size is ❸.

Adding a Value

To figure the list's nodes, use objects with a value (a key or whatever you want to add to the list) and a pointer (ptr) to the following element in the list:

```
const add = (list, position, value) => {
❶ if (isEmpty(list) || position === 0) {
  ❷ list = { value, next: list };
  } else {
  ❸ list.next = add(list.next, position - 1, value);
  }
❹ return list;
};
```

The add(...) recursive function gets a pointer to a list and the position in which to add the new value. If the pointer is null or if the position is zero ❶, the new node goes at the beginning of the list ❷, pointing to whatever was the first element of the list earlier. Otherwise, go down the list recursively to the next node ❸. After the new value is added, return a pointer to the updated list ❹.

Removing a Value

To remove an element from a list, you have two options: remove the first element (in which case the pointer to the first element of the list must be changed) or remove another element in the list (and then modify the pointer in the previous node as mentioned earlier). The following code does all of it:

```
const remove = (list, position) => {
  if (isEmpty(list)) {
  ❶ return list;
```

```
  } else if (position === 0) {
❷ return list.next;
  } else {
❸ list.next = remove(list.next, position - 1);
  return list;
  }
};
```

If the list is null, just return it ❶; you can't do anything else. If it's not null and you want to remove its head, the new list is the list's tail ❷. Finally, if the list isn't null and you don't want to remove its head, advance to the next place in the list to attempt the removal again ❸, but now the position to remove is one less than before. In all cases return a pointer to the list after the removal.

Getting the Value at a Position

You can get the value at a given position in a naturally recursive way by considering several cases. If the list is null, it has no value to return, so you'll return undefined. If the list isn't empty and the position you asked for is 0, you want the first element of the list. If the list isn't empty and you want some element further down the list, advance by one position and apply recursion. The following code does exactly that:

```
const at = (list, position) => {
❶ if (isEmpty(list)) {
    return undefined;
❷ } else if (position === 0) {
    return list.value;
❸ } else {
    return at(list.next, position - 1);
  }
};
```

The logic closely follows the three cases: checking for an empty list ❶, testing for the head of the list ❷, and using recursion to advance down the list ❸.

Searching for a Value

Finally, you can search a list to see whether it includes a given value. This operation isn't as common, but you'll do it anyway to gain more experience with this structure. The general logic is similar to at(...) in the example you just saw. Assume you have a ptr pointer to an element of the list. If the pointer is null, the value isn't in the list. Otherwise, if the object ptr points to has the value you want, you've found it. If the value isn't what you want, keep searching from the next node onward. Here's the recursive logic:

```
const find = (list, value) => {
  if (isEmpty(list)) {
```

```
❶ return false;
  } else {
❷ return list.value === value || find(list.next, value);
  }
};
```

If the list is empty, ❶ return false. Otherwise, if the head of the list is the value you want ❷, return true. If it isn't what you want, search the list's tail. (Note that you're grouping the two tests together by using JavaScript's || operator.)

Considering Performance for Dynamic Memory Lists

To wrap up this discussion of dynamic memory lists, let's analyze their performance (see Table 10-4).

Table 10-4: Performance of Operations for Dynamic Memory Lists

Operation	Performance
Create	$O(1)$
Empty?	$O(1)$
Size	$O(n)$
Add	$O(n)$
Remove	$O(n)$
At	$O(n)$
Find	$O(n)$

As with the array-based implementation, creating a new list and checking whether it's empty are both $O(1)$ operations, but all other operations become $O(n)$! This difference suggests that simply implementing arrays with pointers, as was shown earlier, isn't the best solution. However, some varieties of lists that have a more specific set of operations suited to their particular requirements achieve better performance.

Varieties of Lists

For some tasks, a more specialized ADT is needed than the basic ADT and implementation for common lists that you explored in the previous section. Specifically, we'll consider stacks, queues, deques, and circular lists, including their specific operations and applications.

Stacks

A *stack* is a last-in, first-out (LIFO) data structure, similar to an actual physical stack of plates: imagine you can add a plate only to the top of the pile or remove only the top plate; adding or removing middle plates isn't allowed.

Stacks behave in the same way. You'll add and remove only at the top, and these operations are usually known as *push* and *pop*, respectively. You'll also want to check whether a stack is empty (as with common lists) and learn the value at the top. (Sometimes the pop operation is defined to return the updated stack and also what the top value is. In that case you wouldn't need an operation to get the top value of the stack, since you could just pop it, use it, and push it again.)

Table 10-5 sums up the operations you'll need, and as mentioned previously, you're dealing with a smaller, more specific set of operations here.

Table 10-5: Operations on Stacks

Operation	Signature	Description
Create	$\rightarrow$ S	Create a new stack.
Empty?	S $\rightarrow$ boolean	Determine whether the stack is empty.
Push	S $\times$ value $\rightarrow$ S	Add a value at the top of the stack.
Pop	S $\rightarrow$ S	Remove the value at the top of the stack.
Top	S $\rightarrow$ value	Get the value at the top of the stack.

Stacks are frequently used in applications. For example, to explore how to implement a recursive depth-first process in an iterative fashion by using a stack, check out question 13.3 in Chapter 13. One uncommon place where you'll find a stack is in Hewlett-Packard calculators that provide reverse Polish notation (RPN): you push numbers into a stack, and then operations pop them, do whatever calculation you asked, and push the result back in.

Stacks are also used in programming languages like FORTH or Web Assembly (WASM), as well as page description languages like PostScript. Central processing units (CPUs) use stacks for subroutine calls and interruptions. If code is executing and an interruption comes in, the current status is pushed into a stack, and the interruption is processed; afterward, the normal execution resumes after popping the status from the stack. (Obviously, you could have a new interruption while processing an old one. In that case, the status for the first interruption is also pushed, then the second interruption is processed, and when finished, the status for the first interruption is popped to continue processing it.)

Finally, JavaScript itself implements a stack for calls. Whenever a function calls itself, it's as if the current status and variables were pushed into a stack before starting the recursive call. When returning from a recursive call, the old status is popped from the stack and execution recommences from where it stopped.

Data Structure

Implementing a stack with an array is simple given that you can directly use .pop(...) and .push(...) (see question 10.5). Working with linked memory is also simple, and you'll base the code on the functions you wrote for lists. In this structure you'll have a pointer to the first element, the one at the top of the stack, and each element will have a .next pointer to the element "below" it. The "bottom" element will have a null pointer. Figure 10-4 shows how it works.

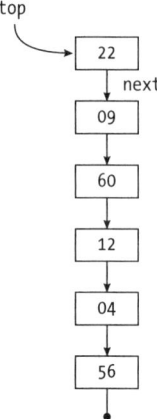

Figure 10-4: A stack implemented with dynamic memory

Pushing a new value onto a nonempty stack just requires adding a new object that points to the old top element and changing the top pointer (see Figure 10-5).

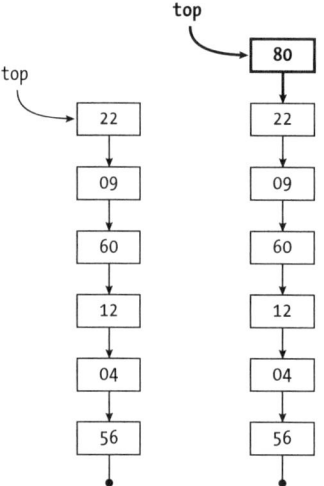

Figure 10-5: Pushing a new element on top of a stack

Popping the top element is even simpler: adjust the top pointer to point to the next one, as shown in Figure 10-6.

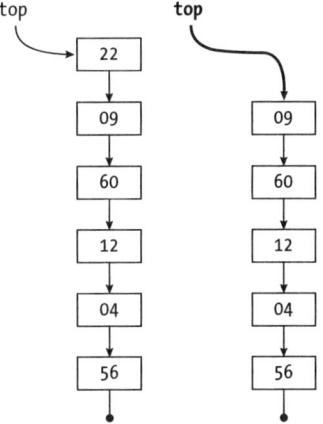

Figure 10-6: Popping the top element from a stack

In both cases (pushing and popping), you need to make simple changes to the logic when dealing with an empty stack.

Implementation

A stack is a list, so creating a stack is exactly the same as creating a generic list, as shown in the previous section; just change the name:

```
const newStack = () => null;
const isEmpty = (stack) => stack === null;
```

Examining the top requires a single line of code (all other operations on stacks are also one-liners):

```
const top = (stack) => (isEmpty(stack) ? undefined : stack.value);
```

For an empty stack, just return undefined; otherwise, stack points to the top element, so stack.value is what you want.

Pushing a value means you'll have a new element on top, which points to the element that was previously at the top:

```
const push = (stack, value) => ({ value, next: stack });
```

This logic also works if the stack is empty. Can you see why?

Finally, popping the top of the stack is also quick:

```
const pop = (stack) => (isEmpty(stack) ? stack : stack.next);
```

If the stack is empty, return it as is. You also could easily change the code, for example, to throw an error. For a nonempty stack, just return the tail of the stack.

Performance for Dynamic Memory–Based Stacks

When considering how this stack implementation performs, the results are much better than with common lists (see Table 10-6).

Table 10-6: Performance of Operations for Dynamic Memory–Based Stacks

Operation	Performance
Create	$O(1)$
Empty?	$O(1)$
Push	$O(1)$
Pop	$O(1)$
Top	$O(1)$

All operations require constant time, and that's optimal. Implementing stacks with arrays, the results would *almost* be the same with an exception: pushing a new value could require moving the array to a new, larger place in memory, and that would make pushing a value an $O(n)$ operation. In comparison to implementing common lists, which brought higher costs for most operations, implementing stacks with dynamic memory is just as good, and in a single case, even better. Now consider other variations on lists that provide similar results.

Queues

Queues are another variant of lists, and they are a first-in, first-out (FIFO) data structure. Queues work the same as a line of people waiting for something. New people enter the queue at the back (nobody may cut in), and the person at the front will exit the queue next. These two operations are *enter* and *exit* (or *enqueue* and *dequeue*), and they mimic what happens in real queues. You'll also want to check whether a queue is empty and be able to get the value of the front of the queue. Table 10-7 shows the operations you'll need.

Table 10-7: Operations on Queues

Operation	Signature	Description
Create	$\rightarrow Q$	Create a new queue.
Empty?	$Q \rightarrow boolean$	Determine whether the queue is empty.
Enter	$Q \times value \rightarrow Q$	Add a value at the back of the queue.
Exit	$Q \rightarrow Q$	Remove the value at the front of the queue.
Front	$Q \rightarrow value$	Get the value at the front of the queue.

An alternative that's sometimes used is that the *exit* operation returns both the updated queue and the value that was removed from the queue, but that's not needed given the *front* operation. You could also have a *rear* operation to access the value at the end of the queue, but that's not common.

Queues are frequently used in situations where things don't have to be (or cannot be) processed immediately and should be attended to in order such as printer queues or call center phone systems that keep you on hold until a representative is free.

Data Structure

Implementing a queue is quite simple using arrays. With linked memory, you need pointers to the first and the last nodes of the queue, so you'll represent a queue with an object that has first and last links. Each element in the queue has a next pointer to the following element, as shown in Figure 10-7. (The next element is actually in the previous place in the queue, so prev could also be the name of the pointer.)

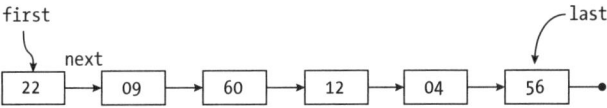

Figure 10-7: A queue implemented with dynamic memory

The first element in the queue (the next to exit) is 22; the following is 9. The last place in the queue is a 56. Adding a new element at the back of the queue simply requires modifying the pointer to the last element and the pointer in the last element itself (see Figure 10-8).

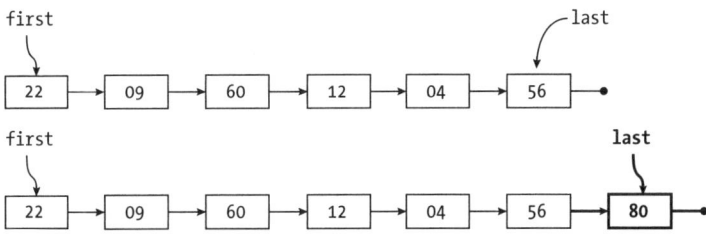

Figure 10-8: Adding an element at the back of a queue

After adding 80, the previous last element, 56, now points to the 80, and so does the last pointer.

Removing the element from the front of the queue is exactly the same as with stacks, as shown in Figure 10-9.

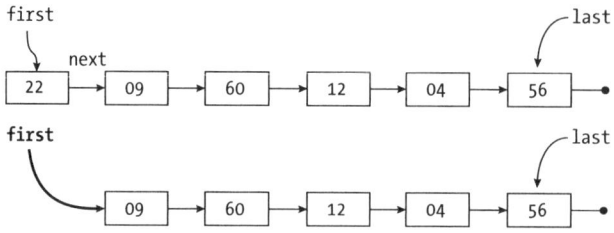

Figure 10-9: Removing an element from the front of a queue

You just have to make the first point at whatever the previous first element pointed at. Now you'll see how to implement all of this.

Implementation

Creating a new queue and checking whether it's empty is simple:

```
❶ const newQueue = () => ({ first: null, last: null });

❷ const isEmpty = () => .first === null;
```

The queue is represented by an object with two pointers ❶ that are initially null. You can tell that the queue is empty ❷ if one of those pointers is null; in fact, either both or none will be null.

Getting the value at the front (first in line) is easy:

```
const front = (queue) => (isEmpty(queue) ? undefined : queue.first.value);
```

If the queue is empty, return undefined; otherwise, queue.first points at the first element of the queue, and you return its value.

Entering a queue at the last place is a short function:

```
const enter = (queue, value) => {
  if (isEmpty(queue)) {
❶   queue.first = queue.last = { value, next: null };
  } else {
❷   queue.last.next = { value, next: null };
❸   queue.last = queue.last.next;
  }
  return queue;
};
```

If the queue was empty ❶, make the first and last pointers point to a new object, with a null pointer to the next node in the queue. Otherwise, make the last element point to a new one ❷, and then make last point to it too ❸.

Finally, exiting the queue is the same as with stacks, but with a special case when the queue becomes empty:

```
const exit = (queue) => {
❶ if (!isEmpty(queue)) {
  ❷ queue.first = queue.first.next;
  ❸ if (queue.first === null) {
     queue.last === null;
    }
  }
  return queue;
};
```

If the queue isn't empty ❶, you just have to make the first pointer ❷ point to the next element in the queue, but if the queue was emptied ❸, you also have to fix the last pointer.

Performance for Dynamic Memory–Based Queues

Given the similarity of queues and stacks (the only difference is that pop removes the first element of the stack, but exit removes the last element of the queue), it's no surprise that performance is the same, as shown in Table 10-8.

Table 10-8: Performance of Operations for Dynamic Memory–Based Queues

Operation	Performance
Create	$O(1)$
Empty?	$O(1)$
Enter	$O(1)$
Exit	$O(1)$
Front	$O(1)$

Again, all operations require constant time; using an array wouldn't be as good (see question 10.9).

Deques

The next variation on lists doesn't really have very many applications (stacks and queues are far more common), but their implementation introduces the interesting concept of double (forward and backward) linking. Assume a queue where entering or exiting is allowed at both ends. (Think of a train with several cars; new cars can be added only at the ends, and cars can be removed only from the ends.) This type of list is called a *deque* (pronounced like "deck"), which stands for "double-ended queue."

Table 10-9 shows operations necessary for deques.

Table 10-9: Operations on Deques

Operation	Signature	Description
Create	→ D	Create a new deque.
Empty?	D → boolean	Determine whether deque is empty.
Enter at front	D × value → D	Add a value at the front of the deque.
Enter at back	D × value → D	Add a value at the back of the deque.
Exit from front	D → D	Remove the value at the front of the deque.
Exit from back	D → D	Remove the value at the back of the deque.
Front	D → value	Get the value at the front of the deque.
Back	D → value	Get the value at the back of the deque.

Basically a dequeue is the same as a queue, except that you enter at or exit from both ends. Similarly, you also need operations to get the values at both extremes; for queues, you looked at only the first (front) item.

Data Structure

Is it possible to implement deques with linked memory? Since you now have full symmetry for all operations, you need links that go in both directions. Figure 10-10 shows how it works: if you were to drop all left-pointing links (or all right-pointing links), you'd be left with a common queue. In this structure, you'll again have first and last pointers to the extremes of the deque, and each node will have next and prev (previous) pointers to the contiguous nodes.

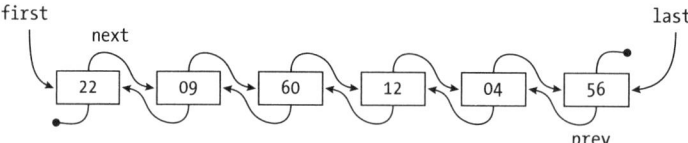

Figure 10-10: Implementing a deque requires two pointers at each node.

Because of the symmetry, operations on one end are totally analogous to the same operation at the other end, so let's just work at the end of the deque (see Figure 10-11).

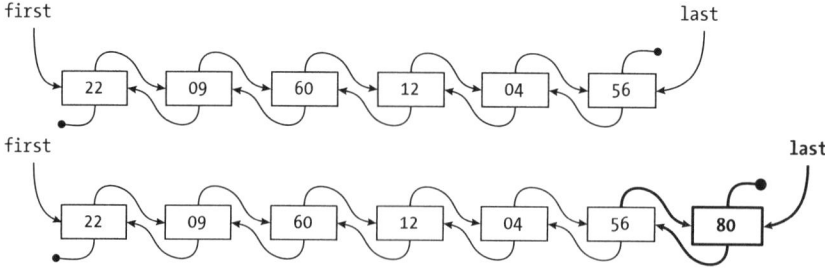

Figure 10-11: Adding an element at one extreme of a deque

Adding a value at the end is the same as for a queue, with the addition that the newly added node must point to the node that was previously at the end of the deque. (Working at the other end is exactly the same, so we'll skip it.)

Removing an element from the end of the deque is the same as shown in Figure 10-11, but from the bottom up; see Figure 10-12.

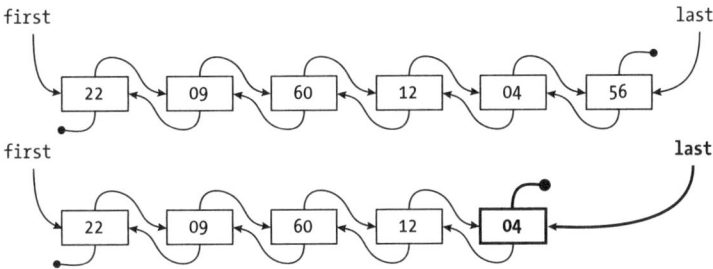

Figure 10-12: Removing an element from one extreme of a deque

When removing a value from the back, modify the corresponding pointer (last) and the next pointer of the new extreme of the deque; working at the other extreme entails modifying first and a prev pointer.

Deletions are simple in doubly linked lists. If you have a pointer to some element and want to remove it (say, the 60 in the list shown in Figure 10-13), doing so is easy.

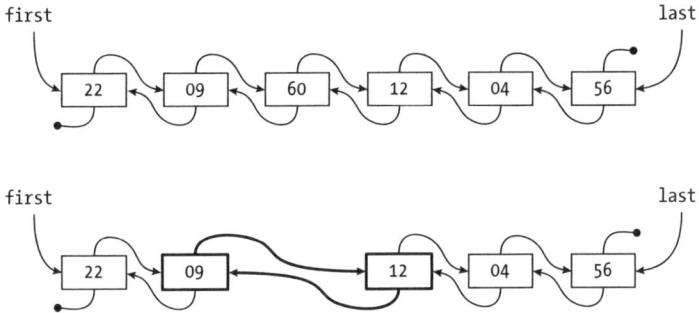

Figure 10-13: Removing an element somewhere in a deque

The key is that all nodes have pointers to both neighbors, so you have to do something along the lines of the following code, assuming that ptr points to the node to be removed:

```
ptr.prev.next = ptr.next;
ptr.next.prev = ptr.prev;
```

This kind of pointer work is common, but it can be jarring the first time you see it, so it merits careful study. The code works for elements in the middle of the deque. For elements at both ends, you need to make minor changes, as well as adjust at least one of (and possibly both) the first and last elements. Even if you don't ever use deques, the concept of double links and the ease of extracting any element from the middle is the key takeaway from this section. You'll use this for circular lists later in this chapter and in future chapters as well.

Implementation

Creating a deque and checking whether it's empty are exactly the same as with a queue, since you have the same first and last pointers:

```
const newDeque = () => ({ first: null, last: null });
const isEmpty = (deque) => deque.first === null;
```

Adding a new element to a deque is the same as entering a queue; the only difference is that you can add it at either extreme, subtly changing what pointers you modify:

```
❶ const newNode = (value, prev = null, next = null) => ({ value, prev, next });

const enterFront = (deque, value) => {
  if (deque.first === null) {
❷   deque.first = deque.last = newNode(value, null, null);
  } else {
❸   const newValue = newNode(value, deque.first, null);
    deque.first.next = newValue;
    deque.first = newValue;
  }
};

❹ const enterBack = (deque, value) => {
  if (deque.last === null) {
    deque.first = deque.last = newNode(value, null, null);
  } else {
    const newValue = newNode(value, null, deque.last);
    deque.last.prev = newValue;
    deque.last = newValue;
  }
};
```

You use an auxiliary function to create a new node with its pair of pointers ❶. Entering at the front requires changing both first and last if the deque is empty ❷. Otherwise, use the same kind of pointer work as for queues ❸. The code for entering a deque at the back is exactly the same, in symmetrical fashion: just change last to first and prev to next ❹.

Similarly, removing an element from the front or back of a deque is the same as exiting from a queue; both algorithms are symmetrical:

```
const removeFront = (deque) => {
❶ if (!isEmpty(deque)) {
❷   deque.first = deque.first.next;
```

```
❸ if (deque.first === null) {
      deque.last === null;
    }
  }
};

❹ const removeBack = (deque) => {
    if (!isEmpty(deque)) {
      deque.last = deque.last.prev;
      if (deque.last === null) {
        deque.first === null;
      }
    }
  };
```

If the deque is empty ❶, there's nothing to do. Otherwise, to remove the front element, advance to the next element of the deque ❷, and if that element is null ❸, you also adjust the last element. A bit of symmetry produces exactly the same "remove last" operation ❹.

Performance for Dynamic Memory–Based Deques

A deque is essentially a queue that goes both ways: half of its operations are exactly the same as for queues, and the rest are symmetrical, but with the same style of code, so the results are not unexpected (see Table 10-10).

Table 10-10: Performance of Operations for Dynamic Memory–Based Deques

Operation	Performance
Create	O(1)
Empty?	O(1)
Enter at front (or at back)	O(1)
Exit from front (or from back)	O(1)
Front (or back)	O(1)

All the operations of deques perform the same as those of queues. Everything is $O(1)$.

Circular Lists

Circular lists are useful for "round-robin"–style processing. For example, PCs place apps in a list and cycle through them, and after the last completes, processing returns to the first. (You'll see another example of this when looking at Fibonacci heaps in Chapter 15.) Instead of an open-ended list, a circular list joins the first and last elements together. This kind of ADT allows for continuous processing, with a "current" element and the possibility of advancing to the next, but cyclically. Table 10-11 shows the operations we'll need.

Table 10-11: Operations on Circular Lists

Operation	Signature	Description
Create	→ C	Create a new circular list.
Empty?	C → boolean	Determine whether the circular list is empty.
Add	C × value → C	Add a new value before the current one and make it current.
Remove	C → C	Remove the current value from the list and advance.
Current	C → value	Get the current value from the circular list.
Advance	C → C	Advance to the next value in the list cyclically.

Some variations and changes are possible. You could require a "go back" (retreat) operation that performs in the opposite direction as an "advance." You could also use an "add after current," but you could achieve that by first advancing and then using the add operation. These changes aren't significant, and the structure is useful as shown. The work you did with deques, however, will help in implementing it.

Data Structure

Circular lists can be singly or doubly linked, but the latter is the most useful version. Basically, you just want a list that has no first or last element. Instead, the elements form a circle, and you'll have a pointer to the element that's being processed currently. Figure 10-14 shows such a list; the nodes have next and prev pointers, as with deques.

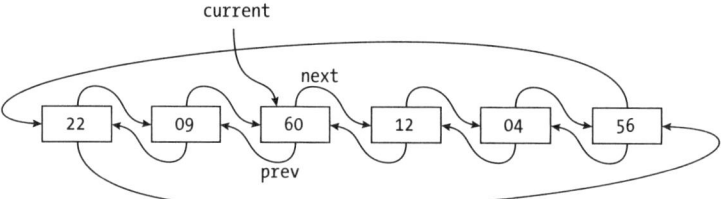

Figure 10-14: A circular list also needs two pointers at every node.

The "advance to the next" operation simply requires following the next link (see Figure 10-15).

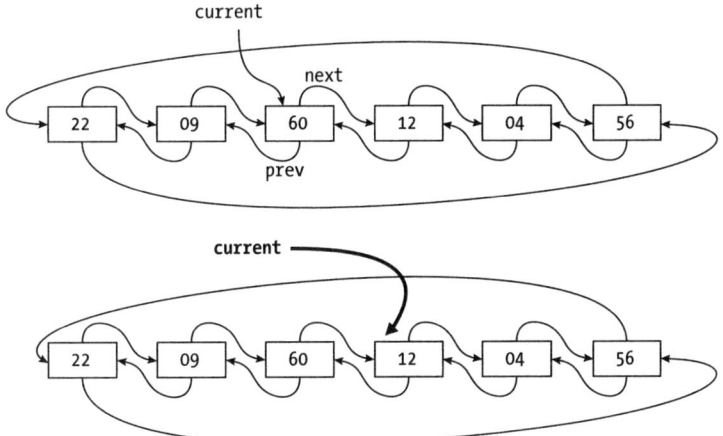

Figure 10-15: Moving along the list is possible in both directions.

Adding a new element before the current one is also a matter of dealing with several pointers (see Figure 10-16).

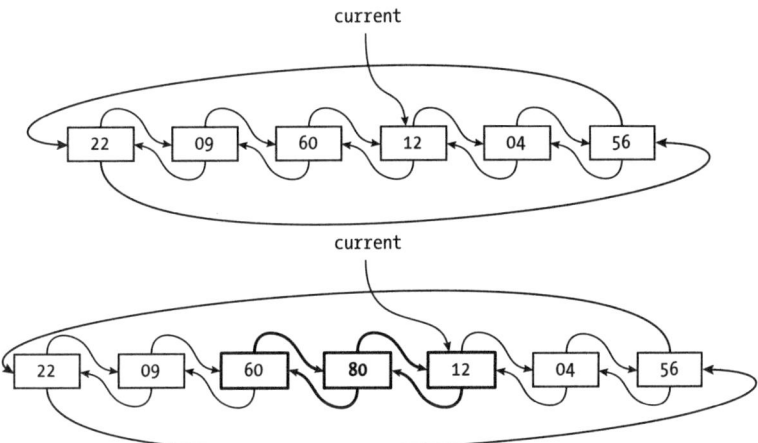

Figure 10-16: Adding an element to a circular list is done by changing a few pointers.

Removing the current element requires some juggling with pointers, but as with deques, having links in both directions makes it easy (see Figure 10-17).

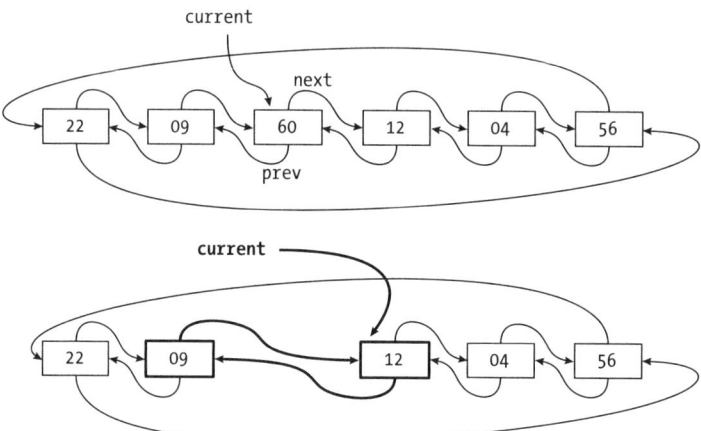

Figure 10-17: Removing an element from a circular list also requires just a few pointer changes.

Operations in a circular list require essentially the same kind of logic that you've already explored. Now consider an actual implementation.

Implementation

Creating a circular list is the same as for common lists, and so is testing whether such a list is empty. The only difference is the naming:

```
const newCircularList = () => null; // current
const isEmpty = (circ) => circ === null;
```

In a stack, you had a pointer to the top element. Here you have a pointer to some element in the list, the current one.

Adding a new node merely involves more work with pointers:

```
const add = (circ, valueToAdd) => {
  const newNode = { value: valueToAdd };
  if (isEmpty(circ)) {
❶  newNode.next = newNode;
    newNode.prev = newNode;
  } else {
❷  newNode.next = circ;
    newNode.prev = circ.prev;
    circ.prev.next = newNode;
    circ.prev = newNode;
  }
❸ return newNode;
};
```

If the list is empty, it consists of a single node ❶ whose next and prev links point to itself. Otherwise, the new node is between the nodes that circ (the current node) and circ.prev (the previous one) point to. Fix the four involved pointers so that the new node lies in its correct place ❷. At the end, return the new node ❸.

Removing the current element is a tad shorter:

```
const remove = (circ) => {
  if (isEmpty(circ)) {
❶  return circ;
  } else if (circ.next === circ) {
❷  return newCircularList();
  } else {
❸  circ.prev.next = circ.next;
    circ.next.prev = circ.prev;
    return circ.next;
  }
};
```

You have three distinct cases to consider. If the circular list is empty, do nothing ❶. If the list consists of a single element (and in that case both its next and prev links point to itself), return a new, empty list ❷. Finally, if the list isn't empty, make the nodes at circ.prev and circ.next (the ones that surround the current node) point to each other ❸.

Finally, getting the current value and advancing to the next one are both one-liners:

```
❶ const current = (circ) => (isEmpty(circ) ? undefined : circ.value);
❷ const advance = (circ) => (isEmpty(circ) ? circ : circ.next);
```

The current element of an empty list is just undefined ❶; otherwise, circ.value gives its value. Advancing the current element to the next position, for a nonempty circular list, is just a matter of going to the next node ❷.

Performance for Circular Lists

Checking all the implemented functions, none of them require loops or recursion, so as with other data structures in this chapter, the performance is constant (see Table 10-12).

Table 10-12: Performance of Operations for Circular Lists

Operation	Performance
Create	O(1)
Empty?	O(1)
Add	O(1)
Remove	O(1)
Current	O(1)
Advance	O(1)

You could use arrays, of course, but the performance for some operations, such as adding a new value, would suffer because of the possible need to move the whole array to a new place in memory: $O(n)$.

Summary

In this chapter we examined several linear structures and a circular one that are based on linked memory, and you'll have the opportunity to reuse them in later chapters. Linked memory is key for all the dynamic structures we'll explore in this book, and in upcoming chapters, we'll work with more complex structures to enable better performance for more complex operations.

Questions

10.1 Iterating Through Lists

All the examples in the "Implementing Lists with Dynamic Memory" section on page 180 were written using recursion, but they are often coded in iterative fashion. Can you rewrite them in that way?

10.2 Going the Other Way

Implement a reverse(list) algorithm that given a list will reverse it, meaning the first element becomes the last, the second element the next to last, and so on.

10.3 Joining Forces

Implement an append(list1, list2) function that given two lists will append the second one to the first one.

10.4 Unloop the Loop

Imagine you are given a list that may or may not have a loop; in other words, instead of eventually finishing with a null pointer, there may be an element that points back to some previous element, so the list has a loop. Can you write a hasALoop(list) function that given a list will determine whether it has a loop? Your solution should use constant extra memory; don't assume anything about the length of the list, because it may be incredibly long.

10.5 Arrays for Stacks

Since JavaScript provides operations on arrays like .pop(...) and .push(...), implementing a stack with an array should be pretty straightforward. Can you write appropriate code?

10.6 Stack Printing

Can you write code that prints out a stack's contents in top-to-bottom order? Could you print it in reverse (bottom-to-top) order?

10.7 Height of a Stack

Suppose you need to know how many elements are in a stack. How could you implement this?

10.8 Maximum Stack

Suppose that you need a stack for some process, but you also need to know, after each push or pop, the maximum value in the stack. How can you implement this efficiently without having to go through the whole stack every time?

10.9 Queued Arrays

In a previous question, you saw that JavaScript provided operations that made it simple to emulate a stack with arrays. Is the same true for queues? How would you emulate queues with arrays? What would the performance of such an implementation be?

10.10 Queue Length

Write a function that given a queue will count how many values are in it; in other words, find the queue's length.

10.11 Queueing for Sorting

In Chapter 6 you implemented radix sort with arrays, but using queues and linked memory is more efficient. Can you rewrite the algorithm accordingly?

10.12 Stacked Queues

Imagine that you needed to use a queue for some program, but all you had was a library that implemented stacks. With some trickery, you can simulate a queue by using a pair of stacks; can you see how? (You'll explore this strategy in Chapter 18.)

10.13 Palindrome Detection

How could you use a deque to decide whether a string is a palindrome? Palindromes are words that can be read the same way forward or backward, like "Hannah" or "radar," or ignoring spaces and punctuation "Step on no pets" or "A man, a plan, a canal: Panama."

10.14 Circular Listing

Implement a function to list all the contents of a circular list; take care not to go into a loop.

10.15 Joining Circles

Suppose you have two circular lists. How could you join them into a single, larger list? For simplicity, assume neither of the lists is empty.

11

BAGS, SETS, AND MAPS

In this chapter we'll consider some widely used abstract data types (ADTs): bags, sets, and maps. A *bag* is just a collection of values (repeated or not), a *set* is a collection of *distinct* values, and a *map* is a set of key + data pairs. We'll consider some new ways of implementing these ADTs, starting with JavaScript's own objects, and then move on to bitmaps, lists, and *hashing*, a new method we haven't yet explored.

Introducing Bags, Sets, and Maps

In Chapter 3 we defined the ADT for bags with the set operations shown in Table 11-1. (The use of the word *set* in this context is fully correct according to its mathematical definition.) When you need to store many (possibly repeated) values, you need a bag.

Table 11-1: Operations on Bags

Operation	Signature	Description
Create	→ bag	Create a new bag.
Empty?	bag → boolean	Given a bag, determine whether it is empty.
Add	bag × value → bag	Given a new value, add it to the bag.
Remove	bag × value → bag	Given a value, remove it from the bag.
Find	bag × value → boolean	Given a value, check whether it exists in the bag.

In Chapter 3 we had an extra operation to retrieve the greatest value from the bag, but that won't be considered here because it's not standard. You could also have an operation to find the current size of the bag, and possibly some more, but these are enough.

In some cases you want an actual set, so you don't want to allow for repeated values, and that restriction calls for a slightly different set of operations, as shown in Table 11-2.

Table 11-2: Operations on Sets

Operation	Signature	Description
Create	→ set	Create a new set.
Empty?	set → boolean	Given a set, determine whether it is empty.
Add	set × value → set | error	Given a new value, add it to the set.
Remove	set × value → set	Given a value, remove it from the set.
Find	set × value → boolean	Given a value, check whether it exists in the set.

All operations are the same, except that when you try to add a new value and find that it is already there, you'll do something different. One possibility is just to ignore the situation (after all, if you want to include a value in the set and the value is already there, everything is fine), or you could throw an error or perform some other action. You could always test beforehand whether the value to be added is in the set already, but it's usually more efficient to do it when adding.

Finally, in some cases you'll want to store key + data pairs. For example, for an application that uses information about countries, the key might be an ISO 3166 country code (such as CH for Switzerland, TV for Tuvalu, or UY for Uruguay), and the data could be the country name, its population, and so on. Having implemented sets, implementing maps is simple. Instead of storing single values, you would store objects with key + data and make changes so *find* and *remove* work with just keys; the former returns the data if found instead of a boolean. See Table 11-3 for all operations.

Table 11-3: Operations on Maps

Operation	Signature	Description
Create	→ map	Create a new map.
Empty?	map → boolean	Given a map, determine if it is empty.
Add	map × (key + data) → map | error	Given a new key + data, add it to the map.
Remove	map × key → map	Given a key, remove it from the map.
Find	map × key → data | undefined	Given a key, check whether it exists in the map, and if found, return the data or undefined instead.

All of these changes are fairly straightforward to do, so we'll work with plain bags and sets. Let's now consider specific implementations, starting with JavaScript's own.

JavaScript's Solutions for Sets

You learned how to implement a bag in Chapter 3 using several different methods. With a few changes, you can implement sets instead of bags; all you need to do is before adding a new value, check whether it was already present. In this section we'll consider two more ways to implement sets in JavaScript: using plain objects (which is not the best way) and using standard set objects.

Objects as Sets

Even if objects aren't designed to be used as sets (or maps), many developers use plain objects for them. If you can use values as attributes (mostly strings or numbers converted to strings), you can use them as property names:

```
❶ const mySet = {};
❷ mySet.one = 1;
  mySet.two = 2;
```

In plain JavaScript, creating an object means assigning an empty object ❶ and adding values to it ❷. Here, you now have a set with two keys: one and two. (If you want a map, the values associated with those keys are the data.)

You can test whether a key is in the object with the in operator:

```
"two" in mySet;   // true
"three" in mySet; // false
```

Finally, you can use delete to remove a key:

```
delete mySet.two;
```

As extra operations, you can get a list of all an object's attributes with `Object.keys(...)`, and you can even iterate over them with `for...in`.

Using plain JavaScript objects obviously works, but you probably will want to make your intentions clearer for more understandable code and use a proper set, which, after all, directly represents the data structure you want.

Set Objects

Sets are objects that let you store unique values. Creating a new JavaScript set and adding a couple of values is straightforward; try redoing the examples from the previous section:

```
❶ const mySet = new Set();
❷ mySet.add("one");
  mySet.add("two");
```

Create a set by producing a new instance of the `Set` class ❶ and add values to it with its `.add(...)` method ❷. By the way, you can chain calls, so you can write those two additions on a single line:

```
mySet.add("one").add("two");
```

To test whether a value is in the set, use a `.has(...)` method:

```
mySet.has("two");   // true
mySet.has("three"); // false
```

Finally, you can remove values with the `.delete(...)` method:

```
mySet.delete("two");
```

JavaScript's sets have a couple of extra interesting methods. To remove all values, you can use `set.clear()`. You also can find the number of elements in a set using the `.size` property.

Bitmaps

In some cases you can implement sets by making do with bitmaps (recall the bitmap sort from Chapter 6). If the values to store are numbers with a restricted range of values, an array of boolean flags will suffice.

We won't see the code here, as it's directly based upon the sorting you learned in Chapter 6. The main idea is to set up an array filled with `false` values. The index to that array is the value itself. To add a value, set its flag to `true`; to remove it, reset its flag to `false`. Finally, to test whether a value is in the set, check the corresponding flag. You can't do any better.

Using Lists

We discussed lists in Chapter 10, and you can adapt them to work as bags or sets. Consider three distinct possibilities:

Ordered lists   Plain lists that keep their values in ascending order

Skip lists   Two-dimensional structures with fast searches

Self-organizing lists   Interesting applications that work for caches and similar situations

Note that some of the solutions you'll consider in the "Hashing" section on page 218 will also use lists.

Ordered Lists

As mentioned in Chapter 10, the concept of an ordered list is simple: instead of always adding values at one extreme or the other, you'll add them so the values remain in order. This practice slows down insertions (you have to look for the right place to add the new value), but it makes for faster searches on average (when you reach a higher value than the one you wanted, you can stop the search). Take a look at the implementation.

Searching for a Value in an Ordered List

The logic for searching is as direct as it gets: start at the beginning and follow the links until you reach the value or learn that the value isn't there, because you either reached the end of the list or found a greater value than the one you wanted. Figure 11-1 shows how you'd (successfully) look for value 22 in an ordered list.

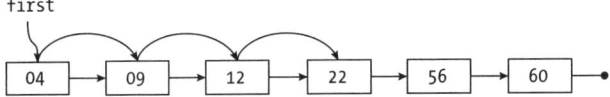

Figure 11-1: A successful search for a value (22) in a list

This works the same way as linear searching (Chapter 9). Start at the head of the list and keep going. In this case you found the value you wanted, so you succeeded. If you were searching for 20 instead, at this same point you'd have given up the search. If you reach a value that's higher than what you wanted, the search has failed.

The other possibility for failure is searching past the end of the list; see Figure 11-2, which shows a search for 86.

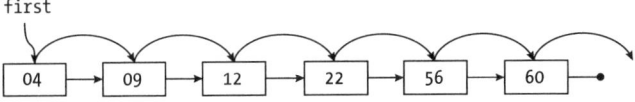

Figure 11-2: A failed search for a value (86) in a list

The code for a linear search is as follows:

```
const find = (list, valueToFind) => {
❶ if (isEmpty(list) || valueToFind < list.value) {
    return false;
❷ } else if (valueToFind === list.value) {
    return true;
❸ } else {
    // valueToRemove > list.value
    return find(list.next, valueToFind);
  }
};
```

If you got an empty list—either it was empty from the beginning or you traveled down it and reached its end (you saw the code for this in Chapter 10)—you know the value isn't there. It also isn't there if the list isn't empty but its first value is greater than the value you're seeking ❶. If the list isn't empty and its first element matches the value you want ❷, the value was found. Finally, if the list isn't empty and the value you want is greater than the first element of the list ❸, continue the search starting at the next node of the list.

Adding a New Value to an Ordered List

To add a new value, first do a search until you reach the place where the new value should go (meaning between a node with a smaller value and a node with a greater one), and then change a couple of pointers to include the new value in the list. Figure 11-3 shows how to add 20 to the list.

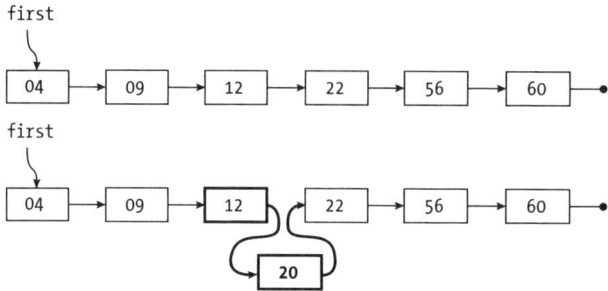

Figure 11-3: Adding a new value (20) to an ordered list

If you want to add a value that's smaller than the one at the head of the list, you'll need to change the pointer to the list itself. Another border case is adding a value greater than the last one in the list; you have to be careful when going down the list. You can use recursion to implement all of these cases more easily:

```
const add = (list, valueToAdd) => {
❶ if (isEmpty(list) || valueToAdd < list.value) {
    list = { value: valueToAdd, next: list };
```

```
❷ } else {
  ❸ list.next = add(list.next, valueToAdd);
  }
  return list;
};
```

The logic is similar to what you saw for the linear search. If the list is empty, or if it's not empty but the first value is greater than the one you want to add ❶, create a single node with the new value whose next pointer points to the list you had. (This covers the case of adding a value past the end; can you see how?) If you want to make a set, add a test ❷, because if you find the value you want to add, you would throw an error or otherwise reject the operation. If you're making a bag and the value to add is greater than the first of the list ❸, add it using recursion, after the first element.

Removing a Value from an Ordered List

Removing a value is a matter of finding it (which you already know how to do) and then modifying its predecessor's link to point at the next value, as shown in Figure 11-4.

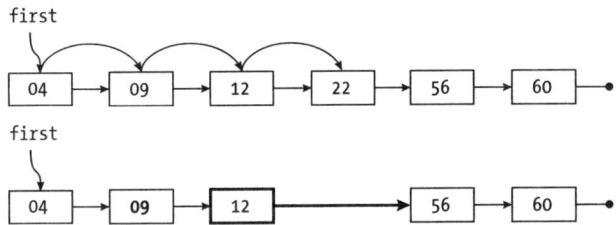

Figure 11-4: Removing the value (22) from an ordered list

As mentioned, do a search, and if it's successful, skip the value to be removed. The only unusual case is when you delete the head of the list, you need to modify the pointer to the list:

```
const remove = (list, valueToRemove) => {
❶ if (isEmpty(list) || valueToRemove < list.value) {
    return list;
❷ } else if (valueToRemove === list.value) {
    return list.next;
❸ } else {
    // valueToRemove > list.value
    list.next = remove(list.next, valueToRemove);
    return list;
  }
};
```

The logic fully matches the search, case by case. It's also logical; you must first find the value before you can remove it from the list. If the list is empty or if its first value is greater than the value ❶, return the list as-is, because there's nothing to remove. If the value to remove is at the head of the list ❷, return the list's tail, skipping the value to be removed. Finally, if the value to be removed is greater than the head of the list ❸, proceed recursively to delete the value from the tail of the list and return the (updated) list.

Considering Performance for Ordered Lists

There's no way to speed up any of the processes, and if the list has n nodes, all functions are $O(n)$. On average, all operations will visit half the nodes of the list. This implementation is good enough for small values of n, but for larger values, you'll need something that allows you to move faster through the list—and in the next section we'll see just that.

Skip Lists

As just mentioned, searching a list is an $O(n)$ process, because there's no way to speed things up and move faster. However, you can take a tip from the jump search method (Chapter 9). What would happen if you could take long jumps to skip many places quickly, and when you get closer to the needed value, start doing smaller jumps, and then even smaller ones, until you finish with a one-by-one search? In this section we'll consider *skip lists*, which allow you to traverse a list much more quickly by providing ways to skip ahead faster.

Consider an ordered list as shown in Figure 11-5 (for clarity, I haven't included the arrows; all go from left to right).

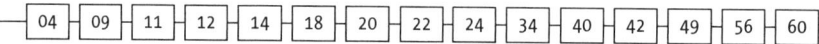

Figure 11-5: Searching a long list is logically slower.

As is, you can't jump quickly as with a jump search, but with an auxiliary second list, doing so is possible, as shown in Figure 11-6 (the vertical lines indicate pointers from top to bottom).

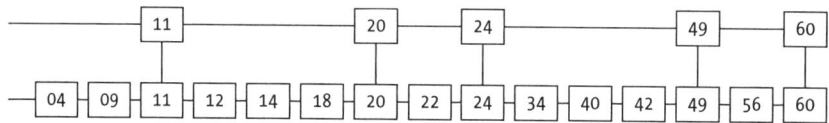

Figure 11-6: Adding a second list to advance faster when searching the list

If you wanted to search for 42, you'd start at the topmost list and move right until you get past 42; then you'd go back and down and continue the search (see Figure 11-7).

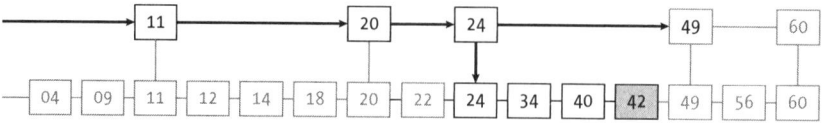

Figure 11-7: Searching for a value (42) with the aid of the second list

The topmost list, which includes only a few values of the bottom list, allows you to make longer jumps, so the searches are speedier. Of course, for even faster processes, you could have three or more levels, as shown in Figure 11-8.

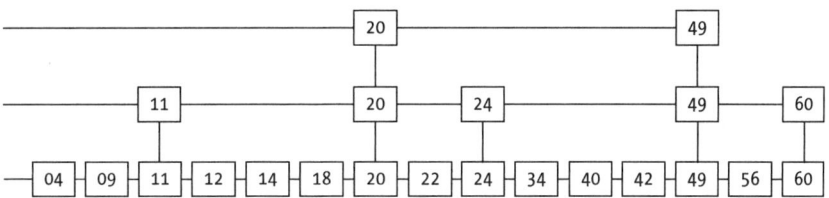

Figure 11-8: A third list helps speed up the search even more.

This works well and provides better expected performance of $O(\log n)$. However, having all of these lists with repeated values everywhere isn't good in practice. It would be better to have each value only once in a node with several pointers, according to the different levels, as shown in Figure 11-9. (In reality, all nodes could have the same number of pointers, but Figure 11-9 shows only the used ones.)

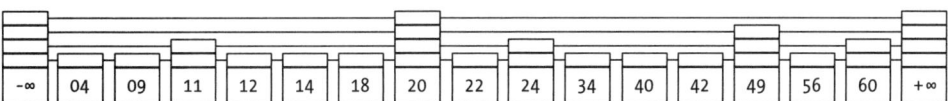

Figure 11-9: An actual implementation of skip lists, with several pointers per node

We'll use this style of search to simplify our work. We'll also add some sentinel nodes at the beginning and end of the lists to simplify all logic. You won't ever have cases like "adding at the beginning" or "adding at the end," because no value can be lower than the first sentinel or higher than the last one. Also, you won't have to deal with empty lists (at least the sentinels will be there), and you'll never go past the last item of a list.

Creating a Skip List

An empty list will consist of a node with two sentinels: a minus infinity value and a single level with a plus infinity value. Implementing it is as follows:

```
const newSkipList = () => ({
  value: -Infinity,
  next: [{ value: Infinity, next: [null] }]
});
```

You have a list with just two values: -Infinity first and Infinity last. (You are working with numeric values here; for strings, you'd have to use appropriate low and high strings.) The skip list is flat, having just one level. (No next array has more than one element.) Knowing this, testing whether a skip list is empty is a tad more complex:

```
const isEmpty = (sl) => sl.next[0].next[0] === null;
```

If a skip list doesn't have any values added to it, you'll have the initial configuration, so your sentinels will be on the bottom level. In this data structure, the only value that has a null pointer to the next is the +Infinity sentinel; all other nodes have non-null pointers.

You can find how many levels a skip list has by simply looking at the length of the next arrays. Another function that will come in handy just returns the last index of the array of pointers:

```
const _level = (sl) => sl.next.length - 1;
```

You have to subtract 1 because the array of pointers is zero indexed, as usual.

Searching for a Value in a Skip List

You saw the general idea for searches earlier, but now consider how they'd work with the actual implementation. Searches begin at the topmost level and advance to the right unless they go past the searched-for value, in which case they go down a level. If there are no more levels, the search is a failure.

The code isn't very long, but dealing with multiple levels requires care:

```
❶ const _find = (node, currLevel, valueToFind) => {
❷   if (currLevel < 0) {
      return false;
❸   } else if (valueToFind === node.value) {
      return true;
❹   } else if (valueToFind >= node.next[currLevel].value) {
      return _find(node.next[currLevel], currLevel, valueToFind);
❺   } else {
      return _find(node, currLevel - 1, valueToFind);
    }
  };

❻ const find = (sl, valueToFind) => _find(sl, _level(sl), valueToFind);
```

We'll use an auxiliary recursive function for the search. This function has three arguments: a node (somewhere in the skip list), the level at which it is searching, and the value to find ❶. If you try to go below level 0 ❷, you've failed, as you were at the bottom level and couldn't find the value there. If the node has the value that you want ❸, the search succeeded. If the value you want is greater than or equal to the next value at the same level ❹, keep going without changing level. Otherwise, if you've already reached a higher value ❺, go down to the next level. The implementation of a general search ❻ starts at the first node at the top level.

Adding a Value to a Skip List

We haven't really discussed how to decide what values go at which levels. The solution we'll use is based on random numbers. Obviously, all values will be at the bottom level, but they all won't be at the other levels. We'll decide whether a new value goes up one level by "flipping a coin"; we want approximately 50 percent of values to be in the next level. We'll keep deciding randomly whether to move the value up one more level until the flip fails or you get a maximum level. In the code that follows, set MAX_LEVEL to 32, which implies that, on average, one value out of 2^{32} will go that high—a really big structure!

We'll require an auxiliary function to add a value at a certain level and all levels below it. An obvious question is why do you first add the value at a higher level and then at lower levels? Lists that are higher up have fewer elements, so insertions are faster there. Here's the code:

```
const _add = (currNode, currLevel, newNode, newLevel) => {
❶ if (newNode.value > currNode.next[currLevel].value) {
    _add(currNode.next[currLevel], currLevel, newNode, newLevel);
  } else {
  ❷ if (currLevel <= newLevel) {
    ❸ newNode.next[currLevel] = currNode.next[currLevel];
      currNode.next[currLevel] = newNode;
    }
  ❹ if (currLevel > 0) {
    ❺ _add(currNode, currLevel - 1, newNode, newLevel);
    }
  }
};
```

If the new value is greater than the next value at this level ❶, you must advance; you'll be able to add the new value when it lies between two consecutive values in the list. If you're at a level lower than or equal to the maximum new level ❷, add the value and adjust the pointers to include the new value in the list ❸. Finally, if you haven't reached bottom yet ❹, use recursion to add the value one level down ❺.

With this function, adding a value is as follows:

```
const add = (sl, valueToAdd) => {
❶ let newLevel = 0;
  while (newLevel < MAX_LEVEL && Math.random() > 0.5) {
    newLevel++;
  }
❷ const newNode = { value: valueToAdd, next: new Array(newLevel) };

  let currLevel = _level(sl);
❸ while (newLevel >= currLevel) {
  ❹ sl.next[currLevel].next.push(null);
    sl.next.push(sl.next[currLevel]);
    currLevel++;
  }
```

```
❺ _add(sl, currLevel, newNode, newLevel);
  return sl;
};
```

First, decide up to which level you'll find the new node ❶. Going up one more level will depend on the "coin flip." After deciding that ❷, create a node with the value and an array with the right number of pointers. There's the possibility that you're going "higher" than before ❸ and that the skip list will be taller. If so ❹, you'll have to add new pointers to the rightmost value. After this is taken care of ❺, use the auxiliary function to add the value to all the corresponding lists.

Removing a Value from a Skip List

Removing a value requires two steps: first, remove it from all the lists it's in, which you'll do with an auxiliary function, and than possibly make the skip list "shorter" because by removing the value, it may not be as tall as before.

Here's the logic for actually removing the value:

```
const _remove = (currNode, currLevel, valueToRemove) => {
❶ if (valueToRemove > currNode.next[currLevel].value) {
    _remove(currNode.next[currLevel], currLevel, valueToRemove);
  } else {
  ❷ if (valueToRemove === currNode.next[currLevel].value) {
    ❸ currNode.next[currLevel] = currNode.next[currLevel].next[currLevel];
    }
  ❹ if (currLevel > 0) {
      _remove(currNode, currLevel - 1, valueToRemove);
    }
  }
};
```

Advance down the list ❶ until you find where the value should be. If you actually found it ❷ (the user may be asking to remove a value that simply isn't in the list), fix the pointers ❸. Then keep doing removals until you reach the bottom level ❹.

Removing the value is the first step, as described; you may have to restructure multiple levels after that:

```
const remove = (sl, valueToRemove) => {
❶ _remove(sl, _level(sl), valueToRemove);
  for (
  ❷ let level = _level(sl) - 1;
  ❸ level > 0 && sl.next[level].next[level] === null;
    level--
  ) {
  ❹ sl.next[level].next.splice(level, 1);
    sl.next.splice(level, 1);
  }
  return sl;
};
```

After removing the value ❶, go down all levels from the top ❷, and while the lists are basically empty (only the sentinels) ❸, you've made the list shorter ❹.

Considering Performance for Skip Lists

Skip lists are probabilistic in nature, and the average performance can be shown to be logarithmic (see Table 11-4).

Table 11-4: Performance of Operations for Skip Lists

Operation	Average performance	Worst case
Create	$O(1)$	$O(1)$
Add	$O(\log n)$	$O(n)$
Remove	$O(\log n)$	$O(n)$
Find	$O(\log n)$	$O(n)$

There is a quite low probability that the structure will behave badly (maybe having just a single level or having most values at all levels), but that's not likely.

As with hashing (which you'll explore later in this chapter) and other structures, you can solve performance problems by restructuring the skip list; see question 11.4 for a possible idea. You can also modify the list to allow retrieving a value by position; see question 11.5.

Self-Organizing Lists

There's a particular case for which you can use an "auto-modified" bag successfully. Think about a cache with a limited maximum size. It's not infrequent that one may require a given element several times in a short interval and then for a long time may not require it at all. In that case you can use a self-organizing list that places the elements required most often near the beginning (for quicker searches) and the ones required less often near the end (allowing a slower search).

As an example, think of a mapping (Global Positioning System [GPS]–style) application. You can't hold every street name in memory, but to optimize speed, you could have a small cache of street names. Traveling in a certain zone, one often requires a certain group of street names, and it's more unlikely one will need to find streets much farther away. The idea of a self-organizing list is to always add new values at the front, and if you search for a value and find it, move it to the front with the idea that if you soon require it again, you'll get to it in a few steps.

Searching for a Value in a Self-Organizing List

Searching an unordered list is not hard; you just have to keep going until you either find the value or reach the end of the list. You saw how to do this kind of search in Chapter 10 (see the section "Implementing Lists with

Dynamic Memory" on page 180). The important detail is what to do if you find the element. Make it the head of the list and take it out of its original place. Figure 11-10 shows an example search for 12.

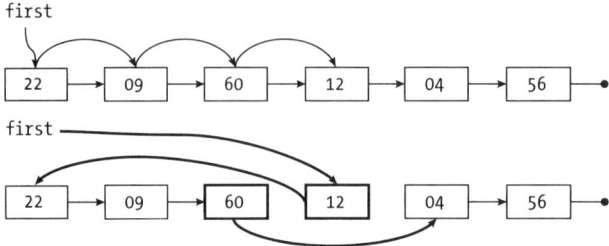

Figure 11-10: After a successful search in a self-organizing list, the found node is brought to the head of the list.

First, do a search for 12 (this is nothing new), but after finding it, there are some changes. Since the value wasn't already at the head of the list, you'll restructure the list so 12 is at the head, pointing to the old head of the list. If in the future you need to search for it again, those searches will be quite fast, because the value will be either at the head of the list or very close.

The code is as follows:

```
const findMTF = (list, keyToFind) => {
❶ if (isEmpty(list)) {
    return [list, false];
❷ } else if (list.value === keyToFind) {
    return [list, true];
  } else {
  ❸ let [prev, curr] = [list, list.next];
  ❹ while (!isEmpty(curr) && curr.value !== keyToFind) {
      [prev, curr] = [curr, curr.next];
    }

  ❺ if (isEmpty(curr)) {
      return [list, false];
  ❻ } else {
      [prev.next, curr.next] = [curr.next, list];
      return [curr, true];
    }
  }
};
```

When you search now you're also modifying the list, so you'll have to return two values: the list itself (which was possibly updated if the value was found) and a boolean value with the result of the search. (Too much bother? See question 11.3.) If the list is empty ❶, return the list and false, because trivially the value isn't there. If the list isn't empty and the value you want is at the list's head ❷, you don't have to change the list, so you return it as is, plus true because you succeeded. If the list isn't empty and the value you want isn't at the list's head ❸, set up a loop in which `prev` and `curr` will point to consecutive nodes in the list. That loop will finish when either you get to the list's end or you find the value you want ❹. In the case of the former condition ❺, return the list and a `false` value, just as when the list is empty ❶; if the latter ❻, change pointers and return the current node as the new list's head, plus `true`.

Adding a Value to a Self-Organizing List

Since these lists are unordered, you can add values anywhere, using the simple logic you've used before:

```
const add = (list, valueToAdd) => {
❶ list = { value: valueToAdd, next: list };
  return list;
};
```

Putting the new value at the top of the list, as its head, is as simple as can be. Make a new node that points to the old list ❶, and the new list has that node as its head. This code is functionally equivalent to the `push(...)` method that you wrote for stacks in Chapter 10.

Removing a Value from a Self-Organizing List

Earlier we saw how to remove a value from an ordered list. Doing the same with an unordered list is not very different, except that you may always have to go to the end of the list, because there's no way to stop the search earlier. You already saw how to do the search in an iterative way, so now do this recursively:

```
const remove = (list, valueToRemove) => {
❶ if (isEmpty(list)) {
    return list;
❷ } else if (valueToRemove === list.value) {
    return list.next;
  } else {
❸ list.next = remove(list.next, valueToRemove);
    return list;
  }
};
```

If the list is empty ❶, just return it, because the value to be removed isn't there. If the value you want to remove is the one the list points at ❷, returning the tail of the list (which `list.next` points to) causes the removal. Finally, if the head of the list doesn't have the value you want ❸, make that node point to the result of removing the value from the tail of the list.

Considering Performance and Variants for Self-Organizing Lists

This structure's performance is $O(n)$, as with common lists, and a search looks at $n/2$ elements on average. However, in actual experience with clustered requirements, it behaves much better, with far fewer looks at elements. It's not a theoretical advantage, but a fully empirical, pragmatic one, and in the worst case, you are no worse off.

There are other variants with similar performance. The "move to front" (MTF) solution is not the only possible one. Another possibility is "swap with previous" in which instead of moving the found element to the head of the list, you just swap it with the one before, making it closer to the head. If you make many searches for a given value, it eventually reaches the front of the list, but if the search was just a one-off case, then it stays around where it was.

Another variant is to add a count of references to each value, increment it by 1 every time a value is searched for and found, and move it nearer the head of the list so that the values are in descending order of counts.

Hashing

In this section we'll move on to a different concept that potentially provides the fastest possible searches: *hashing*. The idea of hashing is somewhat related to bitmaps. If the values to be stored in the set are taken from a small range, you can use a bitmap, which provides $O(1)$ searches, as you saw. However, if the values are from a very large range (for instance, US Social Security Numbers, nine digits long, with 1,000,000,000 possible values), a bitmap becomes prohibitive because of the needed space. In addition, it's most likely that you'll be dealing with a very small percentage of all possible keys. The idea is to first use an array of *slots* to store values, but then instead of using the key as an index (as in bitmaps), you'll compute a hash of the value and use that hash as the index.

In reference to a hash, Ambrose Bierce said, "There's no definition for this word—nobody knows what [a] hash is." For us, a hash is any function that transforms a value—numeric, string, and so on—into a number in a given range. For the Social Security number example, to get a hash between 000 and 999, you could just take the three final numbers. To get a hash between 0 and a top number K, dividing values by K and taking the remainder would do. There are many ways to compute hashes, but we'll use the remainder function, like the following:

```
const hash = (ht, value) => value % ht.slots.length;
```

When using hashing to decide where to store (or look for) a value, we first compute the hash and then go to the corresponding slot in the array (see Figure 11-11). It's quite similar to what we did with bitmaps, but in that case, we used the key as an index; here we assume that the number of possible keys is exceedingly large, so we apply hashing to reduce it to a manageable value.

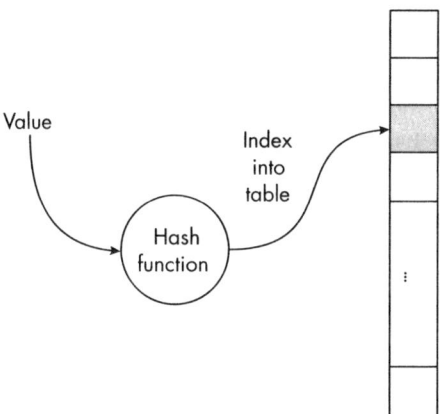

Figure 11-11: In hashing, a hash function is used to decide where a value should be stored in a table.

If the value you want occupies the slot, you've found it. If the slot is free, you know for sure that the value isn't in the set. But how do you deal with different values that produce the same hash, so that all should go in the same slot? This situation is called a *collision*, and you must specify how to solve it. (If you think this isn't likely, try searching for "Birthday Paradox" online; you'll be surprised!) Different hashing strategies differ in how they handle collisions. This chapter will discuss three distinct strategies: buckets with chaining, open addressing, and double hashing. The implementations will be bags, but we'll consider how to do sets in the questions at the end of the chapter.

Hashing with Chaining

A first solution to collisions is to consider each slot as a bucket into which you may place multiple values. The simplest way to implement this is by taking advantage of the ordered lists you saw earlier in the chapter, and most of the work will already be done for you. All values that go into the slot are placed in a list, and you'll work with a small set of numbers for simplicity. Figure 11-12 shows slots at the left (slot #3 is unoccupied) and the lists at the right.

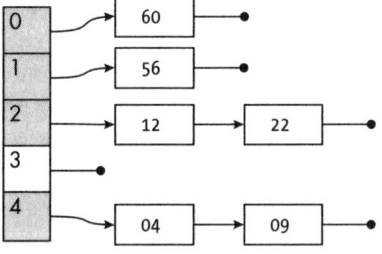

Figure 11-12: Hashing with chaining uses lists for values that hash to the same value.

To search for a value, first compute in which slot to look for it (which in this case means finding the remainder of the value divided by 5, the table's length) and then search the corresponding list. If the value is in the set, you should find it in the list. This implementation isn't hard, considering you already saw how to use lists for bags or sets, so here are the details.

Creating a Chained Hash Table

To create a new hash table for use with chaining, create an empty array and fill it with new lists (you'll use ordered lists, as shown earlier in this chapter, as a refresher):

```
const newHashTable = (n = 100) => ({
  slots: new Array(n).fill(0).map(() => newList())
});
```

You are creating an object instead of an array in case of other solutions that will require extra fields—for instance, to keep track of how many slots are used or are free. (See question 11.6 for a common mistake.)

Adding a Value to a Chained Hash Table

You can add a new value to this hash table easily; the code is as follows:

```
const add = (ht, value) => {
❶ const i = hash(ht, value);
❷ ht.slots[i] = addToList(ht.slots[i], value);
  return ht;
};
```

You just have to compute into which slot the new value should go ❶ and then add it to the corresponding list ❷.

Searching for a Value in a Chained Hash Table

Searching for a value is also simple: after deciding in which slot the value should be, search the corresponding list. You could write the search in a single line, but it's clearer as follows:

```
const find = (ht, value) => {
❶ const i = hash(ht, value);
❷ return findInList(ht.slots[i], value);
};
```

Calculate the corresponding slot ❶ and then search for it ❷. Note that you had to rename the find(...) method from lists to findInList(...) to avoid recursively calling the wrong function. Another possibility would be writing something like List.find(...).

Removing a Value from a Chained Hash Table

Again, removing a value is easy because of all the code you developed earlier:

```
const remove = (ht, value) => {
❶ const i = hash(ht, value);
❷ ht.slots[i] = removeFromList(ht.slots[i], value);
  return ht;
};
```

As when adding a value, first compute the correct slot ❶ and then remove the value from the list ❷ by using the removeFromList(...) method from lists, which is also renamed to avoid conflicts.

Considering Performance for Chaining

The performance of the worst case of hashing with chaining is, obviously, $O(n)$ if all values map to the very same slot. If the situation isn't as extreme and there are s slots, each chain should be around n/s values long, so searches are $O(n/s)$, which is actually $O(n)$ but with a better expected constant. The more slots you have, the shorter the chains and the better the performance.

You could keep track of how many values are in the table (or of the lengths of the individual chains), and should those numbers exceed some limit, you could re-create the table with a larger number of slots to improve performance. We'll study this kind of process in the next sections.

Hashing with Open Addressing

Another common solution to dealing with collisions is this: if the slot to use is already occupied, try the following place (and, if needed, the one after that, and the next, and so on, returning to the beginning cyclically after reaching the end of the table) until you find an empty slot. To do a search, apply the same scheme: first check the corresponding hash, and if the slot is empty, the search failed. If the slot is occupied and it's the value you wanted, the search succeeded; otherwise, proceed to the (cyclically) next place and try again. You can see how this works with a simple example. Start with an empty hash table into which just 22 was added; see Figure 11-13.

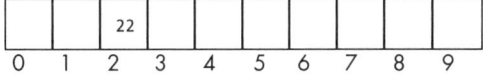

Figure 11-13: A hash table with just one element

You could add 04, 75, 09, and 60, and each would go into its corresponding slot, as shown in Figure 11-14.

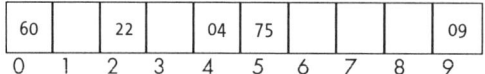

Figure 11-14: Four more elements were added, with no collisions so far.

The problem appears if you try to add 12, because the corresponding slot, the second one, is already occupied. You must start advancing, so 12 ends up in slot 3, as shown in Figure 11-15.

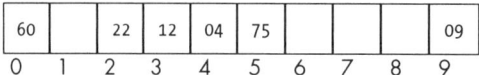

Figure 11-15: A collision occurs when you try to add a value (12) to an occupied slot.

As the table becomes more and more full, it's more likely that new values will end up far from their correct slot; for instance, if you add 63, it ends up at slot 6, as shown in Figure 11-16.

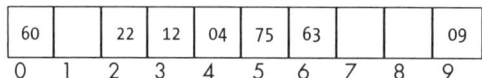

Figure 11-16: In a fuller table, values end up far from their corresponding slot.

Of course, as defined, it becomes apparent that if the table becomes full, you'll be in an infinite loop. The *load factor* is defined as the ratio between occupied slots and total slots. An empty hash table has a zero load, and a totally full one has a load factor of 1. The result is intuitive, but you can show mathematically that as the load factor grows, insertions and searches will progressively become slower. As a rule of thumb, if the load factor gets above 0.75, you should move to a larger hash table.

```
const load = (ht) => ht.used / ht.slots.length;
```

When conducting searches, the process is exactly the same as for insertions. If you are looking for 63, you'd start at slot 3, and if it's not there, advance by one until finding it at slot 6. If you are looking for 73 instead, you'd advance until slot 7, which is empty, and then decide that 73 isn't in the table.

Deletions are not a straightforward process. Consider removing 22, as shown in Figure 11-17.

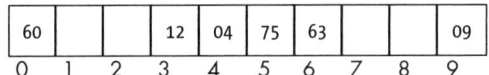

60			12	04	75	63			09
0	1	2	3	4	5	6	7	8	9

Figure 11-17: The wrong way to delete a value (22) messes up other searches.

Now, if you want to search for 12, what happens? Finding that slot 2 is empty, you'd decide that 12 isn't there, which is bad. We'll have to do lazy deletions. Instead of actually emptying a slot when removing a value, we'll mark it as available. We'll treat deleted locations as empty when adding new values, but as occupied when searching. Deleting 12 would get the result shown in Figure 11-18.

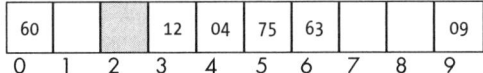

60			12	04	75	63			09
0	1	2	3	4	5	6	7	8	9

Figure 11-18: The right way to delete a value (22) just marks the slot (#2) as "used but available."

For searches, slot 2 is considered to be occupied, so when looking for 12, you won't stop at slot 2, but will instead keep advancing. For insertions (suppose at a later time you wanted to add 42 to the table), slot 2 is considered available, so you could use it, as shown in Figure 11-19.

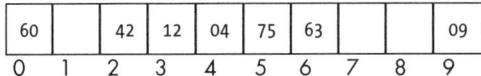

60		42	12	04	75	63			09
0	1	2	3	4	5	6	7	8	9

Figure 11-19: A "used but available" slot (#2) may be used for new insertions.

Now that you've explored how a hash table works with the important detail of how to deal with deletions, consider the actual code.

Creating an Open-Addressed Hash Table

For open addressing, all you need is a table, but you'll also keep track of how many slots have been used in order to compute the load factor. You'll start by defining, for ease of coding, a couple of constants:

```
const EMPTY = undefined;
const AVAILABLE = null;
```

The EMPTY value will be assigned to all as yet unused slots, and AVAILABLE will be used for slots that were occupied before but that are now available because you removed the original value that was there.

A new hash table will be an object with 100 slots by default:

```
const newHashTable = (n = 100) => ({
❶ slots: new Array(n).fill(EMPTY),
❷ used: 0
});
```

Given the desired size for the hash table (defaulting to 100), create an empty array of that size, filled with the EMPTY value ❶, and set the initial count of used slots to 0 ❷.

Adding a Value to an Open-Addressed Hash Table

You've already seen the logic for additions, and the code is somewhat long:

```
const add = (ht, value) => {
❶ let i = hash(ht, value);
❷ while (ht.slots[i] !== EMPTY && ht.slots[i] !== AVAILABLE) {
❸   i = (i + 1) % ht.slots.length;
  }

❹ if (ht.slots[i] === EMPTY) {
    ht.used++;
  }
  ht.slots[i] = value;
  return ht;
};
```

Start by calculating in which slot the value should go ❶. Then start a linear search ❷ that stops when you get to an EMPTY or AVAILABLE slot; note that using the modulus operation ❸ makes the search wrap to the beginning. After the search for an open slot succeeds, if the slot was empty ❹, add one to the count of used slots. If you're wondering why you don't do it if the slot was AVAILABLE instead, you'll understand why when you look at removing a value. An important detail is that you assume the table has some free space. You'll see how that works when you look at performance for hashing with open addresses.

You're actually implementing a bag here. To do a set instead, see question 11.7.

Searching for a Value in an Open-Addressed Hash Table

As described previously, the search is similar to the insertion process. You'll do the same kind of process as if you were looking to insert a new value, but you'll skip the actual insertion. The code is as follows:

```
const find = (ht, value) => {
❶ let i = hash(ht, value);
❷ while (ht.slots[i] !== EMPTY && ht.slots[i] !== value) {
    i = (i + 1) % ht.slots.length;
  }
```

```
❸ return ht.slots[i] === value;
};
```

As when inserting, first decide in which slot the value should be found ❶, and if needed, loop ❷ until you either get the value or reach an EMPTY slot. Depending on how the loop ends, return true or false ❸. You're ignoring AVAILABLE slots, for reasons that will become apparent.

Removing a Value from an Open-Addressed Hash Table

Removing a value is strongly based on how you do searches. The key issue is that upon finding the value to be deleted, you'll mark the slot as AVAILABLE, implying that the slot is free now for future insertions, but it isn't really free, so for searches, consider it as filled and keep going.

Here's the code:

```
const remove = (ht, value) => {
❶ let i = hash(ht, value);
  while (ht.slots[i] !== EMPTY && ht.slots[i] !== value) {
    i = (i + 1) % ht.slots.length;
  }

❷ if (ht.slots[i] === value) {
    ht.slots[i] = AVAILABLE;
  }
  return ht;
};
```

The first part ❶ is the same code as in the search function. The only difference with that code is that after exiting the loop, if the search succeeded, set the slot to AVAILABLE ❷.

An important question is why you didn't decrement the used count. A border case shows the problem: imagine you entered n values, from 1 to n, into a hash table of size n, then removed them all, and finally tried to add any new value. What would happen? An infinite loop when checking if the value was already in the table! The load factor considers all slots that are or were occupied; you'll see what to do when that load becomes too high in the next section.

Considering Performance for Hashing with Open Addressing

As mentioned, the search performance degrades considerably when the load factor tends to 1, eventually becoming $O(n)$. The worst case is always $O(n)$; an example of this (not the only one, for sure) would be if all keys hash to the same slot. See Table 11-5.

Table 11-5: Performance of Operations for Open-Addressed Hash Tables

Operation	Average performance	Worst case
Create	$O(1)$	$O(1)$
Add	$O(1)$	$O(n)$
Remove	$O(1)$	$O(n)$
Find	$O(1)$	$O(n)$

If you keep the load factor limited, you can expect good performance, but as you add more values to the table, it will degrade. There's no way out of this, but you can modify the add(...) logic to produce a larger table automatically to avoid high loads. You just have to change the final return ht from the code for adding a value to the following:

```
❶ if (load(ht) > 0.75) {
  ❷ let newHT = newHashTable(ht.slots.length * 2);
  ❸ ht.slots.forEach((v) => {
    ❹ if (v !== EMPTY && v !== AVAILABLE) {
      ❺ newHT = add(newHT, v);
      }
    });
  ❻ return newHT;
  } else {
  ❼ return ht;
  }
};
```

If the load factor has exceeded the recommended 0.75 threshold ❶, create a new hash table, double the size ❷, and go through the original table slot by slot ❸. Every value you find ❹ will be added to the new table ❺. At the end, instead of returning the original table as you did before, you'll return the new larger table ❻. Had the load factor been acceptable ❼, you'd return the original table as before. For an alternative technique, see question 11.9.

This sort of logic also comes in handy for the versions of hashing tables that we'll consider next.

Double Hashing

The logic you've been applying doesn't really help much with collisions. Should two values coincide at a slot, they'll also coincide at the following slot, and the next, and so on. This scheme is likely to produce long lengths of adjacent occupied slots that will slow down searches and insertions.

An idea that helps in that situation is not always trying the next slot, but rather skip a number of slots and make that number depend on the value, so different values skip different numbers of slots. The concept of *double hashing* works this way: a first hashing function finds the first slot to try, but if that's occupied, a second hashing function determines what size steps to take, instead of always jumping to the immediate next slot.

If you return to the example from the "Hashing with Open Addressing" section on page 221, while there are no collisions, everything works the same way, so after the first five insertions, you'd have the situation shown in Figure 11-20.

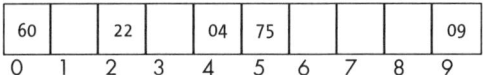

Figure 11-20: A hash table using double hashing, with no collisions so far

Now you want to add 12, and the #2 slot is occupied. In the previous section you used open addressing, so you tried slot #3, and if that also was occupied, you would have tried slots #4, #5, and so on, in succession, until finding an empty one. With double hashing, you'll use a second function to decide how far to jump. Use the remainder of dividing the value by 9, plus 1, which is guaranteed to be a number between 1 and 9 inclusive. For value 12, the step would be 4, so the next attempt would be at slot #6, as shown in Figure 11-21.

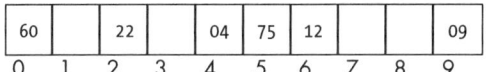

Figure 11-21: Double hashing uses a second function to work with collisions.

If slot #6 had been occupied, you'd have advanced four slots again (cyclically) and tried slot #0, and after that slot #4, and so on. There's a possible problem here. If you wanted to add value 130 to the table in Figure 11-21, what would happen? The first attempt would be at slot #0, and that fails. The second attempt would be at slot #5 (because the step for 130 would be 5), but that also fails. The third attempt would be at slot #0 again (because of the cyclical search), and you'd be in a loop.

You'll have to take care of detecting (and solving) these loops; there are two different ways to achieve this.

Creating a Table That Uses Double Hashing

Creating a table for double hashing is exactly the same as for open addressing, so we don't need special code here. Here's the needed logic again for easier reference:

```
const EMPTY = undefined;
const AVAILABLE = null;

const newHashTable = (n = 100) => ({
  slots: new Array(n).fill(EMPTY),
  used: 0
});
```

The key differences are in how you add, search for, and remove values.

Adding a Value to a Table That Uses Double Hashing

You saw the procedure for adding a value in a double hash earlier in this section, so let's get to the code. You'll use two hashing functions: a first one to determine the initial slot to try and a second one to skip in the search. It's very important that the second function must never return a zero value.

```
const hash1 = (ht, value) => value % ht.slots.length;
const hash2 = (ht, value) => 1 + (value % (ht.slots.length - 1));
```

The logic to add a value needs careful attention to avoid infinite looping:

```
const add = (ht, value) => {
❶ let i = hash1(ht, value);
❷ if (ht.slots[i] !== EMPTY && ht.slots[i] !== AVAILABLE) {
  ❸ const step = hash2(ht, value);
  ❹ let i0 = i;
  ❺ while (ht.slots[i] !== EMPTY && ht.slots[i] !== AVAILABLE) {
    ❻ i = (i + step) % ht.slots.length;
    ❼ if (i === i0) {
        i = (i + 1) % ht.slots.length;
        i0 = i;
      }
    }
  }

❽ if (ht.slots[i] === EMPTY) {
    ht.used++;
  }
  ht.slots[i] = value;
  return ht;
};
```

Start by using the first hash function ❶ to get the initial slot. If that slot isn't empty ❷, use the second hash function ❸ to see how far to jump at every step. You'll save the initial slot at i0 ❹ to detect a loop, and then start looking for an empty or available slot ❺. At each pass of the loop, advance step places ❻, and if you detect that you're back at the initial i0 place, then advance just one place and save the new initial slot ❼. After finding where to put the new value ❽, the logic is as for the previous hash methods: update the count of used slots and save the value. You should regenerate the table in case of a high load factor, in the same way as in the open addressing section.

Searching for a Value in a Table That Uses Double Hashing

The logic for searching for a value matches the way you add new values:

```
const find = (ht, value) => {
❶ let i = hash1(ht, value);
❷ const step = hash2(ht, value);
```

```
❸ let i0 = i;
❹ while (ht.slots[i] !== EMPTY && ht.slots[i] !== value) {
    i = (i + step) % ht.slots.length;
  ❺ if (i0 === i) {
      i = (i + 1) % ht.slots.length;
      i0 = i;
    }
  }

❻ return ht.slots[i] === value;
};
```

Start by deciding on the initial slot to test ❶ and the step amount to jump ❷. Then save the initial spot to detect a loop ❸ and start jumping until you find an empty slot or the desired value ❹; the logic for jumps is exactly the same as for insertions, including the loop detection ❺. At the end, return true or false ❻ depending on where you stopped the search.

Removing a Value from a Table That Uses Double Hashing

I won't repeat the explanation here, but to remove a value, you'll use the same technique as for open addressing. You won't mark removed values as EMPTY, but rather as AVAILABLE. The code is as follows:

```
const remove = (ht, value) => {
  let i = hash1(ht, value);
  let i0 = i;
  const step = hash2(ht, value);
  while (ht.slots[i] !== EMPTY && ht.slots[i] !== value) {
    i = (i + step) % ht.slots.length;
    if (i0 === i) {
      i = (i + 1) % ht.slots.length;
      i0 = i;
    }
  }

❶ if (ht.slots[i] === value) {
    ht.slots[i] = AVAILABLE;
  }
  return ht;
};
```

The code is exactly the same as for the open addressing search, and you change only what to do after finishing the loop. If you found the value ❶, mark the slot as AVAILABLE; if it's not found, don't do anything.

The logic for double hashing works well, but you can easily make it more streamlined, as we'll see next.

Double Hashing with Prime Lengths

The only problem with the logic that you saw with double hashing operations is the need to deal with possible loops. You'll get a loop whenever the step size you choose happens to have some common factor with the table's length. For instance, if the table size is 18 and the step is 12, after three steps you'll be back where you started. If you could choose a table length that doesn't have any common factors with all possible steps, the logic would be simpler. There's an easy way to do that: if the table length is a prime number (divisible only by itself or by 1), no loops are possible, because a prime number has no common divisors with any lower number. Also, if the step is 1, everything is fine, because before returning to the initial slot, you'll have gone through the complete array.

Creating a Table That Uses Double Hashing with Prime Lengths

You can create a new hash table exactly as before, except you must ensure that its length is a prime number. First you need to check whether a number is prime:

```
const isPrime = (n) => {
❶ if (n <= 3) {
    return true;
❷ } else if (n % 2 === 0) {
    return false;
  }

❸ for (let d = 3, q = n; d < q; d += 2) {
    q = n / d;
  ❹ if (Math.floor(q) === q) {
      return false;
    }
  }
❺ return true;
};
```

Small numbers are prime (and for the purposes here 1 works as a prime number, no matter what mathematicians may say) ❶. Even numbers (you excluded 2 in the previous if) aren't prime ❷, so those cases are simple. For other numbers, test all odd possible divisors starting at 3 ❸ and stop when you find an exact division ❹ or when the tested possible divisor exceeds the square root of the number, in which case the number is prime ❺.

Next you need a simple function to find the first prime number greater than a given value, and you can write it simply by using the isPrime(...) function:

```
const findNextPrime = (n) => {
❶ while (!isPrime(n)) {
  ❷ n++;
  }
  return n;
};
```

The logic is straightforward: given a number, if it's not prime ❶, add 1 to it ❷ until the number becomes a prime.

Now you can create a table. The logic is the same as before, except you make sure that the length of the table is a prime number:

```
const newHashTable = (n = 100) => ({
  slots: new Array(findNextPrime(n)).fill(EMPTY),
  used: 0
});
```

Whatever size you get, you find the next higher prime and use it as the table length.

Adding a Value to a Table That Uses Double Hashing with Prime Lengths

To add a value, work the same way as with the double hashing code, except you don't have to test for loops; prime numbers have this covered:

```
const add = (ht, value) => {
  let i = hash1(ht, value);
  if (ht.slots[i] !== EMPTY) {
    const step = hash2(ht, value);
    while (ht.slots[i] !== EMPTY && ht.slots[i] !== AVAILABLE) {
      i = (i + step) % ht.slots.length;
    }
  }

  if (ht.slots[i] === EMPTY) {
    ht.used++;
  }
  ht.slots[i] = value;
  return ht;
};
```

You'll advance as before, but all the code related to i0 (which you used for loop detection) is now gone.

Searching for a Value in a Table That Uses Double Hashing with Prime Lengths

Searching is also simpler:

```
const find = (ht, value) => {
  let i = hash1(ht, value);
  const step = hash2(ht, value);
  while (ht.slots[i] !== EMPTY && ht.slots[i] !== value) {
    i = (i + step) % ht.slots.length;
  }

  return ht.slots[i] === value;
};
```

Again comparing this code with the code for common double hashing, the key difference is that you did away with all the loop detection and prevention.

Removing a Value from a Table That Uses Double Hashing with Prime Lengths

Finally, as expected, removing a value is also easier:

```
const remove = (ht, value) => {
  let i = hash1(ht, value);
  const step = hash2(ht, value);
  while (ht.slots[i] !== EMPTY && ht.slots[i] !== value) {
    i = (i + step) % ht.slots.length;
  }

  if (ht.slots[i] === value) {
    ht.slots[i] = AVAILABLE;
  }
  return ht;
};
```

Once more, the code is the same as for common double hashing, but without checking for loops; it's faster, simpler code.

Summary

In this chapter we've considered several ways to implement bags and sets, including the hashing technique that, when properly applied, can provide the fastest possible search times. The structures considered here were basically linear; in the next chapter we'll start considering nonlinear ones such as trees to explore further implementations of bags and sets.

Questions

11.1 Sentinels for Searches

Show how an ordered list could benefit from a final +Infinity sentinel value for simpler code.

11.2 More Sentinels?

Would adding an initial -Infinity sentinel help with ordered lists?

11.3 A Simpler Search?

Can you simplify the code to avoid having to return two values when searching a self-organizing list? A tip: if the search was successful, the list won't be empty and its head will have the searched value.

11.4 Re-skipping Lists

Can you sketch out an algorithm that will restructure a skip list to make sure it is well balanced?

11.5 Skip to an Index

In the earlier definition you just wanted to search for a value, but what if you had an index i and wanted the ith value of the list? Can you think of a way to modify skip lists in order to find a value by index in an efficient way?

11.6 Simpler Filling

Why wouldn't the following code work to create a hashing table with chaining?

```
const newHashTable = (n = 100) => ({
  slots: new Array(n).fill(newList())
});
```

11.7 A Hashed Set

In the hash table code for insertions you allowed repeated values, so you're doing bags instead of sets. Can you modify the code as efficiently as possible to implement sets instead? Obviously, you could start by doing a search, but if that search failed, you'd be redoing a lot of work for the addition.

11.8 Wrong Seating

This puzzle will remind you of hashing. Imagine that 100 people were given tickets for a show. The theater has 100 seats, and each ticket is assigned to a different seat. There was a problem, though. The first person to arrive at the theater didn't pay attention and sat in a random seat. All the other people tried to go to their seats, and if their seats were already occupied, they also took random seats. What's the probability that the last person (the 100th one) will find that their seat is unoccupied?

11.9 Progressive Resizing

Doing a whole resizing operation (with the corresponding delay in time) may not be acceptable for some systems, so you need some kind of progressive resizing solution. Can you sketch a way to do the rehashing gradually, somehow working with two tables (an old one and a new one) but not rehashing the entire old table at once?

12

BINARY TREES

We've considered linear structures in previous chapters, and now we'll start working with more complex structures—in particular, binary trees and some variants. (We'll explore more general trees in the next chapter.) Binary trees find their way into lots of places, including data compression algorithms, video games, cryptographic techniques, compilers, and more, so they are a structure well worth knowing.

A special variety, binary search trees, can be quite efficient for implementing the bags or sets we explored in the previous chapter. However, since those kinds of trees can, on occasion, provide not-so-good performance, we'll also consider some variants, such as assuredly balanced binary search trees (AVL trees) and probabilistically balanced trees (randomized binary search trees and splay trees).

What Are Trees?

Trees allow you to represent hierarchical data structures. They differ from linear structures because a node can be connected to several other nodes, albeit with some restrictions. Organizational charts (or *organigrams*) are well-known examples of trees where a section of an enterprise can have several subsections, which themselves may have sub-subsections, and so on, in a recursive fashion, such as shown Figure 12-1.

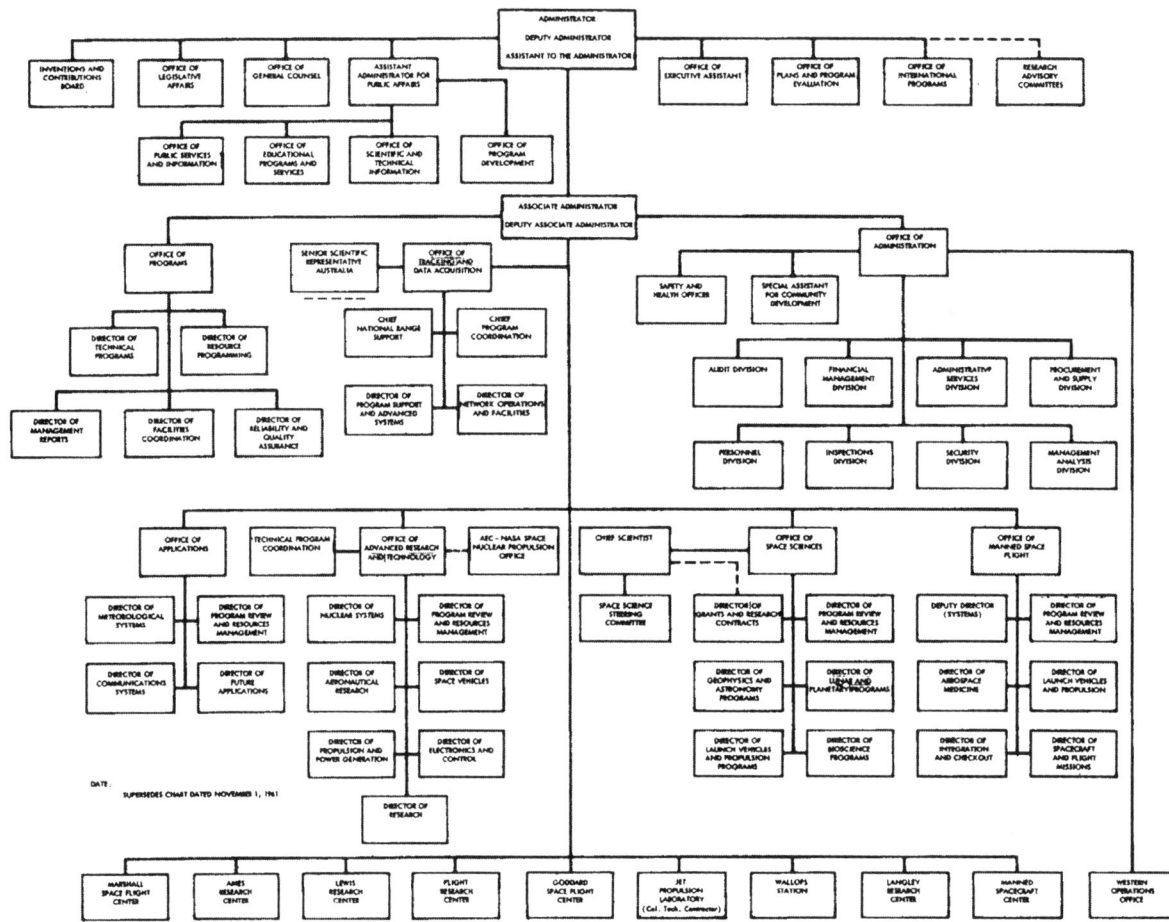

Figure 12-1: An organigram from NASA, November 1961. Don't worry if the text is unreadable; the structure is the important part, not the labels. (From https://commons.wikimedia.org/wiki/File:NASA_Organizational_Chart_November_1,_1961.jpg.)

HyperText Markup Language (HTML) also has a treelike structure. An HTML element can contain several other elements, which may themselves include further elements. Directories on your computer employ a tree structure as well. A directory has files and more directories, which themselves have files and more directories, and so on. (See question 12.2 at the end of this chapter for an exception.)

General Trees

A tree can be empty or consist of a node (called the *root* of the tree) that has several subtrees, each of which may be empty, of course. The root is the *parent* of its subtrees, and the roots of those subtrees are *children* of the root. The *nodes* that form a tree are connected by *edges* or *arcs*. Nodes with both parents and children are called *internal nodes*, and nodes without children are called *external nodes* or (more appropriately for the tree motif) *leaves*. Given a node, its children, and the children of those children, and so on are called its *descendants*. Similarly, the parent of a node, and the parent of that parent, and so on are called the node's *ancestors*. The *level* of the tree's root is 1, its children are level 2, the children of those children are level 3, and so on. The number of a node's nonempty children is called its *degree*. Finally, for any tree or subtree, its *size* is the number of its nodes, and its *height* is the number of nodes in the longest path from the root to a leaf.

NOTE *From the previous definitions, it follows that there can be at most one path between any two nodes; trees cannot have any cycles or loops anywhere. Another property is that given any two nodes, either one is an ancestor of the other or they both have a common ancestor.*

Trees are usually represented with the root at the top and the leaves at the bottom. Even if this goes against biology, we'll follow that style. A possible tree could look like Figure 12-2 (with the root at the top and all links going downward).

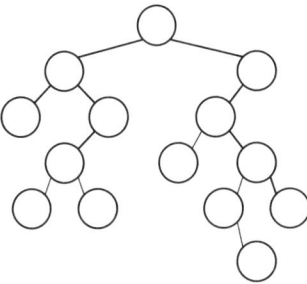

Figure 12-2: Trees are usually shown with the root at the top, branching down to the leaves, despite what biology teaches.

We'll consider general trees (that is, those with any number of subtrees in any node) in the following chapter, so here we'll focus on the most common version, binary trees.

Binary Trees

A *binary tree* is either empty or has exactly two subtrees. We'll see some additional definitions later, so the tree in Figure 12-2 actually could be a binary tree. A binary tree is *full* if every node either is a leaf or has two nonempty children. Figure 12-3 shows a possible case.

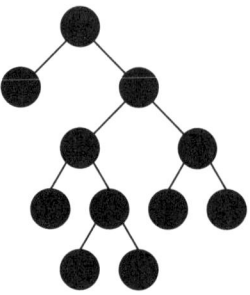

Figure 12-3: A full binary tree: all nodes have zero or two children.

Full binary trees aren't really that interesting unless they also satisfy some other properties. For example, if a tree is full and all leaves are at the same level, it's called a *perfect* binary tree, which implies that nodes are packed as tightly as possible, as shown in Figure 12-4.

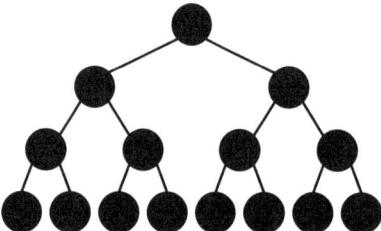

Figure 12-4: A perfect binary tree is full and has all leaves at the same level.

An interesting property, which a little math easily proves, is that the size of a perfect binary tree of h height is $2^h - 1$, so adding a new level approximately doubles the tree's size. Conversely, the height of a perfect tree with n nodes is log n, rounded up. (We are using logarithms in base 2.)

Finally, if you have a tree of h height that becomes perfect if you eliminate all nodes at level h (with the allowed exception of the last level), it's called a *complete* tree. We'll look at some of those structures later in Chapter 14, when we study heaps. Figure 12-5 shows a complete tree, because if you were to remove all the nodes at the bottom level, you'd be left with a perfect tree.

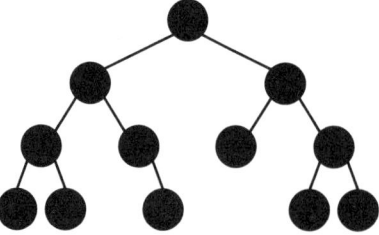

Figure 12-5: A complete tree would become a full tree if the bottom leaves were removed.

With the generic definitions out of the way, let's get started with binary trees. We'll include a key in each node, and you can add more data attributes if needed. We'll also have left and right pointers to the subtrees: each may either be null or point to another binary tree. Let's start writing a binary tree module (you'll reuse some of these methods here for binary tree variants later):

❶ ```
const newTree = () => null;
```

❷ ```
const newNode = (key, left = null, right = null) => ({
  key,
  left,
  right
});
```

❸ ```
const isEmpty = (tree) => tree === null;
```

There's not much to this code: newTree() ❶ builds an initially empty tree; newNode() ❷ creates a new node with a given key and (by default null) subtrees; and isEmpty() ❸ detects whether a tree is empty (no surprise here).

## Binary Search Trees

For the remainder of this chapter, we'll use *binary search trees*, which are a variant of binary trees, to implement the *bag* or *set* abstract data type (ADT), because they provide very efficient searching for keys. (Remember, the difference is that bags allow repeated values, but sets don't.) For these trees, each node will be an object with a key, plus some links to point at its children; in practice, you could also include an extra data field in a node for other usages. Table 12-1 describes the ADT.

**Table 12-1:** Operations on Sets

| Operation | Signature | Description |
| --- | --- | --- |
| Create | → set | Create a new set. |
| Empty? | set → boolean | Given a set, determine if it is empty. |
| Add | set × value → set \| error | Given a new value, add it to the set. |
| Remove | set × value → set | Given a value, remove it from the set. |
| Find | set × value → boolean | Given a value, check if it exists in the set. |

What's the difference between a binary tree and a binary search tree? Binary search trees satisfy the following property: for all nodes, the left subtree has only smaller keys and the right subtree has only greater keys. If you decide to allow duplicate keys, you need to amend the condition to say that the left subtree has smaller or equal keys and the right subtree has greater or equal keys. In Figure 12-6, one of the trees is a binary search tree, but the other is not because of a single unlucky detail. Can you tell which is which?
```

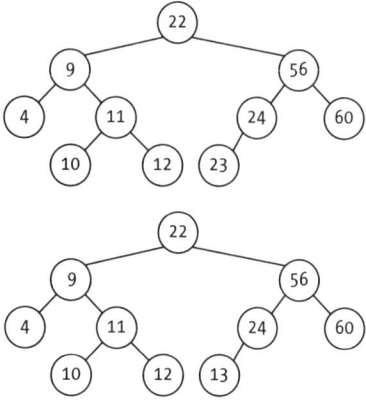

Figure 12-6: Two binary trees, but only one is a binary search tree. Which is it?

The bottom tree isn't a binary search tree, because the 13 key is to the right of key 22, and it should be to its left. Can you figure out exactly where it should go?

This property regarding keys of roots and subtrees is what allows you to use binary search trees as sets.

Finding a Key in a Binary Search Tree

The recursive property regarding the relation of keys (which also applies to every subtree) provides a simple searching method. If you are looking for a given value in a binary search tree, one of three situations must happen:

- If the value is the key at the root of the tree, you're done.

- Otherwise, if the value is smaller than the key at the root, the value (if present) must be in the left subtree.

- Finally, if the value is greater than the key at the root, the value must be in the right subtree.

You can test this. Figure 12-7 shows a successful search for 12, highlighting the path that was taken and all the visited nodes.

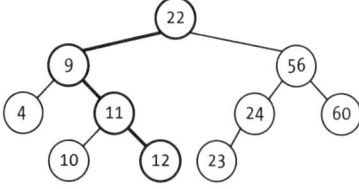

Figure 12-7: A successful search for key 12 in a binary search tree

The search starts at the root. Since 12 < 22, it moves to the left subtree. There, since 12 > 9, it proceeds to the right subtree. Then, as 12 > 11, it again goes to the right subtree, and the key is found. If you had been look-ing for 34 instead, the search would have failed, as shown in Figure 12-8.

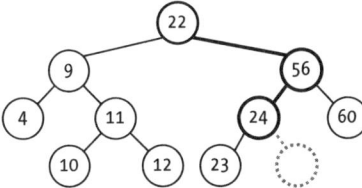

Figure 12-8: A failed search for key 34 in a binary search tree

Since 34 > 22, the search starts down the root's right subtree; next, as 34 < 56, it goes to the left. Then, as 34 > 24, it tries to go to the right but finds an empty tree (shown with a dotted border), so the search was unsuccessful.

You can code this logic straightaway, even before considering how you would do additions or deletions to a tree:

```
const find = (tree, keyToFind) => {
❶ if (isEmpty(tree)) {
    return false;
❷ } else if (keyToFind === tree.key) {
    return true;
  } else {
  ❸ return find(keyToFind < tree.key ? tree.left : tree.right, keyToFind);
  }
};
```

Since trees are recursive by definition, it should be no surprise that this algorithm (and most others in this chapter) is implemented using recursion. There are two base cases: if the tree is empty ❶, the key isn't in the tree, and if the key matches the value you're looking for ❷, the search succeeds. But how do you keep searching? If the key you're looking for is smaller than the key at the root, you recursively search the left subtree, and the right subtree otherwise ❸.

Adding Values to a Binary Search Tree

How can we add a new key to a tree? Let's work with a bag, and accept repeated keys; you'll see how to do a set too. Be careful not to disturb the relationship between the root key and those of its subtrees—using a recur-sive algorithm is the best way to do this. If the tree is empty, you can simply add a new leaf to it. If the tree isn't empty, apply recursion to go down the left or right subtree, depending on where the new key should be, until you reach an empty tree where you can insert the new key.

The previous section showed a failed search for a 34 key, so now the new key would be added at the place where the search ended, as shown in Figure 12-9.

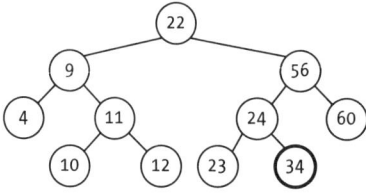

Figure 12-9: Adding a new key to a binary search tree
at the place where it should have been found in a search

The code for this is as follows:

```
const add = (tree, keyToAdd) => {
❶ if (isEmpty(tree)) {
     return newNode(keyToAdd);
  } else {
  ❷ const side = keyToAdd <= tree.key ? "left" : "right";
    tree[side] = add(tree[side], keyToAdd);
    return tree;
  }
};
```

If the tree is empty ❶, create a new node with the key to add, and that will be the root. If the tree isn't empty, decide which of its subtrees must add the new key ❷ and proceed recursively from there. (If implementing a set instead of a bag, you should check whether keyToAdd equals tree.key, and in that case reject the addition; see question 12.16.) This example uses a different coding style from the one in find() just for variety.

Removing Values from a Binary Search Tree

Now let's look at how to remove a key from a binary search tree. Consider the tree shown in Figure 12-10.

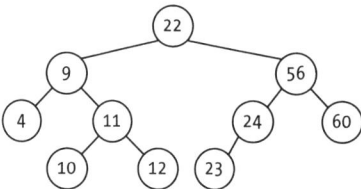

Figure 12-10: A binary search tree before
deleting some keys

If you try to remove a key that you can't find in the tree, you don't need to do anything. Easy.

Another simple case is removing a leaf: just remove its key, which makes it an empty tree. For instance, removing 10 would result in the following situation, where 11 ends up with an empty left subtree, as shown in Figure 12-11.

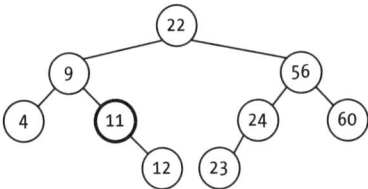

Figure 12-11: Removing a leaf (key 10, in this case) is straightforward.

However, things can get complicated. For example, if you want to remove a node that has at most one child, that's still easy. Just replace it with its child, as in Figure 12-12, where the 24 key was removed by making the 23 key the left child of the 56 key.

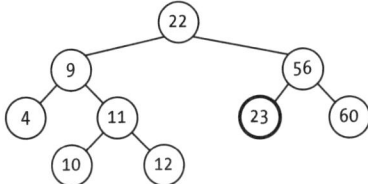

Figure 12-12: Removing a key (in this case, 24) with only one child is also straightforward.

A complex problem is dealing with a node that has two nonempty children. The most common solution is to find the key immediately following it, remove it, and put it in place of the node you wanted to remove. For instance, if you want to remove the 9 key in Figure 12-12, since that node has two subtrees, you would search for the next higher key (in this particular example, 10), remove it, and put it into the 9 key's place, as shown in Figure 12-13.

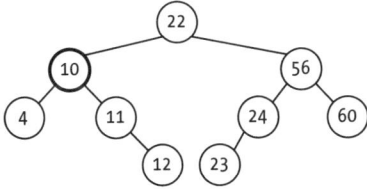

Figure 12-13: Removing a key (here, 9) is the hardest case; you have to put another key in its place to maintain the binary search tree structure.

This method of replacing the removed key doesn't break the search rules. There's a missing step, though—namely, how to find the next higher key. We'll get to that in the next section, but first here's the code to remove a key:

```
const remove = (tree, keyToRemove) => {
❶ if (isEmpty(tree)) {
    // nothing to do
```

```
❷ } else if (keyToRemove < tree.key) {
    tree.left = remove(tree.left, keyToRemove);
❸ } else if (keyToRemove > tree.key) {
    tree.right = remove(tree.right, keyToRemove);
❹ } else if (isEmpty(tree.left) && isEmpty(tree.right)) {
    tree = null;
❺ } else if (isEmpty(tree.left)) {
    tree = tree.right;
❻ } else if (isEmpty(tree.right)) {
    tree = tree.left;
❼ } else {
  ❽ tree.key = minKey(tree.right);
  ❾ tree.right = remove(tree.right, tree.key);
  }

  return tree;
};
```

The first three conditions ❶❷❸ match the find() method: check for an empty tree; if you haven't found the key to delete, proceed to a subtree recursively. The next case ❹ deals with removing a leaf: set the tree to null. The next two conditions ❺❻ deal with nodes that have a single child; set the tree to that child. Finally, in the last case ❼ you must find the key ❽ that follows the one you want to delete and use it to replace that key and then finish by recursively deleting that key from the right subtree ❾. You can complete the algorithm by considering how to find the next higher key.

NOTE *This is not the only way to do a deletion; we'll see more in the sections "Removing a Key from a Randomized Tree" on page 267 and "Removing a Key from a Treap" on page 336.*

Finding the Minimum or Maximum Value in a Binary Search Tree

Because of the relationship between the root and its subtrees, the needed key (the following key) must be the minimum value of the right subtree. (Conversely, the previous key would be the maximum value of the left subtree.) Figure 12-14 shows how to look for the key following the 9 key. You need to go to its right subtree and then go left until you can't move in that direction any longer to find the 10 key.

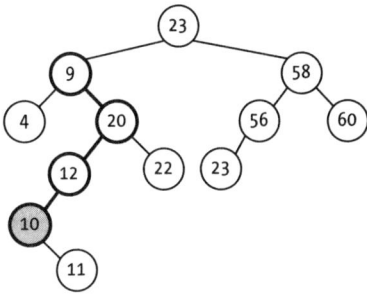

Figure 12-14: Finding the following key; here, you want the minimum key greater than 9.

For a different example, if you wanted to find the *previous* key to 23, you would go to its *left* subtree and then move *right* until reaching the end to find the 22 key. Keep in mind that this logic works only for nodes that have the necessary subtrees. If you want to find the next key of, say, 11, 12, or 22, the logic would fail. Fortunately, this doesn't apply to cases in which you want to find the next higher key.

You can take advantage of similar logic to implement both minKey() and maxKey():

```
❶ const _minMax = (tree, side, defaultValue) => {
❷   if (isEmpty(tree)) {
      return defaultValue;
❸   } else if (isEmpty(tree[side])) {
      return tree.key;
    } else {
❹     return _minMax(tree[side], side, defaultValue);
    }
};

const minKey = (tree) => _minMax(tree, "left", Infinity);
const maxKey = (tree) => _minMax(tree, "right", -Infinity);
```

First look at minKey(), which is what you wanted in this case; maxKey() is analogous. You have an auxiliary _minMax() method that does the actual searching ❶ based on whatever arguments minKey() and maxKey() pass to it. Looking for the minimum requires always going to the left, so that takes care of the second parameter of _minMax(), which will go down that side ❹ until an empty tree is found ❸. Now, if you try to find the minimum of an empty tree ❷, what value should be returned? You'll do the same thing the Math.min() function does; if you call it without any arguments, it returns Infinity (similarly Math.max() === -Infinity), so that's the third parameter of _minMax().

NOTE *If you analyze the removal algorithm, you may decide that it does more work than needed because it travels down the right subtree once to find the next key and then processes the same subtree again to remove the found key. Why not do both things at once? See question 12.17 for this optimization.*

Traversing a Binary Search Tree

Many processes involve accessing all nodes of a tree (also called *traversing* a tree or doing a *tree traversal*) to do something with each—for example, you could have stored words in a binary search tree and want to produce an alphabetically ordered listing of them. This is called *visiting* the nodes. If you don't want to exclude any nodes, three possible scenarios exist for such a general visitation (the pre-, in-, and post- prefixes in these traversal methods are related to when the root is visited):

Preorder   Visit the root of a tree, then traverse its left subtree, and finally traverse its right subtree.

Inorder   Traverse the left subtree first, then visit the root, and finally traverse the right subtree.

Postorder   Traverse the left subtree first, then traverse the right subtree, and finish by visiting the root.

Of course, you traverse an empty tree by doing nothing at all, as visiting applies only to existing keys. Also, note that traversal of subtrees is done by recursively applying the same traversal algorithm.

Here's a basic algorithm where the default visit() method just prints the visited key:

```
const preOrder = (tree, visit = (x) => console.log(x)) => {
  if (!isEmpty(tree)) {
    visit(tree.key);
    preOrder(tree.left, visit);
    preOrder(tree.right, visit);
  }
};

const inOrder = (tree, visit = (x) => console.log(x)) => {
  if (!isEmpty(tree)) {
    inOrder(tree.left, visit);
    visit(tree.key);
    inOrder(tree.right, visit);
  }
};

const postOrder = (tree, visit = (x) => console.log(x)) => {
  if (!isEmpty(tree)) {
    postOrder(tree.left, visit);
    postOrder(tree.right, visit);
    visit(tree.key);
  }
};
```

The code follows the description: for example, preOrder() first visits the root, then traverses the left subtree, and it finally traverses the right subtree.

For debugging purposes, it's useful to be able to print the list of the tree's keys in ascending order. If you have a tree and call inOrder(), all keys are listed in order. It starts at the root and processes all the keys less than the root (listing them in order). Next, it prints the root, and then it processes all the keys greater than the root (also listing them in order), providing the desired result.

NOTE *This algorithm is similar to quicksort from Chapter 6. You have a left set of keys, which you order. Then you have the pivot, and then you have a right set of keys, which you also order, and the result is the complete ordered array.*

Getting a list of keys is fine, but seeing the structure is better, so you want to get a printout of it. Consider the tree in Figure 12-15.

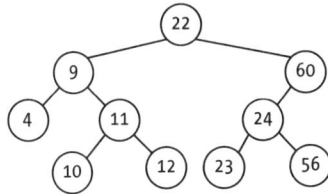

Figure 12-15: A binary search tree for which we want to print out its structure

You could use console.log() for the printout, but that's not too user friendly; console.dir()is a tad better. You could try something like console .log(JSON.stringify(tree)), but that's really hard to read; you get some very unfriendly output:

```
{"key":22, "left":{"key":9,  "left":{"key":4, "left":null, "right":null}, "rig
ht":{"key":11, "left":{"key":10, "left":null,"right":null}, "right":{"key":12,
 "left":null, "right":null}}}, "right":{"key":60, "left":{"key":24, "left":{"k
ey":23, "left":null, "right":null}, "right":{"key":56, "left":null, "right":nu
ll}}, "right":null}}
```

To understand the tree's structure, consider a print() method based on the preorder code. It prints the root first, on one line, followed by its left subtree (with an L: preceding it to signify the left subtree), and then the right subtree (with an R:), indenting children to the right, and children's children even more, and so forth.

The resulting content was similar to the following output:

```
22
 L: 9
 L:  L: 4
 L:  R: 11
 L:  R:  L: 10
 L:  R:  R: 12
 R: 60
 R:  L: 24
 R:  L:  L: 23
 R:  L:  R: 56
```

The root (22) is at the top, followed by L: 9 (and further below, R: 60), showing both of the root's children. For each new key, you also see its children, further indented, so it's clear enough for debugging.

The code to produce this output is as follows:

```
const print = (tree, s = "") => {
  if (!isEmpty(tree)) {
    console.log(s, tree.key);
    print(tree.left, `${s}  L:`);
    print(tree.right, `${s}  R:`);
  }
};
```

If you compare the logic with the earlier preOrder() method, it's the exact same idea: do something with the key first, and then process its left and right subtrees in order.

Considering Performance for Binary Search Trees

Now that we've looked at binary search tree algorithms in detail, what about their performance? Let's start with the *worst* possible case. The most dreadful situation you might get after adding several keys to a tree is some kind of linear structure like the one shown in Figure 12-16, which in searching terms basically is equivalent to simple linked lists with $O(n)$ performance.

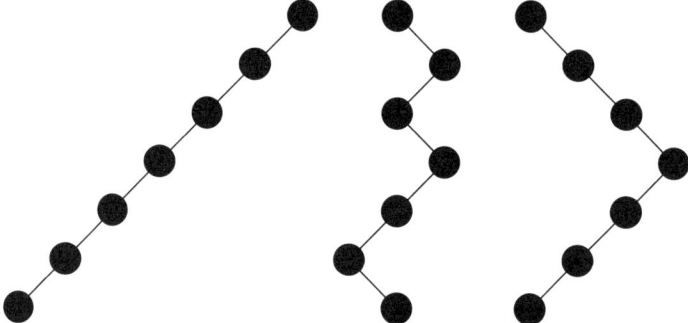

Figure 12-16: Some worst-case binary search trees

Going back to the shapes we've looked at earlier, it's obvious that a tree's shape impacts an algorithm's performance. The perfect tree is best, which would be $O(\log n)$. With linear-like structures, searches would become $O(n)$, and for large trees, that's a huge difference.

In terms of probability, if you take a set of keys in random order, it can be proved that most of the trees will be relatively short in height, and bad cases will be relatively few. While the worst case would still be $O(n)$, on average, we expect $O(\log n)$ performance. Table 12-2 shows average and worst-case performance for the tree.

Table 12-2: Performance of Operations for Binary Search Trees

Operation	Average performance	Worst case
Create	$O(1)$	$O(1)$
Add	$O(\log n)$	$O(n)$
Remove	$O(\log n)$	$O(n)$
Find	$O(\log n)$	$O(n)$
Traverse	$O(n)$	$O(n)$

What can you do about that? We'll look at two options in the following sections that attempt to ensure that the tree never reaches a bad shape and stays as short and balanced as possible.

Assured Balanced Binary Search Trees

As we saw earlier, a tree can become a linear (or almost linear) structure, and its performance will be quite poor. It's possible to ensure that a tree is kept in balance, however, guaranteeing optimum performance. Here are two different ways of dealing with this problem:

- *Assured balanced trees* become efficient because they follow some explicit structural constraint that never lets trees get out of shape, but they imply extra running time and memory usage, needing more complex algorithms—usually add() and remove()—to ensure that the constraints still apply after modifying the tree. These trees offer a consistent performance in an absolute (neither amortized nor probabilistic) way.

- *Probabilistically balanced trees* (or *self-adjusting trees*) are efficient only in an amortized sense. They do not follow any explicit structure rule, but they can be in any possible shape, depending on methods like add() or find() to adjust the structure in such a way that it most likely improves over time.

Height-balanced *AVL trees* do not let trees get out of balance by forcing both subtrees of any node to assuredly have approximately the same height. Weight-balanced trees also offer assured balance, by keeping the weights of both subtrees of any node within a given factor of each other; we'll consider bounded balance (BB[α]) trees later.

AVL Trees

AVL trees, invented by Adelson-Velsky and Landis in 1962, are well balanced by following a simple rule: *for all nodes, the heights of their left and right subtrees must differ at most by one.* This automatically rules out all the badly performing shapes of binary trees.

Figure 12-17 shows a correctly balanced tree and an unbalanced one. Which is which?

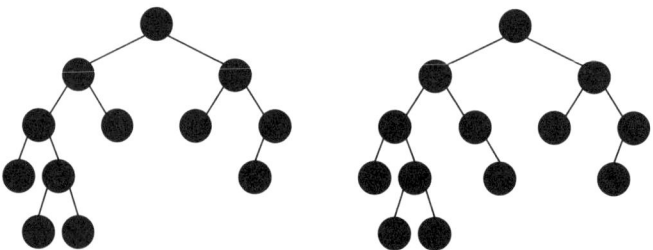

Figure 12-17: Two binary trees, but only one is balanced. Which is it?

The rightmost tree is well balanced, and the leftmost tree is not, because the left child of the root is out of balance: its left subtree has a height of 3, and its right subtree has a height of 1. The balance of a node is the difference in heights between its right subtree and left subtree, so the balances in the correct tree in Figure 12-17 would be as the one shown in Figure 12-18.

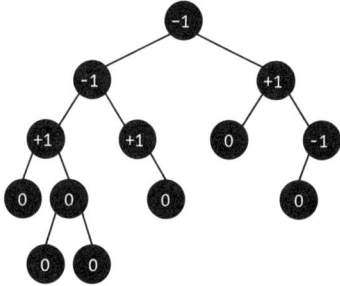

Figure 12-18: The balanced binary tree showing the balances for all nodes

Now that we've looked at the desired shape of AVL trees, you can code them.

Defining an AVL Tree

We'll base the AVL trees on binary search trees. Several of the operations will still work—for instance, finding a key in an AVL tree is exactly the same, so we won't see that code again here. There's a slight difference though: you need to add a _height attribute to each node to help check whether it's in balance, and you need code to access or calculate that attribute. The basic code starts as follows—note that you are reusing some methods of basic binary search trees:

```
const newAvlTree = () => null;

const newNode = (key) => ({
  key,
  left: null,
  right: null,
❶ height: 1
});
```

```
❷ const _getHeight = (tree) => (isEmpty(tree) ? 0 : tree.height);

❸ const _calcHeight = (tree) =>
    isEmpty(tree)
      ? 0
      : 1 + Math.max(_getHeight(tree.left), _getHeight(tree.right));

❹ const _calcBalance = (tree) =>
    isEmpty(tree) ? 0 : _getHeight(tree.right) - _getHeight(tree.left);
```

When constructing a new node, add the new _height attribute ❶, and then have a _getHeight() method to access it; take care so the height of an empty tree is 0 ❷. The new _calcHeight() method ❸ calculates the height of a node; assume both subtrees already have their own heights calculated, and the height of the total tree is one more than the height of its tallest subtree. Finally, calculate the balance of a node ❹ as the difference between the height of its right and left subtrees. That balance can be only –1, 0, or 1; other values imply an unbalanced tree.

Adding a Key to an AVL Tree

To add a new key, the logic is similar to what we already saw, except for a single factor: after deciding where to add the new key, the tree may become out of balance, so you need to move nodes around to restore it. Here's the additional code:

```
const add = (tree, keyToAdd) => {
  if (isEmpty(tree)) {
    tree = newNode(keyToAdd);
  } else {
    const side = keyToAdd <= tree.key ? "left" : "right";
    tree[side] = add(tree[side], keyToAdd);
  }

  return _fixBalance(tree);
};
```

This is exactly the same code as for binary search trees, except it adds a final _fixBalance() call that takes care of balancing the tree if needed. Before getting to that part, let's review how to remove keys, which is also quite similar to what you did previously.

Removing a Key from an AVL Tree

After seeing how to add a new key, removing a key will look familiar:

```
const remove = (tree, keyToRemove) => {
  if (isEmpty(tree)) {
    // nothing to do
  } else if (keyToRemove < tree.key) {
    tree.left = remove(tree.left, keyToRemove);
```

```
  } else if (keyToRemove > tree.key) {
    tree.right = remove(tree.right, keyToRemove);
  } else if (isEmpty(tree.left) && isEmpty(tree.right)) {
    tree = null;
  } else if (isEmpty(tree.left)) {
    tree = tree.right;
  } else if (isEmpty(tree.right)) {
    tree = tree.left;
  } else {
    tree.key = minKey(tree.right);
    tree.right = remove(tree.right, tree.key);
  }

  return _fixBalance(tree);
};
```

As with adding a key, the only difference with the previous code is at the end where you apply the balance fix.

Rotating Nodes in an AVL Tree

Adding or removing nodes essentially uses the same logic as for common binary search trees, but without intervention, the trees are likely to fall out of balance. The solution is to apply *rotations* that won't affect searching but will restore balance.

The two basic tree rotations are symmetrical, as shown in Figure 12-19, where the minus sign represents a smaller key value than the plus sign. After any of the rotations, the tree still allows searching, but the height and balance may change, and this allows you to restore an AVL tree.

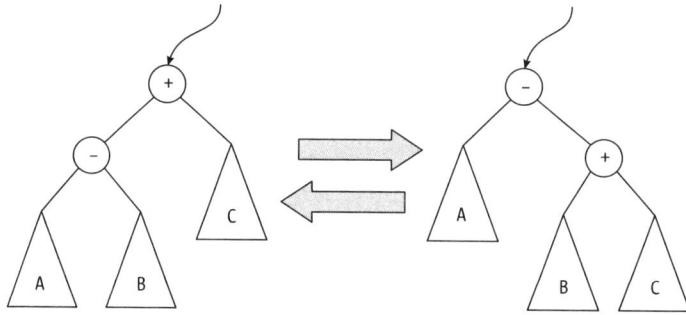

Figure 12-19: The two symmetrical rotations that can be used to solve balance issues

Rotating from left to right is a right rotation and from right to left is a left rotation. To remember which rotation is which, notice the direction the old root moves: in a right rotation, the root becomes its own right subtree, and in a left rotation, the root becomes its own left subtree. Another way of looking at it is in a right rotation, the node that was on the left becomes the root (that is, it moved to the right), and the root becomes a subtree, and in a left rotation, the node on the right becomes the root.

If you search for more information on tree rotations, you'll find many inconsistencies, and in some cases, what we call a right rotation, other sources call a left one, so be careful.

There are two possible cases when rotations are needed: one needs a single rotation, and the other requires two. In the first case (shown in Figure 12-20), the tree was balanced, but a new key was added in subtree A, making it taller, which put the whole tree out of balance. (Alternatively, you could have removed a key from subtree C, making it shorter.) In this case, the problem occurs at the left subtree of the left child of the root or, symmetrically, at the right subtree of the right child. These situations are logically called *left-left* and *right-right*.

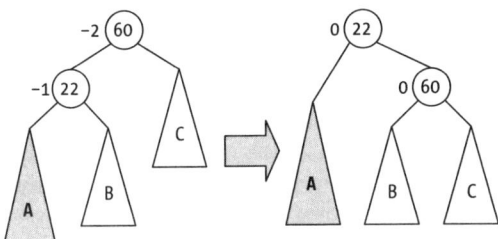

Figure 12-20: Using a right rotation to fix the unbalanced node with key 60

The solution is to apply a right rotation to the root of the left subtree (because the imbalance happened at the left subtree), which results in a balanced situation.

Figure 12-21 shows a more complex scenario. A new key was added at the right subtree of the left subtree of the root, throwing the latter out of balance. This *left-right* case and its mirrored *right-left* case need two rotations to be fixed.

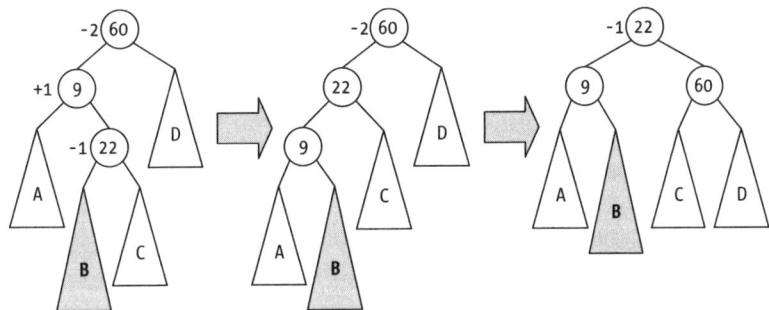

Figure 12-21: Fixing balance in a harder case needs a left rotation first (at the 9 key node) and then a right rotation (at the 60 key node).

A first left rotation brings the lowest key (22, in this case) closer to the root, and now a right rotation takes it all the way up, restoring balance.

Figure 12-21 shows the scenario where the addition was in B; if it had been in C, the solution still would be the same, and it would also apply if instead of an addition, you had removed a key from D, making it shorter.

Now consider the code to rotate a node:

```
const _rotate = (tree, side) => {
❶ const otherSide = side === "left" ? "right" : "left";
❷ const auxTree = tree[side];
❸ tree[side] = auxTree[otherSide];
  auxTree[otherSide] = tree;

❹ tree.height = _calcHeight(tree);
  auxTree.height = _calcHeight(auxTree);
  return auxTree;
};
```

You start by finding the "other" side for the rotation ❶, and you also get a reference to the node on the side of the root ❷ (the one that will become the root of the tree) to make the code briefer. Then, you exchange some pointers ❸ and finish by recalculating the heights of the two involved nodes; it's important to do the "lower" node ❹ first.

NOTE   *If you call _rotate() with a left parameter, it actually does a right rotation, which may be a bit confusing. The idea is you're saying which node should become root. So for a right rotation, the left child moves up to be the root. In some algorithms, you'll see that this is more natural.*

Now, let's finish by providing the missing _fixBalance() method:

```
const _fixBalance = (tree) => {
❶ if (!isEmpty(tree)) {
  ❷ tree.height = _calcHeight(tree);
  ❸ const balance = _calcBalance(tree);
  ❹ if (balance < -1) {
    ❺ if (_calcBalance(tree.left) === 1) {
        tree.left = _rotate(tree.left, "right");
      }
    ❻ tree = _rotate(tree, "left");
  ❼ } else if (balance > 1) {
      if (_calcBalance(tree.right) === -1) {
        tree.right = _rotate(tree.right, "left");
      }
      tree = _rotate(tree, "right");
    }
  }
  return tree;
};
```

If the tree is empty ❶, there's nothing to do. Otherwise, recalculate the root's height ❷ (since the recent addition or removal may have changed it), and also find the node's balance ❸. If the node is imbalanced on the left ❹, check whether an extra rotation is needed ❺ and do it if necessary, ending with a single rotation ❻. The other if is just the symmetrical case ❼, and it does the same things, but the sides are reversed.

Considering Performance for AVL Trees

Given the assured balance that the structure of AVL trees provides, all the operations (adding, removing, finding) have the same logarithmic performance. There's no different worst case, as shown in Table 12-3.

Table 12-3: Performance of Operations for AVL Trees

Operation	Average performance	Worst case
Create	$O(1)$	$O(1)$
Add	$O(\log n)$	$O(\log n)$
Remove	$O(\log n)$	$O(\log n)$
Find	$O(\log n)$	$O(\log n)$
Traverse	$O(n)$	$O(n)$

It can be proved that the height of an AVL tree is bounded by $1.44 \log n$, and that also confirms the performance listed earlier (see question 12.18). In the next chapter, you'll look at *red-black trees*, which have similar restrictions and performance but are based on multiway trees. Searches may be a tad slower (because those trees may be taller) and insertions a bit faster (requiring fewer rotations), but overall, the results are the same.

Weight-Bounded Balanced Trees

Instead of making sure that the heights of both subtrees of any node are within 1 of each other, *weight-bounded balanced (BB[α])* trees maintain a different invariant: that the *weights* (size of the tree plus 1) in the left and right subtrees are in a specific relationship. If a tree has size n and its subtrees have sizes p and q, you then have $(p + 1) \geq \alpha(n + 1)$ and $(q + 1) \geq \alpha(n + 1)$, with $0 < \alpha < 0.5$.

An equivalent way of looking at this is requiring that $(p + 1) / (n + 1) \geq \alpha$ and $(q + 1) / (n + 1) \geq \alpha$. Since both fractions add up to 1 (see question 12.20), this is the same as saying that both subtrees must satisfy $\alpha \leq \text{weight(subtree)} / \text{weight(tree)} \leq 1 - \alpha$. The quotient in the middle is called the *balance* of the subtree. Figure 12-22 shows a BB[0.29289] tree where keys from 1 to 12 were inserted in ascending order; the numbers on the edges show the balance of the corresponding subtree.

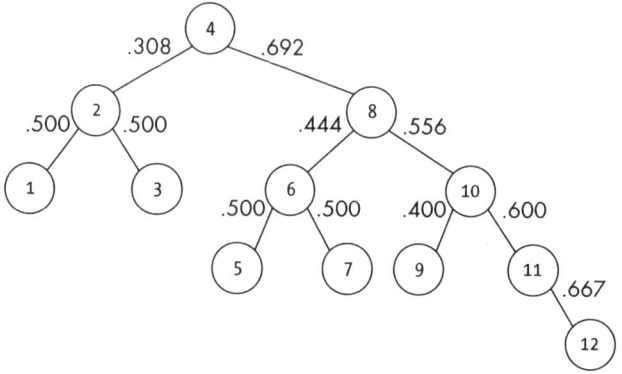

Figure 12-22: A weight-bounded balanced tree (BB[0.29289] in this case) showing the calculated balance for every node with children

The value $\alpha = 0.5$ sounds like a perfect balance (for all nodes, the right and left subtrees would be of equal sizes), but it has been proven that it doesn't really work, and not every value of α does. The α should be between 0.18182 $(= 2/11)$ and 0.29289 $(= 1 - \sqrt{2}/2)$ for balancing to work.

NOTE *When we defined the weight of a node and added 1 to its size, if it weren't for that additional 1, it would be impossible to have a weight-balanced tree of just two nodes. Can you see why?*

A BB[α] tree needs to carry the extra data of its size in every node in order to calculate its weight. This is necessary for balancing (so you can check the balance condition given previously), but it's also useful for other operations, such as finding a key by its rank.

When adding to or removing keys from the tree, if balance is not kept, we apply rotations (as in AVL trees) to restore balance. Since BB[α] trees are binary search trees, the find operation and traversals work without any changes. You need to consider only additions and removals.

Defining a Weight-Bounded Balanced Tree

These new trees and AVL trees share a lot of code. The biggest difference is that instead of including the height of a tree in each node, we include a size attribute and fix the balance by considering sizes instead of heights:

```
const {
  find,
  inOrder,
  isEmpty,
  maxKey,
  minKey,
  postOrder,
  preOrder
} = require("./binary_search_tree.js");
const newBBTree = () => null;
```

```
  const newNode = (key) => ({
    key,
    left: null,
    right: null,
❶ size: 1
  });

❷ const _getSize = (tree) => (isEmpty(tree) ? 0 : tree.size);

❸ const _calcSize = (tree) => 1 + _getSize(tree.left) + _getSize(tree.right);

❹ const _balance = (subtree, tree) =>
    (1 + _getSize(subtree)) / (1 + _getSize(tree));
```

When creating a new node, set its size to 1 ❶ instead of a height attribute. And instead of functions related to getting or calculating heights, you have a function as a getter for the tree's previously calculated size ❷, another function to calculate the size of any tree ❸, and a third one to calculate the balance of a subtree ❹, which you'll need for balance fixing.

Adding and Removing Keys to and from a Weight-Bounded Balanced Tree

I mentioned there would be a surprise, and it's that adding or removing keys is done in exactly the same way as for AVL trees. Look at the code from the previous section. When adding a new key, you did it in the standard way (that is, the same way as for binary search trees), and you finished by calling a function to fix the balance, if needed. The only difference here is that the latter function will be implemented in another way:

```
const add = (tree, keyToAdd) => {
  if (isEmpty(tree)) {
    tree = newNode(keyToAdd);
  } else {
    const side = keyToAdd <= tree.key ? "left" : "right";
    tree[side] = add(tree[side], keyToAdd);
  }

  return _fixBalance(tree);
};
```

Deleting a key worked the same way. You first applied the standard binary search tree algorithm and, at the end, called the same function as with additions to restore balance whenever required.

You'll see the exact same process here; the only difference is how you restore balance.

Fixing Balance in a Weight-Bounded Balanced Tree

The original paper that described BB[α] trees shows (with math that's not included here) that there are two possible cases (plus their symmetrical ones) and that simple or double rotations are enough to restore balance.

Now consider cases where a node has an overweight left subtree; the symmetrical cases are handled the same way.

First, a review of some conditions. The balance of a subtree should be $\alpha \leq \text{balance(subtree)} \leq 1 - \alpha$. Several constants will be used when balancing, but we won't derive the values here:

- α is the underweight limit; if a subtree's balance is below α, the tree is out of balance.
- $\beta = 1 - \alpha$ is the overweight limit; if a subtree's balance is above β, there's also an imbalance.
- $\gamma = \alpha/\beta = \alpha / (1 - \alpha)$ is the underweight limit for a subtree's child.
- $\delta = 1 - \gamma = (1 - 2\alpha) / (1 - \alpha)$ is the overweight limit for a subtree's child.

The following code defines the values (the comments show the approximate value of each constant):

```
const ALPHA = 0.29289;
const BETA = 1 - ALPHA;      // 0.70711
const GAMMA = ALPHA / BETA;  // 0.41421
const DELTA = 1 - GAMMA;     // 0.58579
```

Now, you will fix unbalanced trees. The first situation is shown in Figure 12-23. The left subtree has grown too much (or the right subtree has decreased in size), so the tree is not in balance. You can calculate the balance of the right subtree of the left subtree (B) and find it is below δ, so it's not overweight. In this case, a single rotation to the right (shoving the B subtree to the right, which must have been underweight) rebalances the tree.

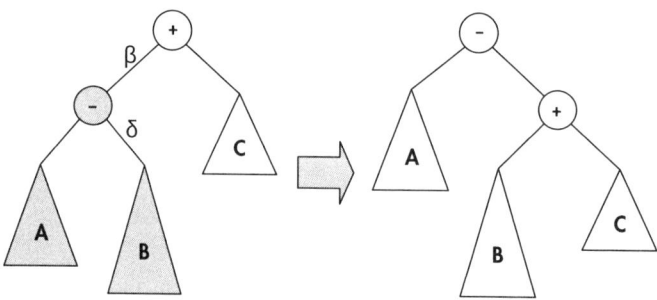

Figure 12-23: A single rotation fixes balance issues in some cases.

The second situation is a bit more complex. The left subtree is overweight, and the balance of its left subtree's right subtree exceeds the δ value. A single rotation wouldn't be enough to restore balance (the tree would still be out of balance), and in this case, a double rotation is needed to bring everything back to normal. Note that part of the overweight subtree is sent to the right (C), and the other part (B) remains on the left, as shown in Figure 12-24.

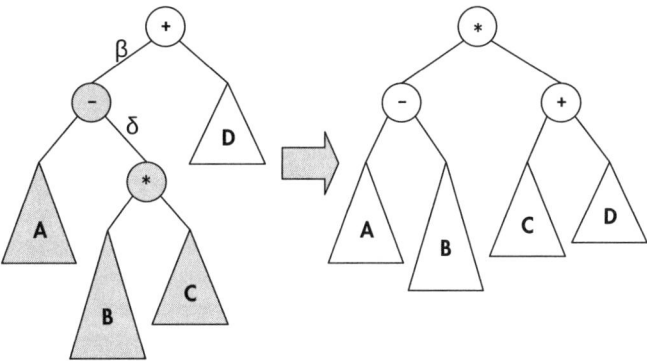

Figure 12-24: Two rotations are needed to fix balance in more complex situations.

So, the logic for deciding whether rotations are needed (and which) is this: first check both children to see if either is overweight (say it's the left one) by comparing its balance to β. Then check whether the other side's grandchild (the right child of the left child) is overweight, but compare its balance to a different limit of δ. Depending on the result of the second check, you'll do one or two rotations:

```
const _fixBalance = (tree) => {
  if (!isEmpty(tree)) {
❶ tree.size = _calcSize(tree);

❷ if (_balance(tree.left, tree) > BETA) {
    ❸ if (_balance(tree.left.right, tree.left) > DELTA) {
      ❹ tree.left = _rotate(tree.left, "right");
      }
    ❺ tree = _rotate(tree, "left");
❻ } else if (_balance(tree.right, tree) > BETA) {
      if (_balance(tree.right.left, tree.right) > DELTA) {
        tree.right = _rotate(tree.right, "left");
      }
      tree = _rotate(tree, "right");
    }
  }

  return tree;
};
```

If the tree isn't empty, start by updating its size ❶. Then, first check whether the left child is overweight ❷; if so, do a second check for the right child of the left child ❸, and if that tree is also overweight, do the first of two rotations ❹; then do the rotation to the right ❺ that will finish the job. If the left child wasn't overweight, check the right child ❻, and the logic is symmetric to the cases noted earlier ❹❺.

You now know how to update the tree by adding or removing keys, but BB[α] trees allow other operations as well, including finding by rank, splitting a tree in two, or joining two trees into one.

Finding an Element by Rank in a Weight-Bounded Balanced Tree

As mentioned previously, having the size of each tree at its root, which is needed by BB[α] trees for balancing purposes, provides an extra benefit, because it allows further operations with good performance. Consider one here: finding an element (in this case, the seventh) by rank, as you saw in Chapter 11. The tree in Figure 12-25 is the same as Figure 12-22, but now the subtrees' sizes are shown next to each node.

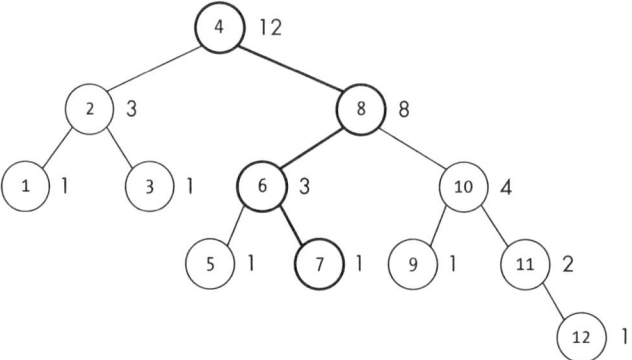

Figure 12-25: Including the size of each subtree at its root allows you to find an element by rank in an efficient way; here you're looking for the seventh key.

The left subtree has three elements, so if you were looking for the fourth element, that would be the root itself, and you'd be done! In this case, however, you are looking for the seventh element, so you need to keep searching. First, decide whether to go left or right: the left subtree has only three elements, so the seventh element must be the third element in the right subtree. You need to discount the three elements of the left subtree and also the root, so that removes four from the count, and you move right.

Now you are at the 8 key root, which has size 8. Its left subtree has three elements, and as you are looking for the third element of that tree, you keep going down to the left. At the 6 key root, repeat the procedure, and this time go right, as you need to discount one element from the left subtree and one from the root, so now you want the first element of the right subtree. Then you arrive at the 7 key, which is what you wanted.

You can easily implement this search with recursion:

```
const findByRank = (tree, rank) => {
❶ if (isEmpty(tree) || rank < 1 || rank > _getSize(tree)) {
    return undefined;
  } else {
❷ if (rank <= _getSize(tree.left)) {
    return findByRank(tree.left, rank);
❸ } else if (rank === _getSize(tree.left) + 1) {
    return tree.key;
❹ } else {
```

```
      return findByRank(tree.right, rank - _getSize(tree.left) - 1);
    }
  }
};
```

First, dismiss all the cases in which the search will fail ❶: an empty tree or asking for a rank outside the size of the tree. If the rank you want isn't greater than the size of the left subtree ❷, continue the search there. Otherwise, if the rank you want is exactly one more than the left subtree's size ❸, the root is the answer, and you are done. Finally, if none of the preceding conditions hold, go right, and you have to discount the left subtree's size and the root to continue the search ❹.

Considering Performance for Weight-Balanced Binary Trees

As in the case of AVL trees, ensuring balance in BB[α] trees makes for constant performance, with no worst cases. For all operations (adding, removing, and finding), the total cost is logarithmic, so weight-balanced binary trees ensure good performance, as shown in Table 12-4.

Table 12-4: Performance of Operations for
Weight-Balanced Binary Trees

Operation	Average performance	Worst case
Create	$O(1)$	$O(1)$
Add	$O(\log n)$	$O(\log n)$
Remove	$O(\log n)$	$O(\log n)$
Find	$O(\log n)$	$O(\log n)$
Find by rank	$O(\log n)$	$O(\log n)$
Traverse	$O(n)$	$O(n)$

Also in comparison with AVL trees, the code isn't very complex, and in both cases, you just depend on a "balance fix" function to be used after additions or removals.

Probabilistic Balance Binary Search Trees

Assured balance trees make operations more complex to ensure that a well-balanced shape will be kept at all times and thus provide a constant performance for operations. The other approach, *probabilistic balanced* trees, are simpler in implementation, require no extra memory usage, and can be as efficient (in an amortized sense) as assured balance trees—but you have to cope with the possible disadvantage of some individual slow operations mixed in with a long series of fast ones.

So, these trees do not ensure balance, but rather, they promise it in a probabilistic sense, and unless you are really unlucky, they will perform quite well. In this chapter, we'll consider two versions of these trees: *randomized binary search trees*, which apply balancing operations in a random manner, and *splay trees*, which restructure trees to make future searches faster. In Chapter 14 we'll consider one more option: treaps.

Randomized Binary Search Trees

Balanced trees guarantee performance by enforcing some constraints. This is an advantage in terms of performance, but it adds an extra level of complexity for operations, plus the need for some bookkeeping information at each node to determine whether restructuring is needed. Another way to avoid bad cases is to use randomized algorithms that provide a guarantee of their *expected* performance in terms of probability for any kind of input data. Depending on the implementation (and also on your particular data), a randomized algorithm may be faster than the corresponding assured balanced version, and it may be better for your needs. For instance, if you add keys in ascending order, balanced trees will have to do frequent balancing operations; if the algorithm works with random-based decisions at some points, fewer balancing operations may be needed; we'll see this more clearly later.

The first such structure we'll look at uses random numbers to decide whether a new addition should be at the root of the tree or go in its usual place. The insertion and deletion algorithms randomly decide how to either add a key to or remove a key from the tree. Both procedures produce a random structure, as if the input values had been shuffled, as you saw in Chapter 10. Remember, we won't need to reconsider how to find a key since we are still dealing with binary search trees and the earlier search logic still applies.

Defining the Randomized Binary Search Tree

Randomized trees will have the same structure as BB[α] trees, including a size attribute, but instead of using it for rebalancing the tree, we'll use it to help determine randomly what action to take. The basic code is as follows, and again we'll be reusing some code from standard binary search trees:

```
const newRandomTree = () => null;

const newNode = (key, left = null, right = null) => ({
  key,
  left,
  right,
  size: 1
});

const _getSize = (tree) => (isEmpty(tree) ? 0 : tree.size);

const _calcSize = (tree) => 1 + _getSize(tree.left) + _getSize(tree.right);
```

This is exactly the same way the BB[α] code started, except here the `newNode()` method lets you provide initial values for the `left` and `right` pointers, otherwise setting them to `null`.

Having a size *attribute means that you'll also be able to find an element by rank quickly, as in BB[α] trees.*

Adding a Key to a Randomized Binary Search Tree

In a standard binary search tree, if you start adding keys to an initially empty tree, the first key you add becomes its root, and it will stay there unless you remove it. This next algorithm acts differently. Each time a key is added, it randomly decides whether it should go at the root or be added as a leaf, wherever that may be, similar to the sampling algorithms described previously. This method ensures that *any* key can be the root, so the order in which you do additions won't matter.

If the algorithm chooses to place the new key at the root, it splits the tree into two subtrees: one with all keys smaller than the future new root and the other with keys greater than it. Otherwise, if the algorithm didn't opt for placing the new key as root, a common insertion logic is applied. Take a look at the basic algorithm first; the details will be filled in later:

```
const add = (tree, keyToAdd) => {
❶ if (isEmpty(tree)) {
    tree = newNode(keyToAdd);
❷ } else if (tree.size * Math.random() < 1) {
  ❸ const newTrees = _split(tree, keyToAdd);
  ❹ tree = newNode(keyToAdd, newTrees.right, newTrees.left);
❺ } else {
    const side = keyToAdd <= tree.key ? "left" : "right";
    tree[side] = add(tree[side], keyToAdd);
  }
❻ tree.size = _calcSize(tree);
  return tree;
};
```

If the tree is empty ❶, set the key at the root with empty subtrees and calculate its size before returning. Since you are moving nodes around, you'll have to recalculate sizes. Just like the sampling algorithm from Chapter 10, you may decide that the new value has to go at the root ❷. In that case, use an auxiliary algorithm to split the tree in two parts ❸: the key that's being added becomes the root, and the two split trees become its subtrees ❹.

As an alternative, if the random test fails, apply the algorithm you know well from binary search trees ❺. (Remember that to implement a set instead of a bag, you check if keyToAdd equals tree.key and reject the new key if so.) Note, however, that in each recursive step, you're also using random numbers to decide whether to split the current tree, so randomness applies not only to the tree's root but also throughout the structure. The last step of add() is to calculate the size of the root ❻, which is done no matter what happens in the earlier steps.

Consider a sample case of this algorithm before dealing with the missing splitting code. Suppose you want to add a key of 20 to the tree shown in Figure 12-26.

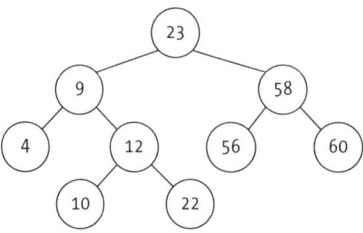

Figure 12-26: A binary search tree before adding a key of 20

Before comparing 20 to 23, generate a random number. Because the tree's size is 9, it's a probability of one in nine that the algorithm will split the tree and set 20 at its root, and in eight out of nine cases, the root will still be 23. Otherwise, you keep working in the usual fashion to add a key in a binary search tree.

Suppose the test passes. Split the tree in two parts and set them as subtrees for 20, which becomes the new root, as shown in Figure 12-27.

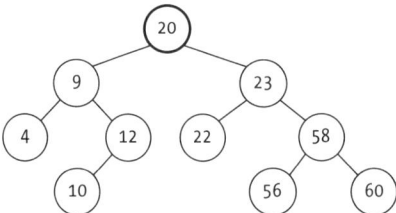

Figure 12-27: If a random test succeeds, the new key becomes the root of the tree.

Now, assume the test fails. Leave the original root in place and move to its left to compare 20 to 9. This time, since the size of the current tree is 5, the random test has a one in five probability of success. If the test succeeds this time, 20 goes in place of 9, splitting the tree rooted at 9 and doing the same kind of job as before.

A third possibility is if the random test fails the first two times. In that case, compare 20 with 12 and do another random test, now with one in three odds, because the original tree rooted at 12 has three nodes. And if that test fails, you still try again, with one in two odds, before comparing 20 to 22. If and *only if* every random test fails, you end by placing 20 exactly where you would place it in a normal binary search tree: in this case, to the left of the 22 key.

Splitting the Randomized Binary Search Tree

The splitting algorithm is reminiscent of the pivoting part of quicksort from Chapter 6. You have a "pivot" key and want to split the structure into two trees, so all keys in the first tree are smaller than the pivot and all keys in the second are greater than it.

Start with the same tree we used before and see how splitting would work with regard to a 20 key, as shown in Figure 12-28.

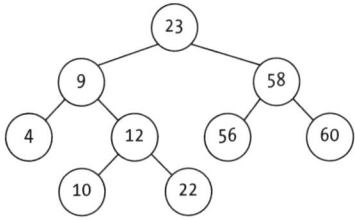

Figure 12-28: The same tree as shown in Figure 12-26 before splitting it in two with regard to the 20 value

First set up two empty trees: one has values less than 20, and the other has values greater than 20. Both start empty. The first step compares 20 with 23. Since 23 is greater, that root and its right subtree go into the tree with greater values. Also, you need to "remember" the left subtree of 23 (now empty), because future values greater than 20 but less than 23 will go there. The two split trees (the one with lesser values, currently empty) would look like the ones shown in Figure 12-29, and you'd go on to process the subtree rooted at 9.

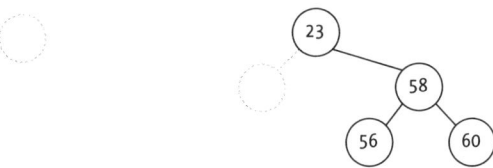

Figure 12-29: The first step: 23 is greater than 20, so part of the tree goes to the right. The dotted circles show where new subtrees will be added.

Now you have 20, which is greater than 9, so 9 and its left subtree go into the "smaller" tree, and you remember the right subtree of 9, which is where any future values greater than 9 but less than 20 will go. The split trees now look like the ones in Figure 12-30, and you can move on to the 12 key.

Figure 12-30: The second step: 9 is less than 20, so part of the tree goes to the left.

This is the same situation: 20 is greater than 12, so connect 12 and its left subtree to the remembered right subtree of the smaller tree, getting the scenario shown in Figure 12-31. Now remember the right subtree of 12 as the possible place to add more values.

Figure 12-31: The third step: 12 is less than 20, so add to the left tree.

You're almost finished: 20 is less than 22, so 22 (and its right subtree, if it has one) goes to the remembered place in the "greater" tree, as shown in Figure 12-32.

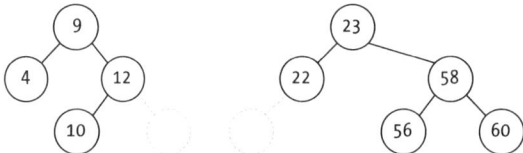

Figure 12-32: The fourth step: 22 is greater than 20, so add to the right tree.

Since there are no more nodes to process, finish by setting the final tree's root to 20, with the "smaller" and "greater" trees as subtrees. The result, shown in Figure 12-33 is what you saw earlier.

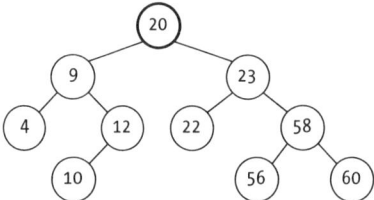

Figure 12-33: The fifth step: the tree was split, and now you set 20 at its root.

Now examine the code, which has the tricky issue of how to remember places in split trees:

```
const _split = (
  tree,
  keyForSplit,
  newTrees = { left: null, right: null },
  lastNodes = { left: newTrees, right: newTrees }
```

```
) => {
❶ if (isEmpty(tree)) {
    return newTrees;
❷ } else {
    const [side, other] =
      keyForSplit <= tree.key ? ["left", "right"] : ["right", "left"];
  ❸ const nextTree = tree[side];
    tree[side] = null;
  ❹ lastNodes[other][side] = tree;
    lastNodes[other] = tree;
  ❺ const newSplit = _split(nextTree, keyForSplit, newTrees, lastNodes);
  ❻ tree.size = _calcSize(tree);
    return newSplit;
  }
};
```

First create two trees, newTrees, as you split the original tree. When you are done with the tree ❶, just return that pair. Otherwise ❷, decide which side to split ❸ and join the split part to the correct new tree; you also have to remember where the next joining will be done ❹ before proceeding recursively down the tree ❺. Finish by calculating the tree size ❻, because you need it for your random tests.

Removing a Key from a Randomized Tree

The algorithm for removing a key is almost the same as before, but with one main difference: what to do if the key to be removed has two children. Here's the basic code:

```
const remove = (tree, keyToRemove) => {
  if (isEmpty(tree)) {
    // nothing to do
  } else if (keyToRemove < tree.key) {
    tree.left = remove(tree.left, keyToRemove);
  ❶ tree.size = _calcSize(tree);
  } else if (keyToRemove > tree.key) {
    tree.right = remove(tree.right, keyToRemove);
  ❷ tree.size = _calcSize(tree);
  } else if (isEmpty(tree.left) && isEmpty(tree.right)) {
    tree = null;
  } else if (isEmpty(tree.left)) {
    tree = tree.right;
  } else if (isEmpty(tree.right)) {
    tree = tree.left;
  } else {
  ❸ tree = _join(tree.left, tree.right);
  }
  return tree;
};
```

The algorithm is pretty standard, and you have seen this code several times now with some small exceptions. When you remove a key from a subtree, you need to update the size attribute ❶❷, but the interesting difference is when you want to remove a key that has two subtrees, you use a joining procedure ❸ to merge the left and right subtrees into a new tree, which then replaced the removed key. (See question 12.22 for more on the deletion algorithm.)

Joining Two Randomized Binary Search Trees

You can build a new tree out of two separate ones by picking one of the subtrees, using its root as the root for the new subtree, and recursively processing the rest of the trees. Consider the sample case shown in Figure 12-34 and try to delete the 20 key added earlier.

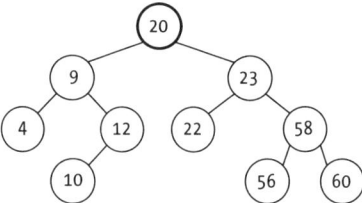

Figure 12-34: Deleting a root (here, 20)
requires joining its subtrees into a single tree.

You need to make a single tree out of both of the root's left and right subtrees and decide how to do it via random selection, so either 9 or 23 will become the new root. Suppose the random choice picks 9. Set 9 as the root of the new tree, along with its left subtree, and at its right subtree, set the result of joining its right subtree with the other subtree, the one rooted at 23.

Now, you have to choose among 12 and 23; suppose you select the latter. You can add 23 and its right subtree to the tree you are building, and then you still have to finish joining the subtrees rooted at 12 and 22. If you randomly pick 12 to be the next root, you'll get the situation shown in Figure 12-35.

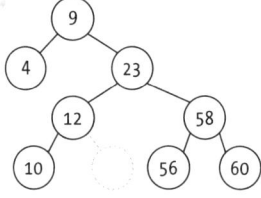

Figure 12-35: The new tree after randomly selecting
9 for its root and 23 for its right subtree

As the last step, you need to join an empty subtree (12's right subtree) and 22's, so the final tree becomes the one shown in Figure 12-36 where you've removed the 20 key using the new style of algorithm.

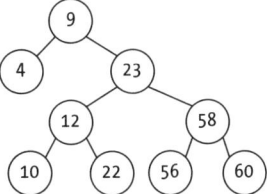

Figure 12-36: The last step, after joining 12's right empty subtree and 22's subtree

Consider the code. To decide from which tree to pick the root, use the same rule when you considered considered sampling: if the subtrees were of sizes 6 and 4 you'd pick the first tree's root with a 6/10 probability and the second tree's root with a 4/10 probability. Here's the algorithm:

```
const _join = (leftTree, rightTree) => {
❶ const leftSize = _getSize(leftTree);
  const rightSize = _getSize(rightTree);
  const totalSize = leftSize + rightSize;

❷ if (totalSize === 0) {
    return null;
❸ } else if (totalSize * Math.random() < leftSize) {
    leftTree.right = _join(leftTree.right, rightTree);
    leftTree.size = _calcSize(leftTree);
    return leftTree;
❹ } else {
    rightTree.left = _join(leftTree, rightTree.left);
    rightTree.size = _calcSize(rightTree);
    return rightTree;
  }
};
```

First, get the sizes ❶ of the trees to join to make the random choice later. If both trees are empty ❷, you're done. If not, randomly decide (based on the trees' own sizes) which one will provide the root ❸. If it's the left one, take its root and its left subtree with no changes and replace its right subtree, with the result of joining it with the other subtree you were working with. Of course, if you picked the right subtree ❹, the logic would be the same, but mirrored.

Considering Performance for Randomized Binary Search Trees

The effects of the randomized addition procedure make the average performance logarithmic. Even if the structure can become out of shape at times, continued operations bring it back to a good shape. The worst cases still are linear in time. There is, after all, a possibility that all random numbers may "work against you" to produce a badly shaped tree, but on average, that doesn't happen; check Table 12-5.

Table 12-5: Performance of Operations for Randomized Binary Search Trees

Operation	Average performance	Worst case
Create	$O(1)$	$O(1)$
Add	$O(\log n)$	$O(n)$
Remove	$O(\log n)$	$O(n)$
Find	$O(\log n)$	$O(n)$
Traverse	$O(n)$	$O(n)$

This structure provides logarithmic performance with high probability: the shape of the search tree will be that of a tree created with a random sequence of keys. Now consider a different structure that will provide amortized logarithmic performance, so a series of operations will have a total time that is logarithmic on average.

Splay Trees

As mentioned previously, binary search trees can have $O(n)$ performance, and although this happens only occasionally, it can be a problem. The balanced trees from the previous sections take preventive actions to avoid that issue, but *splay trees* provide another solution. This version of binary search trees guarantees amortized $O(\log n)$ performance, meaning that a sequence of k successive operations will have $O(k \log n)$ performance, which isn't as good as guaranteed $O(\log n)$, but it's almost as good.

With splay trees, whenever a node is accessed, it's moved to the root by a process called *splaying*, which is a sequence of rotations that brings up the desired node. This doesn't guarantee a well-balanced tree by any means, but over time, splay trees tend to become reasonably well shaped and provide a good alternative to other binary search tree implementations.

Consider the situation in Figure 12-37 where the 12 key is sought.

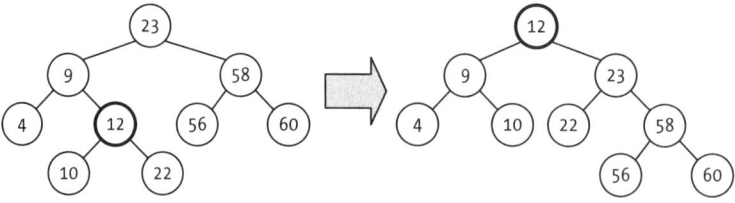

Figure 12-37: In a splay tree, after a search (here, for key 12), the found node becomes the tree's new root.

After finding 12, that key is brought up to the root (you'll see how later), which also causes other paths to change: 23 is pushed down the root's right subtree, 10 moves closer to the root, and 22 moves from left to right. Even if it's possible for the tree to become badly shaped, the sequence of operations usually restructures it for better performance over time. If you frequently require access to a few sets of keys, searches will be quite fast, because those keys will be nearer to the root, which is an advantage for many use cases. An example of this is provided by compilers and their symbol tables: usually when a symbol is defined (say, in a function), there's a good probability you'll be accessing it several times in a short period.

Splaying a Tree

Splay trees have specific rules, with quaint names like *zig-zig* or *zag*, and they are based on simple rotations. Consider the different cases. In Figures 12-38 through 12-40, the key to be moved up is always 1 (highlighted).

Case 1: Left Child of the Root

If the key is the left child of the root, apply a single right rotation to bring the key to the root. This is called a *zig* case, as shown in Figure 12-38.

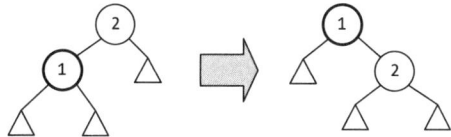

Figure 12-38: A single right rotation moves a left subtree up to the root.

The opposite case is if the key were the right child of the root; then you'd do a rotation to the left, which is called a *zag*.

Case 2: Right Child of the Left Child of the Root

In this *zag-zig* case, first rotate the key to the left and then rotate it to the right to bring it up to the root, as shown in Figure 12-39.

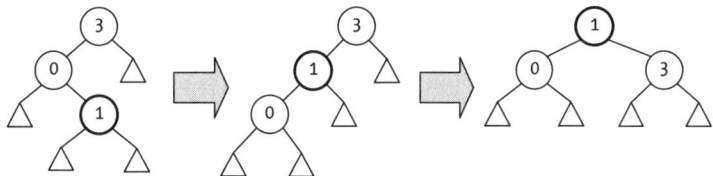

Figure 12-39: Two rotations are needed for the right child of a left child.

In the opposite case (*zig-zag*), first apply a rotation to the right and then to the left.

Case 3: Left Child of the Left Child of the Root

This *zig-zig* case might trip you up, because the order of rotations is altered: first you rotate the *parent* of the bottom key to the right and *then* you rotate the key itself, as shown in Figure 12-40.

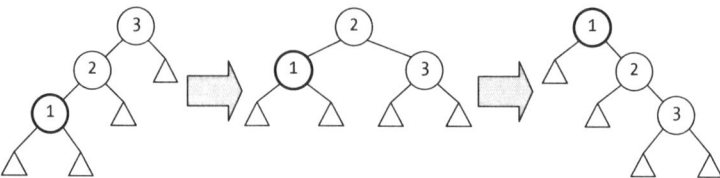

Figure 12-40: Two right rotations are needed for the left child of a left child.

You might think an easier algorithm in this case could rotate the key twice, but the result isn't optimal, and a simple example may convince you of this. Assume you started with a (not very good) tree, as shown in Figure 12-41, and splayed up the 1 key.

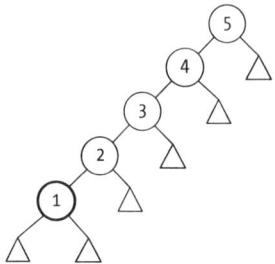

Figure 12-41: Why always rotating the found key isn't very good

You could attempt to use rotations to the right to move the 1 up. At each step, the 1 moves up one place, relocating the original root to its right (first 2; then 3 and 2; then 4, 3, and 2; and so on), and by the time 1 gets up to the root, all the other keys (2–5) are still in the same structure they were before, as shown in Figure 12-42.

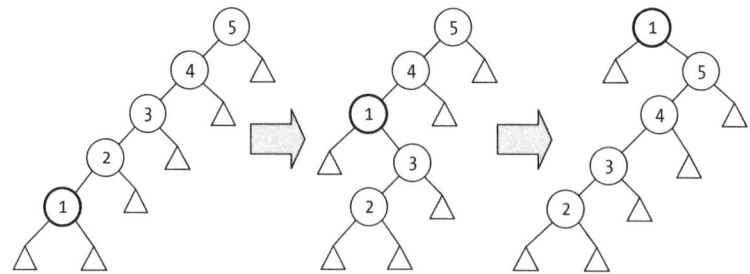

Figure 12-42: After all rotations, the tree's structure becomes worse.

In this splaying algorithm, two zig-zig rotations could handle this case. First, 1 becomes a root with 2 and 3 to its right, and then 1 moves to the tree's top, with 4 and 5 at its right; 2 and 3 are relocated to the left of 4, as shown in Figure 12-43.

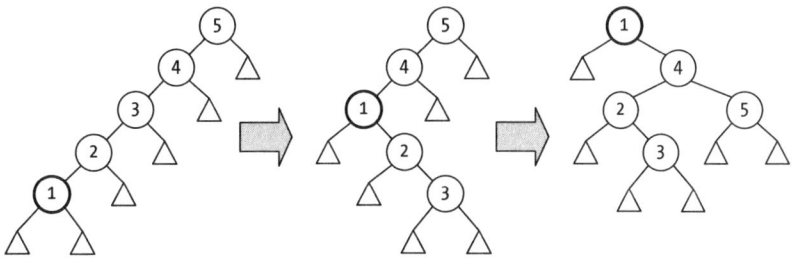

Figure 12-43: The rotations suggested in the text produce a better structure.

The zig-zig logic produces a better balanced tree, with several shorter paths from the root to nodes, and that serves as a justification for using more complex logic.

Now, consider splaying in terms of actual code:

```
const _splay = (tree, keyToUp) => {
❶ if (isEmpty(tree) || keyToUp === tree.key) {
    return tree;
  } else {
  ❷ const side = keyToUp < tree.key ? "left" : "right";
    if (isEmpty(tree[side])) {
      return tree;
  ❸ } else if (keyToUp === tree[side].key) {
      return _rotate(tree, side);
  ❹ } else {
      if (keyToUp <= tree[side].key === keyToUp <= tree.key) {
      ❺ tree[side][side] = _splay(tree[side][side], keyToUp);
      ❻ tree = _rotate(tree, side);
      } else {
      ❼ const other = side === "left" ? "right" : "left";
      ❽ tree[side][other] = _splay(tree[side][other], keyToUp);
        if (!isEmpty(tree[side][other])) {
          tree[side] = _rotate(tree[side], other);
        }
      }
    ❾ return isEmpty(tree[side]) ? tree : _rotate(tree, side);
    }
  }
};
```

Splaying continues until an empty tree is reached or the key you're looking for is found ❶. As long as those conditions are not met, continue. Decide on which subtree you should find the key ❷, and if that subtree is empty, you're also done. (If the tree doesn't contain the key you're looking for, the last key you found is the one that moves up, so *some* restructuring is always done.) If you find the key at the root of the subtree ❸, you have a zig or a zag, and a single rotation suffices. If not, if the key you are searching for is at the same side of the subtree ❹, you have a zig-zig or zag-zag. First recursively splay the lowest subtree ❺, then rotate the root ❻, and finish the last rotation later ❾. The other possibility is either a zig-zag or a zag-zig: splay the lowest subtree ❼ and finish with the two other rotations ❽❾ described earlier.

Finding a Key in a Splayed Tree

This algorithm is simple. Apply the splaying algorithm first, then check whether the value that got to the root is what you were looking for:

```
const find = (tree, keyToFind) => {
❶ if (!isEmpty(tree)) {
    tree = _splay(tree, keyToFind);
  }
❷ return [tree, !isEmpty(tree) && tree.key === keyToFind];
};
```

Unless the tree is empty, splay it. Splaying ❶ is *always* done, whether or not the key exists, so the new key at the root may or may not be what you were looking for, which explains the final test ❷.

Adding a Key to a Splayed Tree

To add a key, first apply splaying to restructure the tree and then add a new root at the top. The tree in Figure 12-44, which is the same tree used earlier when showing how splaying worked, shows how to add an 11 key.

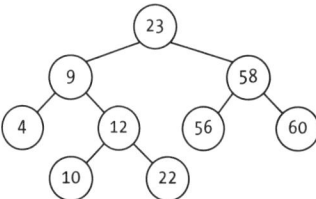

Figure 12-44: A splay tree into which you'll insert an 11 key

The first step is to splay using 11 as the splaying value. This key isn't present in the tree, so the algorithm ends with 10 at the root, as shown in Figure 12-45.

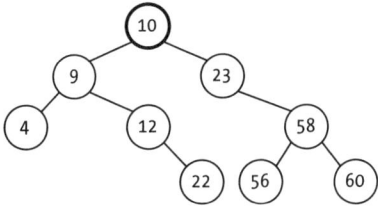

Figure 12-45: The first step: the tree is splayed with regard to the 11 value.

Now it's easy to finish: 11 should become the root, with 10 (the current root) at its left and 23 at the right, as shown in Figure 12-46.

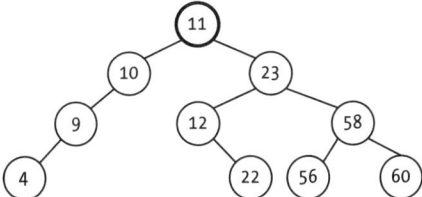

Figure 12-46: The final step: the 11 becomes the new root, with the splayed parts as subtrees.

The code is as follows:

```
const add = (tree, keyToAdd) => {
❶ const newTree = newNode(keyToAdd);
  if (!isEmpty(tree)) {
  ❷ tree = _splay(tree, keyToAdd);
  ❸ const [side, other] =
      keyToAdd <= tree.key ? ["left", "right"] : ["right", "left"];
    newTree[side] = tree[side];
    newTree[other] = tree;
    tree[side] = null;
  }
  return newTree;
};
```

First, create the node that will become the new root ❶. Then, splay the tree ❷, so the root becomes the nearest key to the one added. Then link the new root properly ❸, and the new node will be the tree's root.

Removing a Key from a Splayed Tree

Removing a key starts by splaying the tree, so the root becomes either the key you wanted to remove or a different one, if the key you wanted to remove wasn't present in the tree. If the key was found, do the usual steps. If it has zero children or just one child, removal is simple; if it has two children, find the next key in its right subtree and set it at the root, but splay it as well.

You can see how this works by attempting to remove 12 from the tree shown in Figure 12-47.

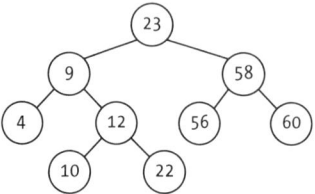

Figure 12-47: A splay tree from which you want to delete the 12 key

The first step, as with adding and searching, is to splay the tree using 12 as the key; you already saw this example, and the result was the updated tree shown in Figure 12-48.

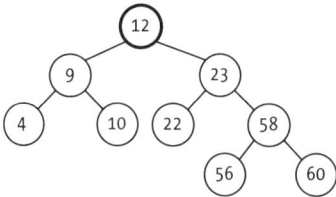

Figure 12-48: The splay tree after splaying, so the 12 becomes the root

Since the 12 was found, you can proceed. In this case, you have two subtrees, so you have to find the key that follows 12 (22) and use that value to splay the root's subtree, getting the new tree shown in Figure 12-49.

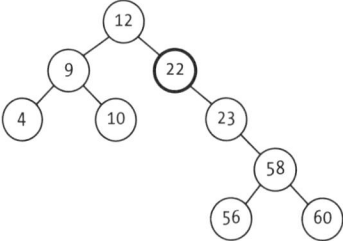

Figure 12-49: The tree after splaying the subtree with 22 at its root

Now you can easily remove 12 by placing 22 in its place, and you have finished the algorithm, as shown in Figure 12-50. Note that the 22 key cannot have a left subtree, because there's no value between 12 and 22.

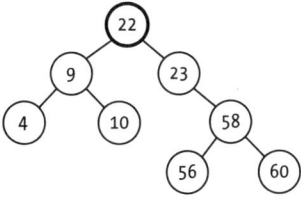

Figure 12-50: The final tree after 22
becomes the root

First look at the code to splay a tree, bringing its minimum value to the top. Remember from earlier algorithms that we find the minimum key going left until we can't go any further. The idea here is to apply rotations so the minimum key ends up at the top:

```
const _splayMinimum = (tree) => {
  if (isEmpty(tree) || isEmpty(tree.left)) {
    return tree;
  } else {
❶ tree.left.left = _splayMinimum(tree.left.left);
❷ tree = _rotate(tree, "left");
❸ return isEmpty(tree.left) ? tree : _rotate(tree, "left");
  }
};
```

The algorithm is basically the same as _splay(), except you always assume you're going left ❶❷❸. Compare the code; it's the same as earlier, except side is replaced with left. (There's another way of deriving the _splayMinimum() code; see question 12.25.) With that out of the way, the code for removal is as follows:

```
const remove = (tree, keyToRemove) => {
  if (!isEmpty(tree)) {
❶ tree = _splay(tree, keyToRemove);
    if (keyToRemove === tree.key) {
❷    if (isEmpty(tree.left) && isEmpty(tree.right)) {
        tree = null;
❸    } else if (isEmpty(tree.left)) {
        tree = tree.right;
❹    } else if (isEmpty(tree.right)) {
        tree = tree.left;
      } else {
❺      const oldLeft = tree.left;
❻      tree = _splayMinimum(tree.right);
❼      tree.left = oldLeft;
      }
    }
  }
  return tree;
};
```

If the tree isn't empty, start by splaying it ❶ and then check whether the key you want to remove is now at the root. If so, you can easily deal with cases where the new root has fewer than two children ❷❸❹. Otherwise, save the left subtree ❺ and then splay the right subtree, bringing its minimum to the top ❻, and the minimum takes the place of the key you are deleting. You just have to fix its left subtree ❼ and you're done.

NOTE *See question 12.24 to verify that you understand an important detail of this algorithm: Why are you only overriding the splayed subtree's left tree in the last steps of the removal process?*

Considering Performance for Splay Trees

Splay trees can, in the worst case, produce a linear tree, so performance would be linear in that case and probably would rule them out in a real-time context when you need absolute guarantees as to processing time. However, the amortized cost of a series of operations is logarithmic, meaning that, on average, a sequence of k operations (additions and removals) would have a total cost $O(k \log n)$, which works to a logarithmic amortized performance; Table 12-6 sums up the results.

Table 12-6: Performance of Operations for Splay Trees

Operation	Amortized performance
Create	$O(1)$
Add	$O(\log n)$
Remove	$O(\log n)$
Find	$O(\log n)$
Traverse	$O(n)$

One interesting feature is that the structure not only self-reorganizes, but it also provides better performance, because frequently accessed keys end up close to the root. This makes splay trees appropriate for implementing caches, for example. The fact that nodes need no extra bookkeeping data (such as the tree's height or size) makes it interesting if lack of memory is a problem, and yet another benefit is that performance is, on average, as efficient as other trees.

Summary

This chapter introduced trees, in particular, binary search trees, which provide a good implementation for the bag and set ADTs, with high-performing *add*, *remove*, and *find* methods. You explored the performance of these trees and saw several variants aimed to ensure good, fast algorithms.

In the following chapter we'll explore more general tree-based structures, and we'll also consider special search-oriented structures that provide quite efficient searches and updates.

Questions

12.1 A Matter of Levels

Can you define the height of a tree in terms of levels?

12.2 Breaking the Rules

Filesystem directories are often said to have a treelike structure, but that's not always true. Can you think of a feature that allows directories to be, say, like circular lists (as seen in Chapter 10) or even possibly like graphs (as you'll see in Chapter 17)? A hint: directory entries can be of many types.

12.3 What's in a Name?

Here are some questions regarding full, perfect, and complete trees: Which term implies another? For example, are full trees also complete? And are complete trees full? What happens with full and perfect trees? What about perfect and complete?

12.4 A find() One-Liner

It's certainly less clear, but can you write the find() method with a single line of code?

12.5 Sizing a Tree

Write a calcSize() function that will find the size of a binary tree.

12.6 Tall as a Tree

Write a calcHeight() function to find the height of a binary tree.

12.7 Copy a Tree

Given a binary tree, write an algorithm that will produce a copy of it. (Hint: you may want to consider using a preorder traversal for this.)

12.8 Do the Math

This problem can pop up if you are writing a compiler or an interpreter. Suppose you have a binary tree whose nodes can have either a number or a mathematical operator (addition, subtraction, multiplication, and division). Such a tree can be used to represent any mathematical expression; for instance, the tree in Figure 12-51 stands for (2+3) * 6.

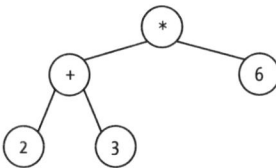

Figure 12-51: Do the math.

Show that you can evaluate such an expression by properly traversing the tree.

12.9 Making It Bad

In what order should you insert keys in a common binary search tree to produce a linear list? And if you have n keys, how many ways can you produce a tree with not one single full node? (Hint: Which values could you pick first to add into the tree? Which values could come next?)

12.10 Rebuild the Tree

If you are given the preorder and inorder listings of the keys in a binary search tree with no duplicate keys, you can rebuild it; write an algorithm to do this. Your input will be two arrays of values: the first will have the tree's keys in preorder, and the second will have them in inorder.

12.11 More Rebuilding?

For the previous question, would you have been able to rebuild the tree out of the inorder and postorder listings? What about out of the preorder and postorder listings?

12.12 Equal Traversals

For which trees are keys visited in the same order with preorder and inorder traversals? What about for inorder and postorder? Or for preorder and postorder?

12.13 Sorting by Traversing

Use the inOrder() traversal to, given a binary search tree, produce an ordered array of its keys.

12.14 Generic Order

Write an anyOrder(tree,order,visit) function that will accept an order parameter that can be "PRE", "IN", or "POST" and will do the corresponding traversal of tree, with the given visit() function.

12.15 No Recursion Traversal

Implement all traversals without using recursion; use a stack instead.

12.16 No Duplicates Allowed

Modify the addition logic for binary search trees to reject attempts to add duplicate keys. After such an attempt, the tree should remain unchanged, and an error should be thrown.

12.17 Get and Delete

Write a _removeMinFromTree(tree) method that will find the least key in a binary search tree, remove it, and return both the key and the updated tree at the same time. Use this new method to optimize _remove() by dropping the need for _findNext().

12.18 AVL Worst

What's the smallest number of nodes that an AVL tree can have related to its height? In other words, if an AVL tree has height 1, 2, 3 . . . , what's the smallest number of nodes it may have?

12.19 Singles Only

Consider a child node with no siblings called *single child*. Can you have more than 50 percent single children in an AVL tree?

12.20 Why One?

In weight-balanced trees, why is it that the balances of the left subtree and the right subtree (that is, the fractions *weight(left subtree) / weight(tree)* and *weight(right subtree) / weight(tree)*) add up to 1?

12.21 Easier Randomizing?

A developer had the following thought:

> Binary search trees behave badly if keys are added in order, but behave well with a random order. What would happen if instead of adding keys to a tree, I hashed them first? The hashed keys are, to all effects, random, and so an ordered sequence of keys would become a totally disordered one, ensuring good performance. Of course, when looking for a key, I'd really need to look for the hashed key, but that's no big deal. Problem solved; binary trees with hashed keys will *always* behave well!

Is this reasoning correct?

12.22 Why Not Decrement?

In the remove() logic for randomized binary trees, why did you use _calcSize() instead of decrementing as in the following?

```
const remove = (tree, keyToRemove) => {
  if (isEmpty(tree)) {
    // nothing to do
  } else if (keyToRemove < tree.key) {
    tree.left = remove(tree.left, keyToRemove);
    tree.size--;
  } else if (keyToRemove > tree.key) {
    tree.right = remove(tree.right, keyToRemove);
    tree.size--;
    return tree;
    ... etc. ...
};
```

12.23 Bad Splay?

You saw earlier that adding keys in ascending or descending order was a bad case for common binary search trees. What happens with splay trees in those cases? And if after those additions you remove a few keys, what tree shape do you get? Is it any better?

12.24 What Left Subtree?

At the end of the remove() method for splay trees, after splaying the root's right subtree, what's the value of this.right.left and why?

12.25 Code Transformation

Show that you can transform _splay() to _splayMinimum() by assuming that keyToUp equals -Infinity. Why should this work?

12.26 Full Rebalance

You've seen several strategies for rebalancing trees, but you may also want to rebalance a common binary search tree. Can you come up with a restructure(tree) function that will balance a binary search tree into as perfect a shape as possible? You should attempt to split nodes as evenly as possible between left and right subtrees, everywhere in the tree.

13

TREES AND FORESTS

In previous chapters we explored binary trees that have the restriction of only two children per node. In this chapter, we'll consider some new structures that go beyond that restriction, such as forests and orchards (when working with a single tree isn't enough). After that, we'll move on to study B-trees and red-black trees for faster searching.

Defining Trees and Forests

Binary trees can be empty or consist of a node (the root) and two children, which are both binary trees themselves. In particular, *binary search trees* are also *ordered* trees, because we define a certain order between children and distinguish the left child from the right child.

Let's expand on those concepts. First, you'll allow a node to have many children, not just two—in other words, nodes can have degrees greater than 2. You may have trees with a specific number of (possibly empty) children per node, as binary or ternary trees, but in general, no restrictions are placed on the degree of a node. Sometimes trees with an unspecified maximum degree are called *multiary* or *multiway*.

Moving beyond single trees, a *forest* is defined as a set of disjointed trees. For instance, you could consider the directory for a given hard drive in your computer to be a tree, but all the different pieces of storage (like hard drives or USB sticks) in your computer would make up a forest.

We can even go further and define an *orchard* as a forest with an ordered relationship between its trees. In a forest, trees are strewn around in a disordered manner, but an orchard has a well-defined layout. Continuing with the computer example, if you assign letters to your drives (C:, D:, and so on, Microsoft Windows style), your storage actually is an orchard. The forestry-related terms do not end here: you can also have *groves*, which are like trees, with the exception that their nodes can have links to other nodes, which would turn the data structure into a directed (and possibly cyclic) graph.

NOTE *To simplify the terminology in this chapter, we'll use the terms* forests *(when no particular order is implied among the trees) and* ordered forests *(rather than orchards), which is the terminology most textbooks use. Also, in this chapter we won't work with groves at all.*

Representing Trees with Arrays

Do you know how to represent a general tree? How about a forest? Let's start with trees, because by doing so we'll discover a tip for dealing with forests. The first solution you'll probably think of is having an array to point at each child, and in JavaScript with varying-length dynamic arrays, that's the simplest solution. Figure 13-1 shows a generic tree where the nodes have different degrees.

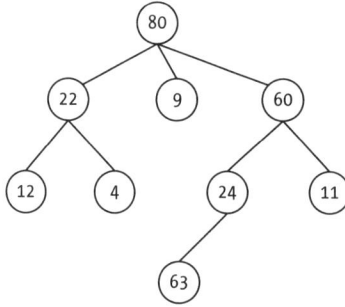

Figure 13-1: A generic tree

To implement such a tree, add an array of subtrees to each node, resulting in something like the following (in terms of JavaScript, this code varies a bit from the style of the book and represents trees with a class, which lets you use a standard interface like the Document Object Model [DOM] node interface):

```
class Tree {
  constructor(rootKey) {
❶ this._key = rootKey;
❷ this._children = [];
  }

  isEmpty() {
❸ return this._key === undefined;
  }

❹ _throwIfEmpty() {
    if (this.isEmpty()) {
      throw new Error("Empty tree");
    }
  }

❺ get key() {
    this._throwIfEmpty();
    return this._key;
  }

❻ set key(v) {
    this._key = v;
  }

❼ get isLeaf() {
    this._throwIfEmpty();
    return this.childNodes.length === 0;
  }

❽ get childNodes() {
    this._throwIfEmpty();
    return this._children;
  }

❾ get firstChild() {
    return this.isLeaf ? null : this.childNodes[0];
  }

❿ get lastChild() {
    return this.isLeaf ? null : this.childNodes[this.childNodes.length - 1];
  }

  // ...more methods...
}
```

A key field ❶, which also doubles as a flag, decides whether a tree is empty ❸, and an array of children ❷ is empty by default. The _throwIfEmpty() method ❹ detects incorrect accesses to empty trees (this is used in several methods). You also add a getter ❺ and a setter ❻ for the tree's key. Then add some getters to check whether a node is a leaf with no children ❼ and, if not, to get its children ❽, in particular, accessing its first ❾ and last ❿ child, mimicking well-known DOM node-related methods.

NOTE *For more details on the DOM node interface, visit* https://developer.cdn.mozilla .net/en-US/docs/Web/API/Node.

You could consider adding more methods to the tree, but you would need some extra fields to reproduce certain methods, such as parentNode or previousSibling. We'll see some ways to achieve those things later in the chapter. Now you can represent general trees and access their nodes, so next take a look at how to add or remove data to update trees.

Adding Nodes to a Tree

First, add a new child in a specific place among its siblings:

```
❶ addChild(keyToAdd, i = this.childNodes.length ) {
❷   this._throwIfEmpty();
❸   if (i < 0 || i > this.childNodes.length) {
      throw new Error("Wrong index at add");
    } else {
❹     const newTree = new this.constructor();
      newTree.key = keyToAdd;
❺     this._children.splice(i, 0, newTree);
❻     return this;
    }
  }
```

To add a new key, all you need to specify is its position among its siblings; by default, you'll add it at the end ❶. If the tree is empty (no root), you throw an error ❷, and you also do that if the index lies beyond the limits of the current array of children ❸. If everything's okay, just create a new tree ❹, place the new key as its root, and place the tree in the correct position among its siblings ❺, ending by allowing chaining, as in other previous cases ❻.

Appending a node is easy:

```
appendChild(keyToAppend) {
  return this.addChild(keyToAppend);
}
```

You just depend on the default parameters for addChild(), which also test whether the tree has a root. No special code needed here.

Removing Nodes from a Tree

To remove a given child, you need only a test and some array manipulation:

```
removeChild(i) {
❶ this._throwIfEmpty();
❷ if (i < 0 || i >= this.childNodes.length) {
    throw new Error("Wrong index at remove");
  } else {
❸ this._children.splice(i, 1);
❹ return this;
  }
}
```

After verifying that the tree has a root and isn't empty ❶, check whether the index of the child to remove is valid or not ❷. If it is valid, do some array manipulation to remove the child from among its siblings ❸. End by enabling chaining as when adding a node ❹.

Representing Trees with Binary Trees

Representing trees with arrays works well, but another way of dealing with trees uses a simpler kind of tree, a binary tree. The trick is to use the left and right pointers in a different way from before: the left one will point to the first child, and the right one will point to the next sibling.

NOTE   *If you are wondering whether this technique is purely academic or if you'll ever use it in reality, we'll be doing so in Chapter 15 when studying binomial heaps and their variants.*

Revisit the tree shown in Figure 13-1. An alternative representation would have each node's left link pointing to the node's first child, and the right links would create a list of the node's siblings. (As with all the other diagrams in this book, left and right null pointers are omitted for clarity.) If you rearrange and rotate the image 45 degrees, so the left pointers actually point down, the scheme becomes clearer, as shown in Figure 13-2.

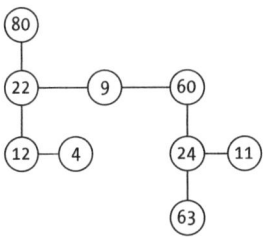

Figure 13-2: A tree represented with the "left-child, right-sibling" style

Many structures use this *left-child, right-sibling* convention, but it's better to rename the left pointer to *down* for clarity; *right* will still point to siblings. As for algorithms (adding or removing values and so on), you won't need to do anything different from what you learned about binary trees in Chapter 12.

Representing Forests

You can extend these methods for representing trees to represent forests. If you use arrays for pointers, you can simply have an array of roots, each pointing to a specific tree (we'll see this concept again in Chapters 14 and 15 when discussing binomial and Fibonacci heaps, so consider Figure 13-3 to be a minor spoiler).

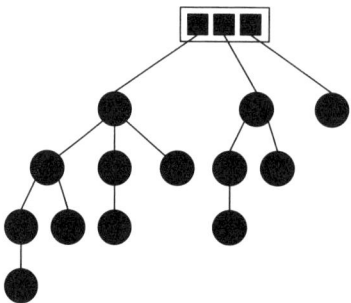

Figure 13-3: A forest, represented with an array of roots

At the top of the forest is an array with a pointer to the root of each individual tree. If you prefer the binary tree representation, you can do two different things: consider that all roots are siblings or add a fictitious "super-root" that has all the forest trees as subtrees. The first is the usual representation, which would give something like the forest in Figure 13-4 (this is the same forest as shown in Figure 13-3).

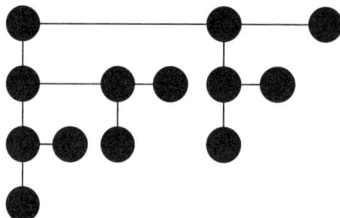

Figure 13-4: An alternative representation of the same forest; the roots are linked to the right.

To access this forest, you need a pointer to the leftmost root; from there, you can access all the trees. You can go even further by making the list of siblings circular and doubly linked; we'll explore this later and see why those enhancements (and complications) are actually needed.

Traversing Trees

When studying binary search trees in Chapter 12, we looked at three different ways of traversing trees by "visiting" all the nodes according to various schemes. With general trees, you don't have all those methods, but we'll add new ones. But first review the traversals we've done before and adapt two of them to the general tree:

Preorder   For binary trees, *preorder* means first visiting the root, then traversing its left subtree, and finally traversing its right subtree. You can adapt this for general trees by first visiting the root and then traversing each of its subtrees in order.

Postorder   The *postorder* method for binary trees is similar to preorder, but it first visits the root's left subtree, then the right, and finally the root itself. The adaptation requires first traversing all the root's children in order and finally visiting the root itself.

Inorder   The *inorder* method doesn't really have an equivalent. For binary trees, it means first traversing the left subtree, then visiting the root, and finally traversing the right subtree. However, for general trees, you don't have any reasonable alternative, so you can forgo this traversal (although for B-trees, discussed later in this chapter, you do have a possible inorder version).

Coding preorder and postorder traversals is fairly straightforward, and the versions used for binary trees require only minor changes (we'll consider their implementations in the questions at the end of this chapter). Two new methods, however, are worth studying and also appear in other types of algorithms, such as game playing or function optimization: *depth-first* and *breadth-first* traversals.

Depth-First Traversal

The possibly most logical traversal to implement requires visiting the root first and then traversing all its children using the same algorithm. In effect, it's equivalent to going as deeply as possible into a branch before moving on to another branch—thus, the name *depth first*. Figure 13-5 shows an example of such a traversal; the numbers in the nodes reflect the order in which the visits occur.

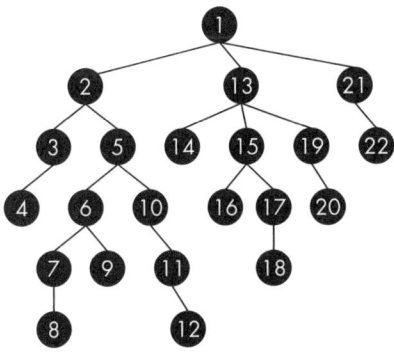

Figure 13-5: A depth-first traversal of a tree

This kind of algorithm is a classic one, generally used for searches or games. For instance, if you're trying to get out of a maze, you'd follow some path until you either exit (and finish) or become blocked, in which case you go back to try another option. (See "Finding a Path in a Maze" on page 69.) Similarly, in games, you consider some sequence of movements as far as you can (because of time limitations), and if you haven't found a winning line, you go back to try another one. The logic looks like this:

```
❶ const depthFirst = (tree, visit = (x) => console.log(x)) => {
❷   if (!isEmpty(tree)) {
❸     visit(tree.key);
❹     tree.children.forEach((v) => depthFirst(v));
    }
};
```

The algorithm is similar to some that you wrote for binary trees in Chapter 12. First, define a default visit() function ❶ that just lists the node's key, and if the tree isn't empty ❷, visit its root ❸. Then proceed to visit all of its children recursively, depth first ❹.

Breadth-First Traversal

The alternative way to traverse trees is breadth first, which is a traversal style that you haven't yet met. (This type of traversal is also called *level order* for reasons that will become apparent.) The idea is that you start at the root; then, you visit all of its children in the next level. Then (and only then), you visit the children's children at the second level of the tree, and so on. You never visit a node until you've already visited all the nodes closer to the root, going down level by level, one by one. Figure 13-6 shows such a traversal for a generic tree. Again, the numbers reflect the order in which the nodes are visited, and you can verify that each level is fully visited before starting the next level.

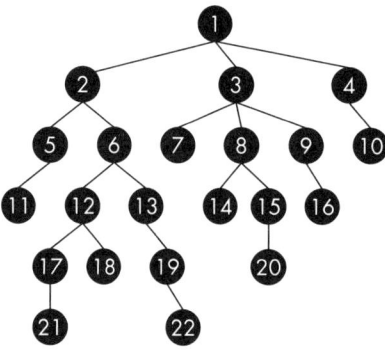

Figure 13-6: A breadth-first traversal of a tree

Implementing this strategy requires using a queue, as discussed in Chapter 10. As you start visiting nodes "horizontally," you need to remember to visit their children later, so a queue with a first in, first out (FIFO) strategy works. You can code this as an independent function just for variety:

```
❶ breadthFirst(visit = (x) => console.log(x)) {
  ❷ if (!this.isEmpty()) {
    ❸ const q = new Queue();
    ❹ q.push(this);
    ❺ while (!q.isEmpty()) {
      ❻ const t = q.pop();
      ❼ visit(t.key);
      ❽ t.childNodes.forEach((v) => q.push(v));
      }
    }
}
```

As with other traversals, a default visit() function ❶ logs the node's key. If the tree to be traversed isn't empty ❷ (in which case you wouldn't have to do anything), initialize a queue ❸ by pushing the tree's root ❹. The rest of the algorithm is straightforward: while the queue hasn't been emptied ❺, you pop its top ❻, visit that node ❼, and push all of its children into the queue for future visits ❽.

This algorithm isn't recursive at all, which isn't common for trees and other recursively defined structures. There's an interesting sort of symmetry here: visiting a tree breadth first without recursion requires using a queue, and visiting it depth first without recursion needs a stack; see question 13.3.

B-trees

B-trees have a self-adjusting tree structure with assured logarithmic performance for additions, removals, and searches, so in that sense, you could consider them an extension of height-balanced binary search trees—and a better-performing one at that. A key characteristic of these trees is that nodes can have more than two children, which allows for wider, shorter trees with faster searches.

NOTE *Nobody really knows what the B in B-tree stands for. This structure was defined in 1972 by Rudolf Bayer and Edward McCreight, but no explanation was given for the term, so you can choose your own interpretation: some proposals have been "balanced," "broad," "bushy," and, obviously, "Bayer."*

The definition (and implementation) for a B-tree varies among different sources and authors, so let's make clear what is used here. A B-tree of order m satisfies the following properties:

- Every node has $p < m$ keys, in ascending order, and $p + 1$ children.
- Every node but the root must have at least $m/2$ (rounded up) children, or in other words, all nodes (except the root) should be at least half full.
- The root should have at least one key.
- All leaves must be at the same level.

The B-tree is structured in a fashion similar to a binary search tree: given any key in a node, all the children to its left will be smaller than the key, and all the children to its right will be greater than it. For instance, you could have a node such as the one shown in Figure 13-7.

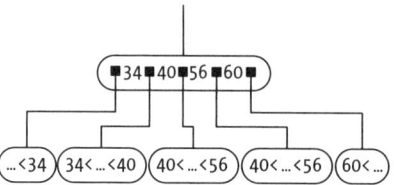

Figure 13-7: A B-tree node, showing where keys are to be found

In this case, the node has four keys, so five children. The first child, to the left of the 34 key, has keys less than 34; the child between 34 and 40 has keys between those two values, and so on, until the last child, the one to the right of the 60 key, has keys greater than that value. (These facts are what we'll use to search in a B-tree; you'll see the algorithm soon.) This structure is similar to binary search trees, except that now instead of a maximum of two children per node, we allow a greater number—and for actual implementations (such as an index for actual files in disk), much larger values are preferred in order to have a shorter height and thus faster access.

Defining B-trees

Let's start in similar fashion as with binary search trees by defining the basic functions we'll need:

```
❶ let ORDER = undefined;

  const newBTree = (order = 3) => {
❷ if (ORDER === undefined) {
     ORDER = order;
   }
   return null;
  };

❸ const newNode = (
    newKeys = [null],
❹ newPtrs = new Array(newKeys.length + 1).fill(null)
  ) => ({
    keys: newKeys,
    ptrs: newPtrs
  });

  const isEmpty = (tree) => tree === null;

❺ const _tooBig = (tree, d = 0) => tree.keys.length + d > ORDER - 1;

❻ const _tooSmall = (tree, d = 0) =>
    tree.keys.length - d < Math.ceil(ORDER / 2) - 1;
```

You can define B-trees of any order, and we'll use a variable ORDER ❶ to store the one you want. The first time you create a B-tree ❷, you'll store the desired order (or 3, by default), so all future B-trees will have that order. (This decision begs the question: What if you want to have B-trees of *different* orders? See question 13.9.) The newNode() function ❸ creates a new node with a single null key by default, flanked by two null pointers; of course, this node will be "too empty," unless ORDER is small. Note, however, that if you provide an array of keys ❹, some JavaScript trickery is used to generate (if needed) a corresponding array of null pointers, with one key more; can you see how this works?

Finally, a couple of auxiliary functions will come in handy. At times, you'll need to test whether a node is oversized (or would be, if d keys were added to it), with more keys than allowed ❺, and _tooBig() will check that. Similarly, _tooSmall() determines whether the node is undersized (or would be, if d keys were removed from it) and doesn't have enough keys ❻. (Be careful not to apply _tooSmall() to the root—the only node that is allowed to be smaller.) You'll use those two methods when adding or removing keys.

Finding a Key in a B-tree

Let's start with the most basic algorithm: searching for a key. In a sense, the algorithm is similar to searching a binary search tree; you look for the key, and if you don't find it in a node, you determine where to continue searching. Consider some examples. Assume you have the following B-tree of order 3; null links are represented with blank boxes, as shown in Figure 13-8.

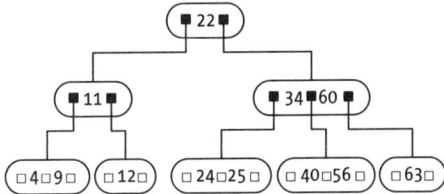

Figure 13-8: A B-tree of order 3

If you were looking for 22, that would be easy: it's in the root, so there's nothing to do. You can make it more complex and look for 56, as shown in Figure 13-9.

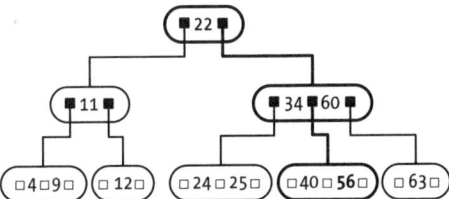

Figure 13-9: The search process for key 56

Start at the root, and since 56 > 22, follow the root's last pointer to a new node. There, you find that 56 should be between 34 and 60, so follow the middle pointer to yet another node. In that one, you finally find 56, so the search is successful.

What about searching for a key that isn't in the tree? If you had been looking for 38, everything would be the same as for 56, but upon not finding 38 in the node with 40 and 56, you would have continued down the first pointer of the node (since 38 < 40), but finding it null, the search would have been unsuccessful.

Now on to the algorithms. Every time you get to a node, you need to see whether the key you want is there; if not, the algorithm tells you what pointer to follow to the next level; there is an auxiliary function for this:

```
const _findIndex = (tree, key) => {
❶ const p = tree.keys.findIndex((k) => k >= key);
  return p === -1 ? tree.keys.length : p;
};
```

This looks for the first element in the keys array that is greater than or equal to the key you are searching ❶. If no key fits, findIndex() returns -1, so in that case, you return the index of the last element of ptrs (you'll see the reason for this tricky code soon enough):

```
const find = (tree, keyToFind) => {
❶ if (isEmpty(tree)) {
    return false;
  } else {
❷ const p = _findIndex(tree, keyToFind);
❸ return tree.keys[p] === keyToFind || find(tree.ptrs[p], keyToFind);
  }
};
```

If you are searching for a key and arrive at an empty node ❶, the key obviously isn't there. Otherwise, the _findIndex() method finds the first key that isn't less than the searched-for key ❷. If the key is actually equal to the value you want to find ❸, you are done; otherwise, continue the search at the corresponding pointer. This is why you had _findIndex() return the last position in the array—because that's the link you need to follow when the key you want to find is greater than all the keys in the node.

It may seem like a backward step to do a linear search when you have already seen better ways of searching; see question 13.7 for some ideas on better ways of searching.

Traversing a B-tree

We can define the equivalent to the inorder traversal for binary trees, which means visiting all keys in ascending order. Since each node has several keys, you need to work carefully. Figure 13-10 shows how to do this.

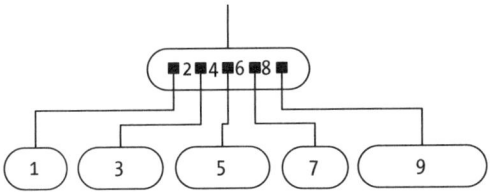

Figure 13-10: A traversal for B-trees, similar to inorder for binary trees

This version of inorder should visit keys in ascending order, so with the node in Figure 15-9, start with the first (leftmost) child, then visit the node's first key, followed by the second child, then the second key, then the third child, then the third key, and so on, ending after having traversed the rightmost child.

Here's a version of it at work:

```
const inOrder = (tree, visit = (x) => console.log(x)) => {
❶ if (!isEmpty(tree)) {
  ❷ tree.ptrs.forEach((p, i) => {
    ❸ inOrder(p, visit);
    ❹ i in tree.keys && visit(tree.keys[i]);
    });
  }
};
```

If the current node is empty ❶, do nothing; otherwise, do a loop ❷, alternating between traversing a child ❸ and visiting a key ❹. For the latter, remember that there's one fewer key than children. This code uses the condition && expression syntax as a shortcut to an if or ternary operator: expression is evaluated if and only if condition is true; in this case, visit a key if and only if the corresponding index is within the array of keys.

Adding a Key to a B-tree

Now consider how to add a key. If you want to add a key to a node that has enough empty space, it's straightforward. The problem is trying to add a key to a node that's already too full to allow another key. Consider both cases using the B-tree of order 3 from earlier as an example (see Figure 13-11).

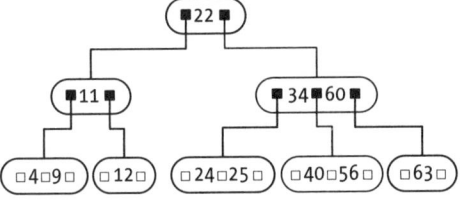

Figure 13-11: A B-tree into which you'll add some new keys

First, try to add a 66 key. After searching, you decide it should go with the 63 key, and since that node has enough space, there's no extra work (see Figure 13-12).

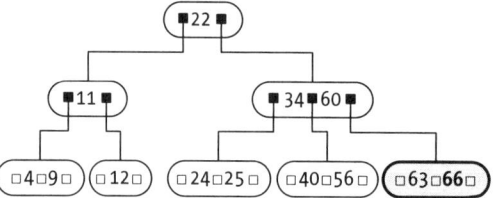

Figure 13-12: Adding a key at a node with space causes no problems.

Now make things more complicated, and add a 10 key. That's a problem, because the bottom-left node has no more space. First, you can let it grow beyond its maximum size (see Figure 13-13).

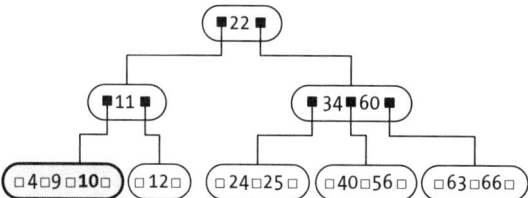

Figure 13-13: Adding a key at a node at limit requires splitting the node and rotating a key up.

Now you need to split the oversized node in two and make its middle key go up to the parent node. Fortunately, that node has space, so you are done (see Figure 13-14).

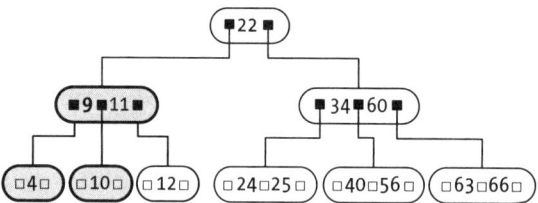

Figure 13-14: After the split, the B-tree structure is okay again.

If you try to add a 78 key, this node would split twice: first, the 66 key would go up to the 34-60 node, and then that node would split, sending its middle key (60) to its parent. This is the only way a B-tree can grow: if its root needs splitting, it adds a new level.

Take a look at the code and see what happens at the root level:

```
const add = (tree, keyToAdd) => {
  if (isEmpty(tree)) {
❶ return newNode([keyToAdd]);
  } else {
❷ _add(tree, keyToAdd);
❸ if (_tooBig(tree)) {
      const m = Math.ceil(ORDER / 2);
    ❹ const left = newNode(tree.keys.slice(0, m - 1), tree.ptrs.slice(0, m));
    ❺ const right = newNode(tree.keys.slice(m), tree.ptrs.slice(m));
    ❻ tree.keys = [tree.keys[m - 1]];
    ❼ tree.ptrs = [left, right];
    }
    return tree;
  }
};
```

The easiest case is if the tree is empty ❶, because then all you need to do is create a new node with the new key and a couple of null pointers. Otherwise, add the key somewhere in the tree ❷ (using an auxiliary recursive _add() method, which will see it immediately), and check whether the root became too big ❸. If so, create two new nodes ❹❺, each with half the keys and pointers, leaving the middle key at the root ❻, which will have the two new nodes as children ❼.

But how is the new key actually added, and what's the _add() method missing earlier? Here's the code:

```
const _add = (tree, keyToAdd) => {
❶ const p = _findIndex(tree, keyToAdd);
❷ if (isEmpty(tree.ptrs[p])) {
    tree.keys.splice(p, 0, keyToAdd);
    tree.ptrs.splice(p, 0, null);
  } else {
❸ const child = tree.ptrs[p];
❹ _add(child, keyToAdd);

❺ if (_tooBig(child)) {
    // Child too big? Split it
    const m = Math.ceil(ORDER / 2);
  ❻ const newChild = newNode(child.keys.slice(m), child.ptrs.slice(m));

  ❼ tree.keys.splice(p, 0, child.keys[m - 1]);
    tree.ptrs.splice(p + 1, 0, newChild);

  ❽ child.keys.length = m - 1;
    child.ptrs.length = m;
    }
  }
};
```

First find in which subtree of the current node to add the new key ❶. If the subtree doesn't exist (the corresponding pointer is null), you are at the bottom level and can simply add the key there ❷; no need to do anything else. (Of course, the bottom node may have grown too large, but that will be checked by its parent, which will fix the situation if needed.) If there is a subtree ❸, recursively add the new key into it ❹ and then check whether the child grew too large ❺. If so, you need to add a new node that will get the second half of the keys and pointers of the outsized node ❻; the middle key and pointer to the new node will go up to the parent ❼, and you'll leave the first half of the keys and pointers in the original child ❽. The code isn't particularly hard to understand, but handling indices and arrays properly requires care.

Removing a Key from a B-tree

Adding a key introduces the complexity that sometimes a node grows too big, and dealing with that requires either rotating or moving a key up. Removing a key also causes difficulties, because nodes may become too empty and need to get keys from other nodes or eventually require a key from the parent, which may make the whole B-tree shorter; in the same way that additions may cause it to grow taller, removals may lower its height.

There are several possible cases to study, but here are two: removing a key that isn't in a leaf and removing a key that is in a leaf.

Removing a Key from a Nonleaf Node

Removing a key in an internal node is easy; it's the same as for binary search trees. Replace the key with the one that follows it in ascending order and then remove *that* key (which will be at a leaf) from the tree. For instance, assume you want to remove the 22 key from the tree in Figure 13-15.

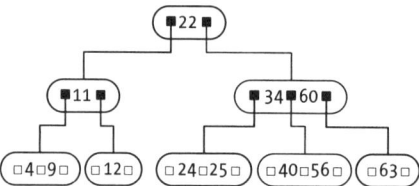

Figure 13-15: A B-tree from which you'll remove a key not from a leaf—in this case, the 22

You first need to locate the key following 22, so follow the link after 22 and then continue following the leftmost link until you get to a leaf to find the 24 key (see Figure 13-16).

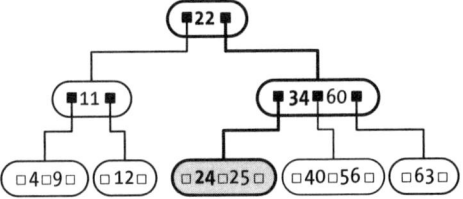

Figure 13-16: To remove 22, first locate the following
(greater) key, which is 24 in this case.

Now, replace the key to be removed (22) with the following key (24)
and proceed with the logic to remove a key from a leaf node (marked in
gray in Figure 13-17).

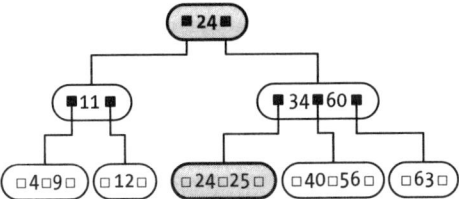

Figure 13-17: After moving 24 to the place of the 22,
you now need to remove the 24 from the leaf node.

With this method, you always have to remove keys from a leaf node.
Here's how to do it.

Removing a Key from a Leaf Node

After finding the key to remove and checking that it's in a leaf node, you
have two possible cases: either the node is "full enough," so removing the
key won't make it too empty, or the node is at its minimum size, which
means removing the key will leave it undersized.

The first case is easy to handle: continuing the example from the previous section, to remove the 24 key, just remove it from the node, which has
enough keys (see Figure 13-18).

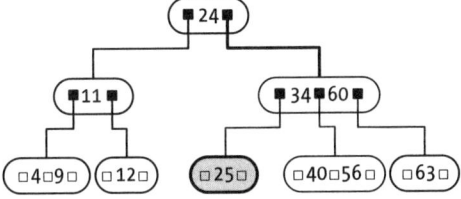

Figure 13-18: If after removal the leaf node still has
enough keys, you're done.

But consider a more complicated case: What happens if you want to remove the 12 key? You have a problem, because the corresponding node would end up with not enough keys, as shown in Figure 13-19. (In this case, the node gets emptied, because you are dealing with a B-tree of order 3; in a B-tree of a higher order, the node would still have some keys, but just not enough.)

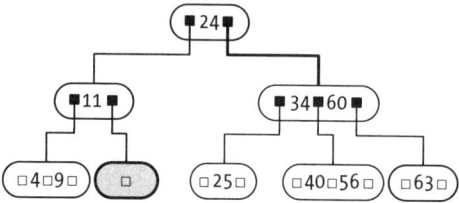

Figure 13-19: An attempt to rotate keys from a sibling to reorganize the B-tree

The solution here depends on the node's siblings. You can try to borrow keys from one of them and check whether it's possible to borrow from the left or right sibling; both siblings are symmetrical. In this case, the left sibling has enough keys (4 and 9), so you borrow one from it. The 9 key goes into the parent node, and the 11 key is rotated down into the leaf (see Figure 13-20).

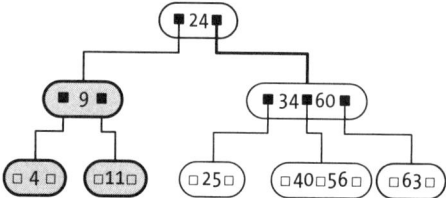

Figure 13-20: A rotation fixed the problem, so you're done.

You have only one pending case to deal with: What happens if no sibling has keys to share? In that scenario, merge the node with a sibling, borrowing one key from the parent. This step may cause it to become undersized as well and need to be fixed. In this example, say you want to remove the 11 key. Merge it with its sibling and borrow the 9 key, reaching the situation shown in Figure 13-21.

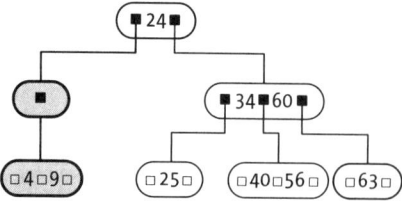

Figure 13-21: The leaf node was fixed, but a problem appears above it.

To solve the new undersized node situation, you again turn to the borrowing concept, rotate keys around, and the final tree would look like Figure 13-22.

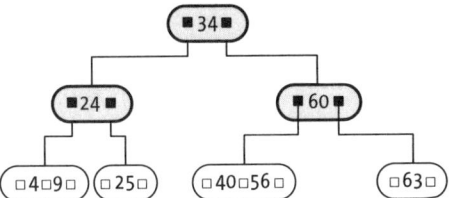

Figure 13-22: A new rotation fixes the problem.

When doing this sharing and joining, it's possible that eventually the tree will grow shorter in height. If you removed the 4 and 9 keys in succession, you'd arrive at the tree in Figure 13-23.

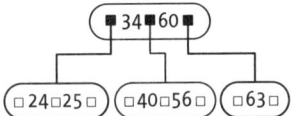

Figure 13-23: Removing several keys can make a B-tree eventually grow shorter.

Can you do the intermediate steps?

Implementing the remove() Method

Now that you've seen all the strategies to apply, here's how to code them. Dealing with a removal at the root level is short:

```
const remove = (tree, keyToRemove) => {
❶ tree = _remove(tree, keyToRemove);
❷ return (isEmpty(tree) || tree.keys.length === 0) && !isEmpty(tree.ptrs[0])
    ? tree.ptrs[0]
    : tree;
};
```

First use a _remove() function to remove the key from the tree recursively ❶. If the tree is empty, or if the root node has been left with no keys (because it had a single key and had to pass it down when its two children got joined), but the node still has one non-null child (the root might already be at the bottom level, without children) ❷, return that child. This means the B-tree has become shorter.

Now start the actual key removal process:

```
const _remove = (tree, keyToRemove) => {
❶ if (!isEmpty(tree)) {
```

```
❷ const p = _findIndex(tree, keyToRemove);
   if (tree.keys[p] === keyToRemove) {
   ❸ if (isEmpty(tree.ptrs[p])) {
      ❹ tree.keys.splice(p, 1);
         tree.ptrs.splice(p, 1);
      } else {
      ❺ const nextKey = _findMin(tree.ptrs[p + 1]);
      ❻ tree.keys[p] = nextKey;
      ❼ _remove(tree.ptrs[p + 1], nextKey);
      ❽ _fixChildIfSmall(tree, p + 1);
      }
   } else {
   ❾ _remove(tree.ptrs[p], keyToRemove);
   ❿ _fixChildIfSmall(tree, p);
   }
 }
 return tree;
};
```

Start as when doing a search: if the tree is empty, you're done ❶. Otherwise, see whether the current node includes the key you want to remove ❷. If you find the key and are at the bottom level ❸, just remove that key and its corresponding pointer ❹; otherwise, if you are at a higher level, find the next key in ascending order ❺ using _findMin() and put it in place of the original key you wanted to remove ❻. Finish by removing the next key from the tree ❼ and fixing its size ❽ if needed (because the child became too small, with not enough keys in it). If the key wasn't in the node, go down to the next level to remove it ❾ and fix the size if necessary ❿.

How do you find the next key? You have seen similar methods before, and for B-trees, the code is also quite short:

```
const _findMin = (tree) =>
  isEmpty(tree.ptrs[0]) ? tree.keys[0] : _findMin(tree.ptrs[0]);
```

If there's no leftmost subtree, return the first key in the node; otherwise, go down to the subtree and look for the minimum there.

The last method is _fixChildIfSmall(), which deals with all the cases mentioned before and properly rebalances nodes. The following includes four distinct cases, but the logic for each of them is short:

```
const _fixChildIfSmall = (tree, p) => {
  const child = tree.ptrs[p];

❶ if (_tooSmall(child)) {
   ❷ if (p > 0 && !_tooSmall(tree.ptrs[p - 1], 1)) {
      ❸ const leftChild = tree.ptrs[p - 1];
         child.keys.unshift(tree.keys[p - 1]);
         child.ptrs.unshift(leftChild.ptrs.pop());
         tree.keys[p - 1] = leftChild.keys.pop();
   ❹ } else if (p < tree.keys.length && !_tooSmall(tree.ptrs[p + 1], 1)) {
      ❺ const rightChild = tree.ptrs[p + 1];
         child.keys.push(tree.keys[p]);
```

```
      child.ptrs.push(rightChild.ptrs.shift());
      tree.keys[p] = rightChild.keys.shift();
❻ } else if (p > 0) {
  ❼ const leftChild = tree.ptrs[p - 1];
      leftChild.keys.push(tree.keys[p - 1], ...child.keys);
      leftChild.ptrs.push(...child.ptrs);
      tree.keys.splice(p - 1, 1);
      tree.ptrs.splice(p, 1);
❽ } else {
  ❾ const rightChild = tree.ptrs[p + 1];
      rightChild.keys.unshift(...child.keys, tree.keys[p]);
      rightChild.ptrs.unshift(...child.ptrs);
      tree.keys.splice(p, 1);
      tree.ptrs.splice(p, 1);
    }
  }
};
```

First, just verify whether the child is still big enough ❶, and if so, nothing needs to be done. Then, check whether the child has a left sibling that won't become too empty if you take one key from it ❷; if this is the case, do a rotation of keys as described earlier ❸. Alternatively, check whether the child has a right sibling with enough keys ❹, and if so, do the rotation with that sibling ❺. If no rotation was possible and if there is a left sibling ❻, join it to the child ❼; otherwise, there must be a right sibling ❽, so join the child with it instead ❾. Again, the cases are not complex, but take care when manipulating indices; it's easy to get things wrong.

Considering Performance for B-trees

A B-tree ensures that every node (apart from the root) will have a minimum number of children, so it grows exponentially as levels are added, meaning that the height is logarithmic; all paths from the root to another key will be $O(\log n)$, so all algorithms turn out to be logarithmic, as shown in Table 13-1.

Table 13-1: Performance of Operations for B-trees

Operation	Performance
Create	$O(1)$
Add	$O(\log n)$
Remove	$O(\log n)$
Find	$O(\log n)$
Traverse	$O(n)$

B-trees ensure good performance, so they're widely used, most notably to create indices for databases; in fact, the B-tree is the default structure for MySQL and PostgreSQL.

Red-Black Trees

B-trees are powerful, but can be a bit complex to implement. However, you can work with them using a binary representation that produces the same results in a different way. In particular, we'll use B-trees of order 3 but we'll represent them in a binary tree style. The resulting *red-black trees* have very good performance and are used, among other places, in the Linux kernel to track directory entries, virtual memory, scheduling, and more. In this section we'll look at left-leaning red-black trees, which are a variant created by Robert Sedgewick and are easier to implement than the original red-black tree.

NOTE *B-trees of order 3 are also known as 2-3 trees, alluding to the fact that their nodes have either two or three children. In the same way, B-trees of order 4 are called 2-3-4 trees or 2-4 trees.*

Consider the nodes in a red-black tree as 2-nodes (with two children) or 3-nodes (with three children). You can represent 2-nodes as common nodes in any binary tree, but here you'll add an extra node to represent a 3-node (see Figure 13-24).

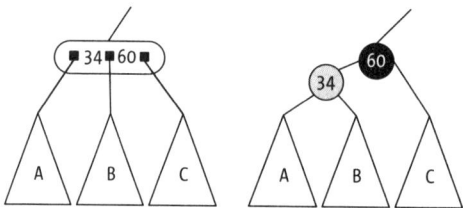

Figure 13-24: A red-black node is actually equivalent to a B-tree of order 3.

The standard nodes are black, and the extra nodes added for 3-nodes are red, which allow you to distinguish 2-nodes from 3-nodes. You could also say that links between nodes are black if they point to a black node or red if they point to a red node.

NOTE *Since this book is black and white, the "red" nodes will be gray with black text and the "black" nodes will be black with white text.*

Because of the way representation was defined, red nodes are always to the left; in addition, a red node can never be connected to another red node (or, alternatively, you can't have two red links in a row). Also, the root is black, and empty trees (leaves at the bottom) are also black.

Now, let's transform the B-tree we worked with earlier into a red-black tree (see Figure 13-25).

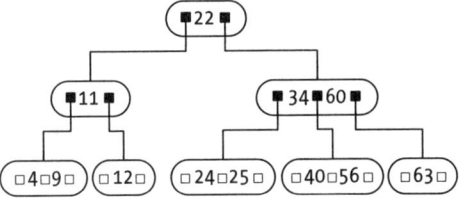

Figure 13-25: A B-tree that will be converted
into a red-black tree

All 2-children nodes become black nodes and 3-children nodes add a
new red node, as shown in Figure 13-26.

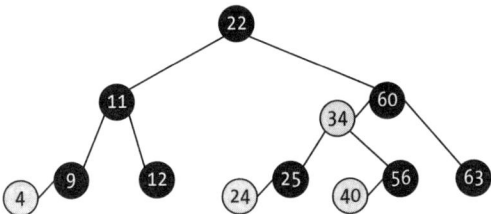

Figure 13-26: The equivalent red-black tree

You now have a binary search tree, which means you can use the earlier
key-searching logic without any changes, but you do need to make adjust-
ments when adding or removing keys.

NOTE *Because of their B-tree origin, red-black trees have another important property. All
paths from the root to a leaf have the same number of black nodes and up to that same
number of red nodes. This property is called* black balance, *and we'll often allude to
it in this book.*

Next, let's look at how to implement these red-black trees, but keep in
mind their equivalence to B-trees of order 3, because algorithms will make
much more sense that way, essentially doing the same type of work from
earlier in this chapter.

Representing Red-Black Trees

Red-black trees are just binary search trees, so you can start with some functions you already have, such as the find() method and others that need no
changes. For these new trees, you need a couple of constants and a method
to flip a node's color (you'll use these frequently):

```
const RED = "RED";
const BLACK = "BLACK";
const flip = (color) => (color === RED ? BLACK : RED);
```

Now start defining the new tree:

```
const newRedBlackTree = () => null;

const newNode = (key) => ({
  key,
  left: null,
  right: null,
❶ color: RED
});

❷ const _isBlack = (node) => isEmpty(node) || node.color === BLACK;
❸ const _isRed = (node) => !_isBlack(node);
```

To represent a node's color, add a color attribute ❶, which is red for new nodes, although that color may be changed to black later. You also add a couple of auxiliary methods to test a node's color ❷❸. Notice you are defining that an empty tree is black.

Adding a Key to a Red-Black Tree

You are essentially just adding a key to a B-tree, which was described earlier. Always add nodes as red, which won't affect the black balance of the tree, but you can possibly fix their color later or make other changes. Also take care that the root is always black and that all the properties of red-black trees are satisfied.

To implement the algorithm, you'll allow (for a while) problems like having right red links or two consecutive red links, but you'll use rotations and color changes to fix those situations before you are done. Just add the key and worry about fixing problems later:

```
❶ const _add = (tree, keyToAdd) => {
    if (isEmpty(tree)) {
      return newNode(keyToAdd);
    } else {
      const side = keyToAdd <= tree.key ? "left" : "right";
      tree[side] = _add(tree[side], keyToAdd);
    ❷ return _fixUp(tree);
    }
  };

❸ const add = (tree, keyToAdd) => {
  ❹ const newRoot = _add(tree, keyToAdd);
  ❺ newRoot.color = BLACK;
    return newRoot;
  };
```

Add the key with an auxiliary _add() method ❶, which is the same algorithm as for common binary search trees with only one innovation, a call to a _fixUp() function ❷ that takes care of restoring the structure if there are any problems in it. The addition itself is done ❸; then first use _add() to add the new key to the tree ❹ and then make sure the root is black ❺.

Restoring a Red-Black Tree Structure

If the red-black structure has been damaged somehow, the trick to restoring it lies in the _fixUp() method. Remember a new node is always colored red. The possible cases when adding a new key depend on whether the new key ends up forming part of a 2-node or a 3-node.

The first situation is simple: if the new child is the left child of a black node, you just turned a 2-node into a 3-node, and since the red child is on the left of the root, everything's okay. Call this case (a). Otherwise, if the new child is to the right of the black root, you can fix it with a rotation. Call this case (b). Figure 13-27 shows both cases; N is the newly added key, and R is the original root for the 2-node.

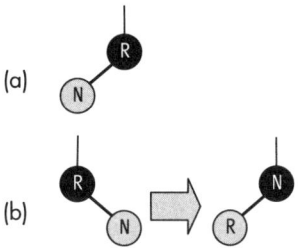

Figure 13-27: A rotation is needed if adding a node to the right.

For case (a), nothing needs to be done, and a rotation to the left solves case (b). In both situations, the black balance of the whole hasn't been affected. You didn't add any black links, so everything's still fine. Also, notice that the N node, originally red, has turned black.

The more complex case happens when you add the new key to an existing 2-node (thus creating a 3-node), because in this case, all situations are wrong. The (relatively) easiest case to fix is when the new key becomes the rightmost key in the 3-node, as shown in Figure 13-28. Call this case (c).

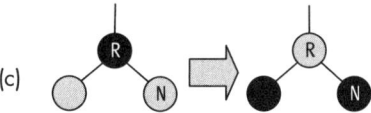

Figure 13-28: You can fix adding a new key to a 2-node tree by flipping colors, but new problems may appear above it.

Here, a quick solution is available: just flip the three nodes' colors. But notice that doing so will send a red link up the tree, which may require further recursive fixing. Also, verify that the black balance of the tree was maintained, so the fix is good.

The next case in terms of complexity is adding the newest key to the left of the leftmost one in a 3-node. Call this case (d); you need two steps to solve it, as shown in Figure 13-29.

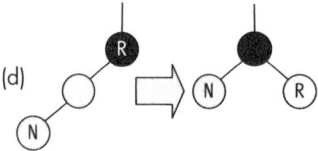

Figure 13-29: Adding a child to the left of a left child can also be solved with rotations.

If the new key is the lowest key in a 3-node, you have two left red children in a row. Start by doing a right rotation at the root, and that will leave you with the previous case (black root, two red children), so a final color flip will again solve the problem, although you may still need more recursive fixes.

The final case (e) is the most complex. Add a new key that ends as the middle key, placed between the two existing keys in a 3-node, as shown in Figure 13-30.

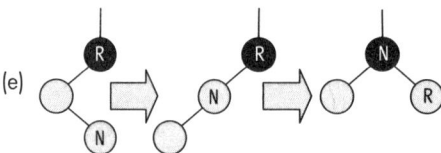

Figure 13-30: You also can fix adding a child to the right of a left child with rotations plus color-flipping.

In this case, you end with a red node to the right of another red node, which is a no-no. You can solve the issue by starting a rotation to the left, which leaves you with a situation you've already dealt with, and finish with another rotation, this time to the right, to get a scenario you've seen twice before (black root with two red children), so a final color flip solves everything.

Take a look at the rotations code, which is exactly the same as when we studied AVL trees in Chapter 12, except we don't need to maintain a height attribute:

```
const _rotate = (tree, side) => {
  const otherSide = side === "left" ? "right" : "left";

  const auxTree = tree[side];
  tree[side] = auxTree[otherSide];
  auxTree[otherSide] = tree;
```

```
auxTree.color = auxTree[otherSide].color;
auxTree[otherSide].color = RED;

  return auxTree;
};
```

The only additions are the two lines in bold that exchange colors.

Now we'll delve into the far more interesting code that applies all the fixes described (note that we'll also use this code for removals; the same logic applies there):

```
const _fixUp = (tree) => {
❶ if (_isRed(tree.right)) {
    tree = _rotate(tree, "right");
  }

❷ if (_isRed(tree.left) && _isRed(tree.left.left)) {
    tree = _rotate(tree, "left");
  }

❸ if (_isRed(tree.left) && _isRed(tree.right)) {
    _flipColors(tree);
  }

  return tree;
};
```

If the node you're looking at has a red right child, do a rotation to the left ❶, which solves case (b) and is the first step for cases (c) and (e), which will be completed later, when you move up recursively. If you have a left red child and a left red grandchild ❷, this is case (d); you could also have arrived here after doing a rotation in case (e). Finally, after the previous changes, either you have fixed everything—if you were originally in cases (a) or (b)—or you still need to flip the color ❸, and you're done.

Consider another takeaway from this algorithm: color flipping and rotations to the left or to the right all retain the black balance in a tree, so if you start with a red-black tree and apply only those transformations, you'll necessarily end with a red-black tree. This concept is important for additions, but you'll also apply it for removals.

Removing a Key from a Red-Black Tree

Removing a key from a red-black tree is probably the most complex algorithm discussed in this book. (Many textbooks and other resources tend to omit it or, at most, only hint at it.) While adding a key is not too complex, being basically the same algorithm as for common binary search trees (plus some logic to ascertain that certain constraints are being kept), deletion requires a more difficult process with changes both up and down the tree.

You need to ensure that the key to be deleted is at the bottom of the tree as part of a 3-node (either the black or red node), because in that case, removing it causes no problem. If the key to be deleted is the red one, just take it off. If you want to delete the black one, put the red key in its place, but change the color to black to maintain balance. Figure 13-31 shows both cases; X marks the key to be removed.

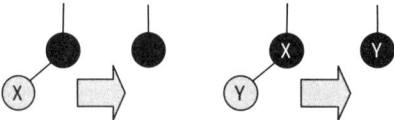

Figure 13-31: Removing a red leaf is direct, and removing the black root is also easy to achieve.

Here's the process. Either do rotations or flip colors as you go down, so at all times, the node at the root either is red or has a red left child, and you'll tolerate (for now) having red right children or black nodes with two red children. When you find the key you want, replace it with the following key (as with binary search trees) and then proceed to remove that key from the tree. When you find it, one of the two cases illustrated in Figure 13-31 will apply, and you'll effectively remove the key. Finally, apply the "fix-up" algorithm to go back to the root and take care of whatever problems might be left.

Remember the invariant to uphold: either the root or its left child must be black. Assume that at some moment in the algorithm, you have to go down to the left. Obviously, if the root is black, you just go down to the left (which is red). The invariant will persist, and now the root will be red. However, if the left child is black, there are two cases, depending on the color of the root's right child's left child. If that child is black, you can just flip colors, as Figure 13-32 shows (it doesn't include other links or subtrees for clarity, so you can focus on the important nodes). The small triangle points to the new red node, which you'll move left for the updated invariant.

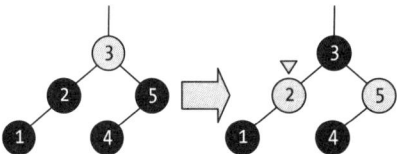

Figure 13-32: Color flipping adjusts this case.

In terms of the equivalent red-black tree, this is like joining nodes to create a 4-node, which you'll need to split later.

If the root's right child's left child had been red, you would have needed more steps: flipping, rotating right, rotating left, and flipping again. But after all those transformations (all of which maintain black balance), you'll be able to go down the left: the root's left child's left child will be red, and again the invariant is maintained. Figure 13-33 shows all the steps.

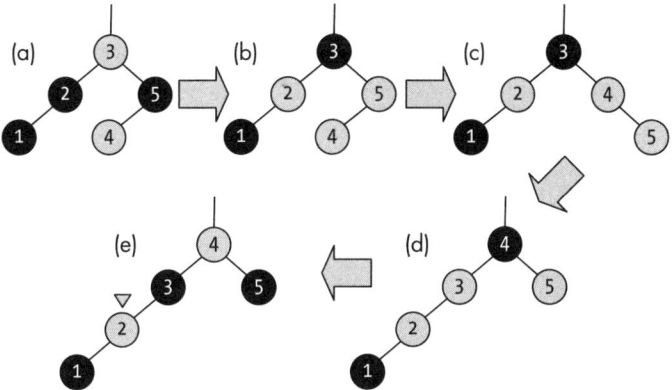

Figure 13-33: The most complex case requires several rotations and color flips.

The initial tree is (a); (b) is the tree after flipping; (c) is after rotating the root's right's left child to the right; (d) is after rotating the root to the left; and finally, (e) is after flipping colors.

Considering the equivalent 2-3 tree, this deletion was like borrowing the 4-key from a 3-node to send the 3-key down to create a 3-node together with the 2-key. As before, you are maintaining black balance and there will be nothing to fix later.

Now consider the other case, which is when you want to move right. This case is similar to the ones you just explored, but it's a tad simpler. If the root is black and its right child is red, just move to it with no fuss. If the root is red, its left child is black, and the root's left child's left child is also black, you can just flip colors. Take a look at the situation shown in Figure 13-34.

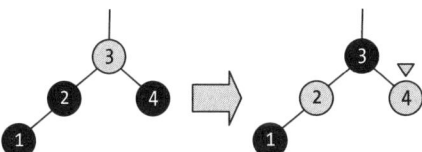

Figure 13-34: Color-flipping also fixes the potential 4-node tree.

As before, this solution is equivalent to joining nodes and creating a 4-node tree, which will need to be fixed later.

The last case occurs when you need to move right and the root's left child's left child is red. You need to flip colors, do a rotation, and flip colors again to re-establish the invariant, as shown in Figure 13-35.

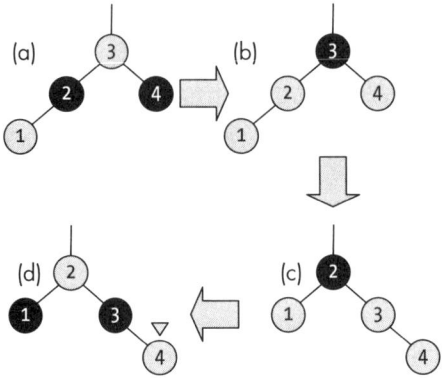

Figure 13-35: This complex case also
requires both color flipping and rotations.

In this scenario, (a) is the initial situation, (b) shows the flipped colors,
(c) is after rotating the root to the right, and (d) is after flipping colors
again.

Again, in terms of the original 2-3 tree, this example is like moving the
2-key up from the 3-node where it was to the place of the 3-key, which joins
the 4-key into a 3-node. However, note that the final situation is not valid
(there's a red child to the right), so this will need to be fixed later.

The complete algorithm is based directly on Sedgewick's own code, but
blame any errors on me. The simple part is this:

```
const remove = (tree, keyToRemove) => {
❶ const newRoot = _remove(tree, keyToRemove);
  if (!isEmpty(newRoot)) {
  ❷ newRoot.color = BLACK;
  }
  return newRoot;
};
```

First apply the algorithm described to actually remove the key from the
tree ❶ and then make sure the root is black ❷ unless, obviously, there's no
key and the tree became empty.

The complex part is the _remove() code:

```
const _remove = (tree, keyToRemove) => {
  if (isEmpty(tree)) {
    return null;
❶ } else if (keyToRemove < tree.key) {
  ❷ if (_isBlack(tree.left) && _isBlack(tree.left.left)) {
      _flipColors(tree);
      if (_isRed(tree.right.left)) {
        tree.right = _rotate(tree.right, "left");
        tree = _rotate(tree, "right");
        _flipColors(tree);
      }
    }
```

```
❸ tree.left = _remove(tree.left, keyToRemove);
❹ } else {
    if (_isRed(tree.left)) {
      tree = _rotate(tree, "left");
    }
❺ if (keyToRemove === tree.key && isEmpty(tree.right)) {
      return null;
    } else {
    ❻ if (_isBlack(tree.right) && _isBlack(tree.right.left)) {
        _flipColors(tree);
        if (_isRed(tree.left.left)) {
          tree = _rotate(tree, "left");
          _flipColors(tree);
        }
      }
      if (keyToRemove === tree.key) {
    ❼ tree.key = minKey(tree.right);
        tree.right = _remove(tree.right, tree.key);
      } else {
    ❽ tree.right = _remove(tree.right, keyToRemove);
      }
    }
  }
}
❾ return _fixUp(tree);
};
```

After verifying that the tree isn't empty, check whether you need to
go left ❶, and if so ❷, you may need to apply the transformations you saw
earlier before actually moving left ❸. If the key you want to delete is greater
than or equal to the root, start by doing a rotation ❹ so that the red node
will be to the right, which you'll need later. If you find the key ❺ and it
has no right child, it must be at the bottom, so you can now delete it. You'll
want to move right, so set things up according to the procedures described
earlier ❻ but don't move just yet. If you find the key but aren't able to delete
it, replace it with the following key in the tree and move right to delete
that value ❼; otherwise, just move right to keep looking for the key to be
removed ❽. At the end, a final fix-up pass ❾ solves any wrong configura-
tions in the tree.

Red-black trees have the shortest code for searching and not very complex code for adding a new key (basically, just adding a fix-up call at the
end), but the deletions are rather more involved. Getting the code right is
difficult (see question 13.10 for a small detail).

Considering Performance for Red-Black Trees

We don't need to analyze the performance of red-black trees, because
they're just another case of B-trees, so you already know that all algorithms
(adding, removing, and searching) are $O(\log n)$, as Table 13-2 shows.

Table 13-2: Performance of Operations for Red-Black Trees

Operation	Performance
Create	$O(1)$
Add	$O(\log n)$
Remove	$O(\log n)$
Find	$O(\log n)$
Traverse	$O(n)$

Of course, because of the possible embedded red links, red-black trees are taller on average than B-trees of higher orders (and not all paths from the root to a leaf have the same length), but that doesn't change the result. Performance will still be logarithmic, even if searches are slower by a (bound) constant factor.

Summary

In this chapter, we moved beyond binary trees and explored two new structures for the dictionary ADT: B-trees and red-black trees (which are derived from B-trees). These structures provide good performance, and they're used often because their implementation is not too complex and their speed is significant.

In the next chapter, we'll study heaps, which are a variant of binary trees, and then in the following chapter, we'll look at extended heaps, which combine heaps and forests to achieve high performance.

Questions

13.1 Missing Test?

In the appendChild() method for trees, shouldn't it include a call to this._throwIfEmpty()?

13.2 Traversing General Trees

Implement the missing preorder and postorder traversals. You might want to do this for both trees represented with arrays of children and for trees with a "left child, right sibling" representation.

13.3 Nonrecursive Visiting

Implement a depth-first traversal of a tree without recursion by using a stack as an auxiliary structure.

13.4 Tree Equality

Implement an equals(tree1, tree2) algorithm that will decide whether two trees are equal—that is, having the same shape and the same keys in the same positions. You may want to "think outside the box" for this. Maybe you won't even need recursion!

13.5 Measuring Trees

Redo the `calcSize()` and `calcHeight()` functions from Chapter 12 (see questions 12.5 and 12.6) to work with multiway trees.

13.6 Sharing More

In a B-tree, instead of siblings sharing just one key, you can achieve a better balance if they share more keys. For example, when adding a key, if a node becomes too full and a sibling has enough space, instead of just passing a single key to it, pass as many as possible until both siblings are about equally full. A similar process would work when removing a key. Implement this optimization.

13.7 Faster Node Searching

This chapter used linear searching in nodes, but since keys are ordered, a better way would be using binary searching. Make this change. Would that make a difference in the order of the B-trees' methods? Why or why not?

13.8 Lowest Order

Would B-trees of order 2 make sense?

13.9 Many Orders of Trees

What would you do if you needed to work with B-trees of different orders? Hint: the problem here is that imported modules are singletons. Look for a way to avoid this behavior.

13.10 Safe to Delete?

In the `remove()` algorithm for red-black trees, when you actually delete a node, are you sure it's possible to remove it without any negative effects?

14

HEAPS

In the last chapter, we worked with binary trees, and we'll continue doing so in this chapter, but with a variant that is stored without the need for dynamic memory: heaps. Heaps allow for easy implementation of a new abstract data type (ADT), a good performance sorting method, and a new structure for an enhanced version of binary search trees. We'll consider implementing heaps (in particular, binary max heaps, but other types too), and we'll look at using heaps for priority queues, sorting arrays with heapsort, and searching with another new structure called treaps.

In the next chapter, we'll follow with another representation for heaps that uses dynamic memory and allows for more freedom and better performance for some new operations.

Binary Heaps

A *binary heap*, usually referred to simply as a *heap*, is a type of binary tree with two particular properties: a *structure property* that determines the shape of the tree and a *heap property* that specifies the relationship between the key of a parent node and those of its children.

The Structure Property

Heaps are a subset of binary trees, and the structure property requires that the tree must be complete and that all leaves on the last level must be located on the left. Consider the trees in Figure 14-1. Only one qualifies as a heap while the other two fail. Can you tell which is which?

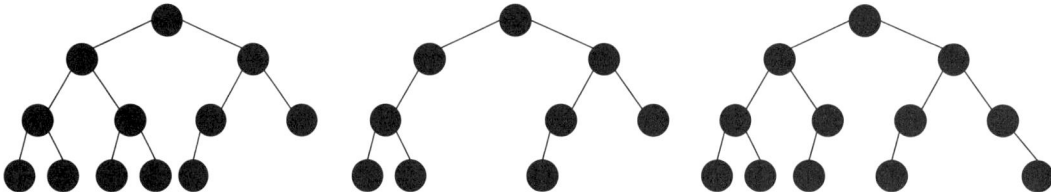

Figure 14-1: Of these three heap candidates, which is the right one?

The tree on the left in Figure 14-1 is the only heap. The tree in the middle has an incomplete middle level, and in the tree on the right, the bottom children are not all on the left.

Given this rule, you can store a heap in a common array without dynamic memory or pointers to simplify implementation. (See question 14.18 to consider an alternative.) Place the root at the first position of the array, followed by the nodes at the second level from left to right, then the nodes at the third level (also from left to right), and so on. Figure 14-2 shows how the array looks for a sample case. The numbers in the nodes correspond to indices in the array.

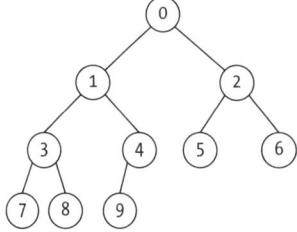

Figure 14-2: Storing a heap's nodes in an array

In this representation, the heap's root is always at position 0. The left and right children of the node at position p are placed in consecutive adjacent positions, 2*p+1 and 2*p+2, respectively—unless they fall beyond the end of the heap, in which case the node has fewer children. The parent of a non-root node at position p is found at the Math.floor((p-1)/2) position.

You can verify those rules with a few examples. The children of the root (at position 0) are at positions 2*0+1=1 and 2*0+2=2. Node 4 has a single child at position 2*4+1=9, because the other child would be beyond the heap size. The parent of node 9 is at position Math.floor((9-1)/2)=4. The parent of node 2 is at position Math.floor((2-1)/2)=1.

These rules let you implement algorithms without needing any pointers; a simple array suffices. As with complete trees in Chapter 12, if a heap has up to $n = 2^{h-1}$ nodes, its height will be h, so its height is bounded by log n, a result that will feature in order calculations.

The Heap Property

The second property of heaps is simple: the key of a node must be greater than or equal to the keys of its children. This is an important contrast from binary search trees: in heaps, there's no difference between a left child and a right child (either could be greater than the other), although they both will be smaller than or equal to their parent. Any tree that follows both the structure property and the heap property is called a *binary max heap* or, more simply, a *heap*.

NOTE *Why do I say the word* heap *means "max heap" by default? The song "New York, New York" may give us a clue: Frank Sinatra describes wanting to be "king of the hill" or "top of the heap." This suggests the root (top) of the heap should be the greatest value, doesn't it?*

You can reverse the condition and specify that the parent's key be smaller than or equal to those of its children, meaning the root would be the minimum value of the heap. This variant is called a *min heap*, and you can use it (among other scenarios) to merge several linked lists, which requires finding the minimum of many elements repeatedly (see question 14.5). The heap shown in Figure 14-3 satisfies both the structure and heap properties.

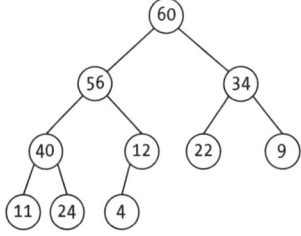

Figure 14-3: A valid heap

As mentioned previously, this tree also could be represented by an array where 60 (the root) is at position 0 of the array, as shown in Figure 14-4.

60	56	34	40	12	22	09	11	24	04
0	1	2	3	4	5	6	7	8	9

Figure 14-4: The same heap shown in Figure 14-3, as stored in an array

The heap property has an immediate consequence: the highest value in the heap will necessarily be at its root; can you see why? Where in the heap will you find its second highest value? The third highest? The fourth? (See question 14.13.) This result will be key for a sorting algorithm called *heapsort*, which we'll study later in this chapter.

Heap Implementation

To implement a heap, you need only a simple array. A heap is a data structure with a few operations, as shown in Table 14-1.

Table 14-1: Operations on Heaps

Operation	Signature	Description
Create	→ H	Create a new heap.
Empty?	H → boolean	Determine whether the heap is empty.
Top	H → key	Given a heap, produce its top value.
Add	H × key → H	Given a new value, add it to the heap.
Remove	H → H × key	Given a heap, extract its top value and update the structure correspondingly.

The functions you'll implement for those operations are:

newHeap()   Creates a new heap

isEmpty(heap)   Determines whether the heap is empty

top(heap)   Gets the value of the topmost (maximum) element of the heap

add(heap, value)   Adds a new element to the heap

remove(heap)   Removes the topmost element of the heap

The first three functions are very short:

```
❶ const newHeap = () => [];

❷ const isEmpty = (heap) => heap.length === 0;

❸ const top = (heap) => {
    if (isEmpty(heap)) {
      return undefined;
    } else {
      return heap[0];
    }
  };
```

Creating a new empty heap is the same as returning an empty array ❶. The size of the heap is heap.length, so checking it for 0 tells you whether the heap is empty ❷. Also, the top of the heap (unless the heap is empty, in which case this code returns undefined) is at the first position of the array ❸, so the implementation details are straightforward.

Adding to a Heap

To add a new value to the heap, follow these steps:

1. Add the new value at the end of the array.
2. If the value is greater than its parent, exchange places with it repeatedly.
3. When the value is smaller than its parent or when it gets to the top of the heap, stop.

Let's see how this works. Start with the heap shown in Figure 14-5.

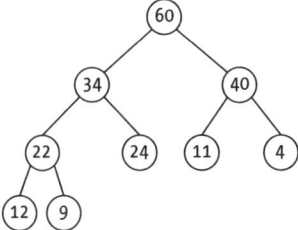

Figure 14-5: An initial heap, before adding a new value

If you want to insert a new value of 56, the first step is to add it at the end of the heap, so you'd get the result shown in Figure 14-6.

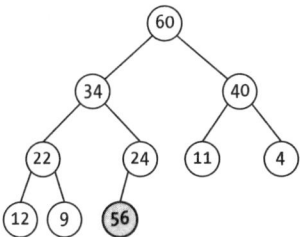

Figure 14-6: The new value (56) starts at the end of the heap.

Let's see if the new value should bubble up. Comparing 56 with its parent (24) shows that they need to be swapped, resulting in a new heap configuration (see Figure 14-7).

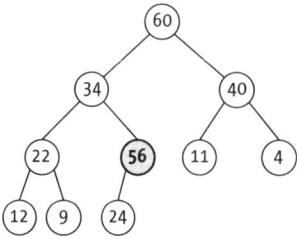

Figure 14-7: If the new value is greater than its parent, it has to "bubble" up.

After bubbling up, keep checking recursively, and a new upward movement is required (see Figure 14-8).

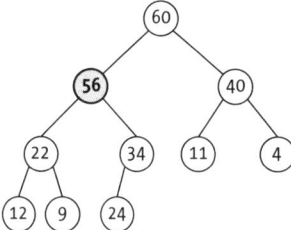

Figure 14-8: Bubbling up continues until the added value is not greater than its parent or at the root of the heap.

The last step led to a situation where the inserted value is not greater than its parent, so the algorithm stops.

Our version of add() is short and to the point:

```
const add = (heap, keyToAdd) => {
❶ heap.push(keyToAdd);
❷ _bubbleUp(heap, heap.length - 1);
  return heap;
};
```

As described, the new value was added at the end of the heap ❶, and it was forced to bubble up to its final position ❷ by using the _bubbleUp() auxiliary function.

As before, a recursive implementation is easiest. If the element has moved up, apply _bubbleUp() recursively to keep it moving:

```
const _bubbleUp = (heap, i) => {
❶ if (i > 0) {
  ❷ const p = Math.floor((i - 1) / 2);
  ❸ if (heap[i] > heap[p]) {
    ❹ [heap[p], heap[i]] = [heap[i], heap[p]];
```

```
  ❺ _bubbleUp(heap, p);
    }
  }
};
```

If the element is not already at the top of the heap ❶, use math (as in "The Structure Property" on page 318) to determine the parent p of position i ❷. If you need to swap elements ❸, destructuring makes it easy ❹, and you can keep bubbling up (if needed) by using recursion ❺.

Removing from a Heap

Next you need the remove() method. Remember that the whole heap must become one element smaller, so what happens after removing the top of the heap? If the heap is empty, there's nothing to remove; throw an exception, and you're done. Otherwise, pick the last element of the heap, place it at the top, and then reduce the heap size by one. If the element doesn't have any children, stop. If the element is greater than the greatest of its children, also stop. Otherwise, exchange the element with its greatest child and keep moving it down.

Here is an example of how this works. Start with the heap in Figure 14-9.

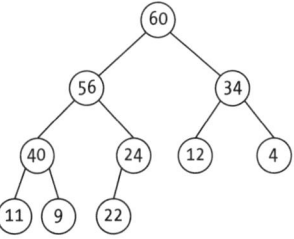

Figure 14-9: An initial heap before removing its top

The first step involves removing the top value (60), replacing it with the last value in the heap (22), and shortening the heap by one, which leads to the situation shown in Figure 14-10. The value that needs to be moved downward to restore the heap is highlighted.

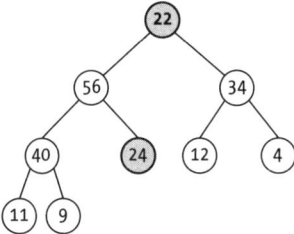

Figure 14-10: After removing the top, replace it with the last element of the heap (22).

Now start sifting down. Comparing 22 with its children, it needs to be swapped with 56, which results in the new situation shown in Figure 14-11.

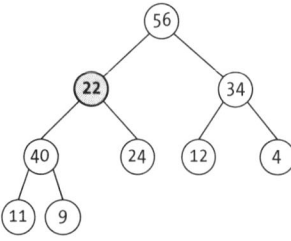

Figure 14-11: If the new top isn't greater
than its children, it must "sink" down.

Recursively, compare 22 with its new children, and again, it needs to
sink down, as shown in Figure 14-12.

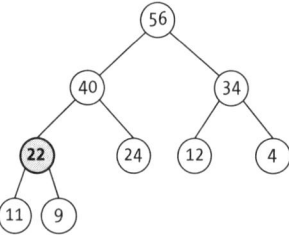

Figure 14-12: Sinking down proceeds
until the value is greater than its
children or reaches a leaf.

In this situation, 22 is now greater than its children, so the sinking-
down procedure ends. If the value 40 had been a 20, the 22 would have
been exchanged with the 24, and the shifting-down procedure would have
finished as well, since 22 would have no children.

Take a look at the following code, which uses a recursive _sinkDown()
auxiliary function to push a value down the heap:

```
const _sinkDown = (heap, i, h) => {
❶ const l = 2 * i + 1;
❷ const r = l + 1;
❸ let g = i;
  if (l < h && heap[l] > heap[g]) {
    g = l;
  }
  if (r < h && heap[r] > heap[g]) {
    g = r;
  }
  if (g !== i) {
❹ [heap[g], heap[i]] = [heap[i], heap[g]];
❺ _sinkDown(heap, g, h);
  }
};
```

Calculate l and r, the children of the parent at i; you can use the formula discussed earlier in "The Structure Property" section on page 318 ❶ and find r just by adding 1 ❷, since r follows l in the array. Use g to determine the greatest value at positions i, l, and r ❸. If the value at i isn't greater than its children, swap it ❹ and keep sinking it down recursively ❺.

At this point, you can finally write the remove() function:

```
const remove = (heap) => {
❶ const topKey = top(heap);
❷ if (!isEmpty(heap)) {
   ❸ heap[0] = heap[heap.length - 1];
   ❹ heap.length--;
   ❺ _sinkDown(heap, 0, heap.length);
   }
❻ return [heap, topKey];
};
```

This code closely follows the description in the previous example. When you get the top (which may be undefined, if the heap is empty) ❶, and if the heap isn't empty ❷, place its last value at the top ❸, reduce the heap's length by one ❹, and sink the new top down ❺. Finally, return the top value and the updated heap ❻.

Considering Performance for Heaps

Table 14-2 shows the performance of the algorithms just explored.

Table 14-2: Performance of Operations for Heaps

Operation	Performance
Create	$O(1)$
Empty?	$O(1)$
Top	$O(1)$
Add	$O(\log n)$
Remove	$O(\log n)$

Three operations work in constant time: creating a heap, testing whether it's empty, and getting the top value. The other two operations, adding and removing, are more complex. Adding an element may make it bubble up from the bottom of the heap all the way up to the top. Since you know that the heap's height is log n, this operation requires logarithmic time. Similarly, removing an element implies placing a new one at the top and possibly sinking it down to the bottom. It's the same number of operations as in adding a value, but in reverse (also logarithmic time).

Let's move on to consider a new ADT, and compare the performance of heaps versus the other structures already discussed.

Priority Queues and Heaps

Priority queues (PQs) are different from the queues discussed in Chapter 10, because each element has an associated priority, which determines what element is removed first. In a PQ, the first element to be removed is the one with the highest priority, not the first one that was added, as in a first in, first out (FIFO) strategy.

NOTE *There's a problem with the English language! The term* priority one *implies the highest priority, but 1 is the lowest-priority number. If you order tasks by priority and the lowest numbered task is the one you should tackle first, then lower numbers have a higher priority. However, some tools (like Microsoft Project) assume that 0 is the lowest priority and higher numbers are higher priority, so there's no clear-cut case. Regardless, if you actually need a min heap instead of a max heap, see question 14.4.*

PQs are used in multiple algorithms and many different situations. Operating system schedulers use priorities to select what process will be the next to run. Discrete event simulations decide the next step to apply based on a timestamp (and in this case, lower timestamps equal higher priorities). Dijkstra's shortest path algorithm (which we'll consider in Chapter 17) requires finding the vertex with the minimum distance to another given vertex. Prim's algorithm for finding a minimum spanning tree for a graph also needs to find the vertex with the smallest (cheapest) connection to another vertex. Huffman's coding algorithm builds a tree and repeatedly needs to find the two nodes with the smallest probabilities to replace them with a new node with the sum of those probabilities. All of these things require PQs.

In terms of an ADT, the description for a PQ then requires the following operations shown in Table 14-3 (other versions that add more operations are considered later).

Table 14-3: Operations on Priority Queues

Operation	Signature	Description
Create	$\rightarrow$ PQ	Create a new PQ.
Empty?	PQ $\rightarrow$ boolean	Determine whether the PQ is empty.
Top	PQ $\rightarrow$ key	Given a PQ, produce its top value.
Add	PQ $\times$ key $\rightarrow$ PQ	Given a new key, add it to a PQ.
Remove	PQ $\rightarrow$ PQ $\times$ key	Given a PQ, extract its top value and update the PQ correspondingly.

In terms of the provided operations, a heap matches the requirements of a PQ, so implementation is straightforward. For the sake of variety, though, take a look at other simple ways of implementing PQs and compare their performances:

- With an unordered array or list, getting the top value would be $O(n)$. Removing it would also be $O(n)$, because you have to go through all the values to find it, and adding a new element would be $O(1)$.

- With an ordered array (the top value at the last position), getting and removing the top would be $O(1)$, but adding a new element would be $O(n)$; after finding where the element goes, which is $O(\log n)$ with binary search, you have to move the elements physically to make space, and that's $O(n)$.

- With an ordered list (the top value in the first place), the results are the same as with an ordered array.

- With a balanced binary search tree, all three operations would be $O(\log n)$. If you have an extra pointer to the maximum value, getting the top value becomes $O(1)$, but insertions and deletions will be a tad slower because they have to maintain the added pointer.

This list of possible implementations for PQs isn't complete, but it should be enough to show how the heap is one of the best ways of implementing them, with extra points for low complexity. In the following chapter, we'll consider some extra operations you may need, which will lead to other implementations of PQs.

Heapsort

Heaps can be used to create a well-performing sorting method. Given a set of values, with a heap, you can easily find the highest value of the set. After removing it and restoring the heap, you can find the second-highest value of the set, and so on. The basic algorithm structure is as follows:

1. Build a heap out of the values to be sorted.

2. Then, until no more values are left, swap the heap's top element with its last, reduce the heap size by one, and restore the heap condition.

Take a look at how this algorithm works and then consider an enhancement.

Williams' Original Heapsort

First, here is an example implementation of the heapsort algorithm, invented by John W. J. Williams in 1964. You can reuse the _bubbleUp() and _sinkDown() functions from earlier without any change (so I won't list them here), and what's added is just the following:

```
function heapsort_original(v) {
❶ for (let i = 1; i < v.length; i++) {
    _bubbleUp(v, i);
  }

❷ for (let i = v.length - 1; i > 0; i--) {
  ❸ [v[i], v[0]] = [v[0], v[i]];
  ❹ _sinkDown(v, 0, i);
  }

  return v;
}
```

The first stage of heapsort goes from the beginning to the end of the array, making each element bubble up to its correct place ❶. The second stage ❷ swaps the top element with the last one of the (current) heap ❸ and sinks it down, using the second argument to _sinkDown() to limit how far it can sink ❹.

Here is the algorithm in action. The building phase goes through the stages shown in Figure 14-13. The highlighted area corresponds to the heap that's being built, and the rest are the values that haven't yet been added to the heap.

09	22	60	34	24	40	04	12	56	11
22	**09**	60	34	24	40	04	12	56	11
60	**09**	**22**	34	24	40	04	12	56	11
60	**34**	**22**	**09**	24	40	04	12	56	11
60	**34**	**22**	**09**	**24**	40	04	12	56	11
60	**34**	**40**	**09**	**24**	**22**	04	12	56	11
60	**34**	**40**	**09**	**24**	**22**	**04**	12	56	11
60	**34**	**40**	**12**	**24**	**22**	**04**	**09**	56	11
60	**56**	**40**	**34**	**24**	**22**	**04**	**09**	**12**	11
60	**56**	**40**	**34**	**24**	**22**	**04**	**09**	**12**	**11**

Figure 14-13: The first phase of heapsort is to build up the heap (highlighted area) one value at a time.

In every step of the algorithm, a new value is added to the heap as it is so far, bubbling up as needed, until you get a heap with one more element than before. After the building phase is done, the second part of the algorithm starts. The top element of the heap is swapped with the last value of the heap, which shrinks down in size by one, and the new value at the top sinks down to restore the heap, as shown in Figure 14-14.

60	56	40	34	24	22	04	09	12	11
56	**34**	**40**	**12**	**24**	**22**	**04**	**09**	**11**	60
40	**34**	**22**	**12**	**24**	**11**	**04**	**09**	56	60
34	**24**	**22**	**12**	**09**	**11**	**04**	40	56	60
24	**12**	**22**	**04**	**09**	**11**	34	40	56	60
22	**12**	**11**	**04**	**09**	24	34	40	56	60
12	**09**	**11**	**04**	22	24	34	40	56	60
11	**09**	**04**	12	22	24	34	40	56	60
09	**04**	11	12	22	24	34	40	56	60
04	09	11	12	22	24	34	40	56	60
04	09	11	12	22	24	34	40	56	60

Figure 14-14: In the second phase of heapsort, the top value is repeatedly removed to build up the sorted array.

In the first step, the top value (60) is exchanged with the last value of the heap (11). The 11 sinks down, and 56 moves up to the top. In the next step, 56 is exchanged with the last value (again 11), and 11 sinks down as 40 goes to the top. This continues step by step, and when the heap is at size 1, the array is sorted.

Heapsort Analysis

What's the order of heapsort? Without going too deep into mathematical details, if you are sorting n items, you call the _bubbleUp() function n times, and each time, an element may bubble up to the top of the heap, which is $\log n$ high, so that's $O(n \log n)$. Similarly, when removing elements from the heap to produce the ordered array, call _sinkDown() n times, and elements may sink down to the bottom of the heap, so that's $O(n \log n)$ again; the conclusion is that the algorithm is $O(n \log n)$.

An interesting property is that this behavior is guaranteed. No set of data will lead to a worse case, as in quicksort, which could become $O(n^2)$. Also, since we've established that $O(n \log n)$ is the best possible order for comparison-based sorting algorithms, you can see that heapsort is a solid algorithm with consistent performance, often used in libraries and in other algorithms that may require sorting.

Finally, heapsort isn't a stable sort in the sense shown in Chapter 6 (see question 14.12 for an example).

Floyd's Heap-Building Enhancement

Williams' version of the algorithm is quite efficient, but a possible enhancement, thanks to Robert Floyd, speeds up the heap-building part to $O(n)$, although we won't go into the math here. The reason for this result is that most of the elements are near the bottom, so sinking them down is much faster than bubbling them up. Very few are near the top, where sinking them is slower than bubbling up, all of which is enough to change the order of heap building. Since the second part of the procedure is still $O(n \log n)$, the algorithm's total order won't change, but it will run faster in any case.

Instead of making each element bubble up to its position, the algorithm builds small heaps, which are then made larger by joining them together, until you get the complete heap. Initially, you can consider all the leaves of the tree to be small one-sized heaps. Then, take two leaves and their parent and reorganize them (if needed) so that those three values form a heap. Keep doing this, and eventually, you'll get to the top of the heap and be done.

Take a look at the code first and then at an example. The code for this new heapsort version will depend on the previous _sinkDown() code, which you'll use unchanged. The rest of the algorithm is as follows:

```
function heapsort_enhanced(v) {
  for (let i = Math.floor((v.length - 1) / 2); i >= 0; i--) {
    _sinkDown(v, i, v.length);
  }
}
```

```
  for (let i = v.length - 1; i > 0; i--) {
    [v[i], v[0]] = [v[0], v[i]];
    _sinkDown(v, 0, i);
  }
  return v;
}
```

The code for the second part of the algorithm (exchanging and restructuring) is the same; the only difference is where you build up the heap using _sinkDown(). Figure 14-15 shows just the heap-building part of this code in action—specifically, how more parts of the array successively become a heap. At each step, the part of the array that becomes a "mini heap" is highlighted.

| 09 22 60 34 24 40 04 12 56 11 |
| 09 22 60 34 **24** 40 04 12 56 **11** |
| 09 22 60 **56** 24 40 04 **12 34** 11 |
| 09 22 **60** 56 24 **40 04** 12 34 11 |
| 09 **56** 60 **34 24** 40 04 **12 22 11** |
| **60 56 40 34 24 09 04 12 22 11** |

Figure 14-15: The enhanced heap-building algorithm creates the heap out of smaller, previously created ones.

To better understand Figure 14-15, look at the heap at different stages. Initially, the array looks like Figure 14-16 and obviously isn't a heap.

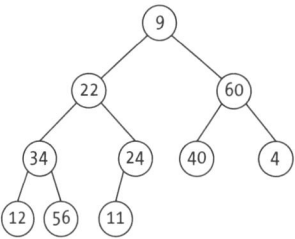

Figure 14-16: An initial array, which doesn't fulfill the heap property

After two steps, subheaps out of 11 and 24 are built (which were left as they were) as well as 12, 34, and 56 (where 34 sifted down and 56 took its place).

Figure 14-17 shows two more steps.

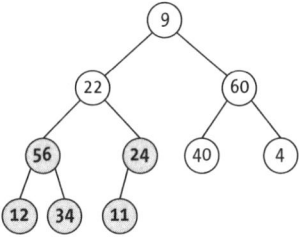

Figure 14-17: After some rotations,
a few subheaps are built.

The heap is practically done with root 56 (the 22 value sank down
and 56 took its place) and another with root 60 (where no changes
were needed). You are just one step away from finishing, as shown in
Figure 14-18.

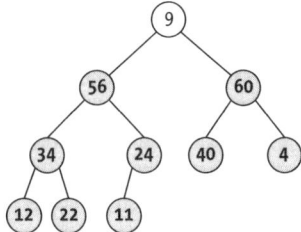

Figure 14-18: More and more subheaps
are built, reaching to the top.

The last step completes the heap; the 9 value sinks down to its place,
being replaced by 60. Figure 14-19 shows the finished heap.

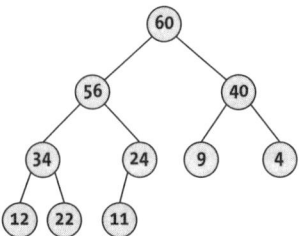

Figure 14-19: Upon reaching the top,
the array has become a heap.

Floyd's enhanced algorithm has two advantages: a faster first phase (the
second phase, which generates the sorted result, is the same) and shorter
code. You can take advantage of this method for your heap logic and modify the newHeap() function (see question 14.8).

Treaps

In Chapter 12, we discussed binary search trees and several ways to try to keep them balanced in order to avoid slow searches. In 1989, a new structure was invented that mixes the characteristics of trees and heaps: treaps. These trees are balanced, although their heights are not guaranteed to be $O(\log n)$; rather, randomization and the heap property are used to maintain balance with high probability.

The (invented) term *treap* is a portmanteau of the words *tree* and *heap*. How does this mix come about? Basically, every key is associated with a random priority, and when you build the binary search tree, take care to satisfy the heap property, so a parent node always will have a priority greater than those of its children. (The structure property will not be satisfied; nodes are linked by pointers, not an array.) Note that instead of using a random number generator, you can apply a hash function to the key, and thus produce its "random" priority. Mathematical analysis of treaps depends on truly random numbers, but the form of randomness generated by hashing also works. And in terms of testing algorithms, hashing has the advantage of determinacy.

Let's think about this a bit more. If you happened to order the keys by priority and insert them in the tree in decreasing order of priority, the resulting tree would satisfy the heap property. (Can you see why?) This means that assigning random priorities is equivalent to taking a random permutation of the keys before inserting them in the tree, which will probabilistically provide a good shape for the resulting tree, with an expected height that is $O(\log n)$, as with balanced trees.

Given a set of distinct keys and their corresponding (also distinct) priorities, the resulting treap is unique, and we can construct a recursive proof for this. First, the root of the treap must be the key with the highest priority. Then, all the smaller keys will go into the root's left subtree and the greater keys will go into the right subtree, and we can apply the same reasoning recursively to prove that those two subtrees will also be unique.

You can also modify the binary search tree algorithms from Chapter 12 to create treaps. The code is simpler than code for AVL or red-black trees, yet it provides competitive performance, often better than that of its more complex alternatives.

Creating and Searching a Treap

Given that treaps are just binary search trees at heart, most of the code discussed in Chapter 12 will still work. Start coding treaps as follows:

```
const {
  find,
  inOrder,
  isEmpty,
  maxKey,
  minKey,
  postOrder,
  preOrder
```

```
❶ } = require("../binary_trees/binary_search_tree.func.js");

const newTreap = () => null;

const newNode = (key) => ({
  key,
  left: null,
  right: null,
❷ priority: Math.random()
});
```

The treap is based on the previous binary search tree ❶, and many of the functions written there are still valid. When creating a new node, add a random priority ❷, but that's all that will change here.

NOTE *If you want to test your code and need deterministic results, you could compute priority as a hash of the key; as long as the results are random enough, it will do.*

Let's move on to adding a new key, which requires some extra coding.

Adding a Key to a Treap

Insertion into a treap requires basically the same logic as for binary search trees, except that after having added the node in its place, you may need to do some rotations to maintain the heap property. Adding new keys to a treap is possible using the _rotate() method introduced in previous chapters. The basic idea is first to place the new node in the tree according to its value and then rotate it, if needed, in a way that the binary search tree condition is kept but that also satisfies the heap condition. Figure 14-20 shows the basic rotations.

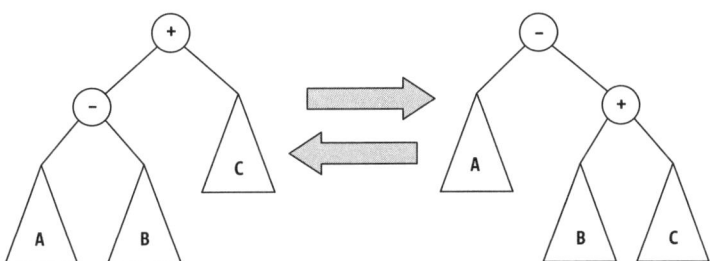

Figure 14-20: Rotations also make the binary search tree become a heap.

The minus sign represents a smaller key value than the plus sign. If the node placed lower (the minus) has a higher priority than its parent (the plus), which is shown in the left side of the figure, you can apply a *right rotation* and produce the situation on the right, which would satisfy the heap property. Conversely, if you have the situation on the right and the lower node (plus) has a higher priority than its parent (minus), you can apply a *left rotation* to get the situation on the left. In both cases, the resulting tree is still a binary search tree, but the nodes are shifted around so that the final parent node has a greater priority than its children.

Here is an example of how the add() algorithm should work. Start with the treap shown in Figure 14-21, where priorities are shown to the right of nodes.

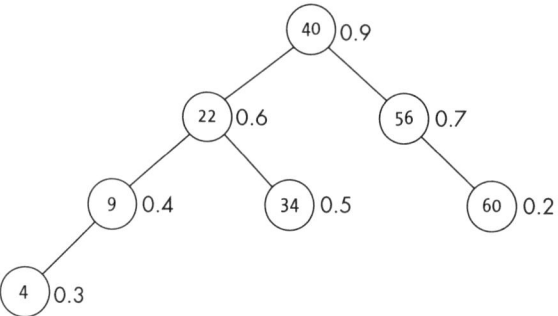

Figure 14-21: In a treap, keys form a binary search tree and priorities form a heap.

If you want to add a 12 node with a (random) priority of 0.8, the first step is to insert the new node, without worrying about priorities and the heap property, which will be sorted out later. This insertion (done in the standard way for binary search trees, as described in Chapter 12) leads to the tree shown in Figure 14-22.

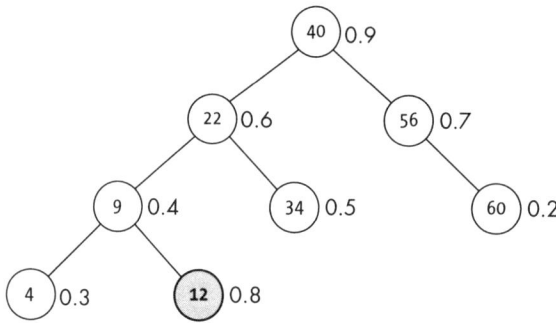

Figure 14-22: After standard insertion, the resulting binary search tree may no longer be a heap.

The treap works for searches, but the heap property isn't satisfied, because the priority for the 12 node is higher than the priority of its parent. You can fix that problem by doing a left rotation, leading to the tree shown in Figure 14-23.

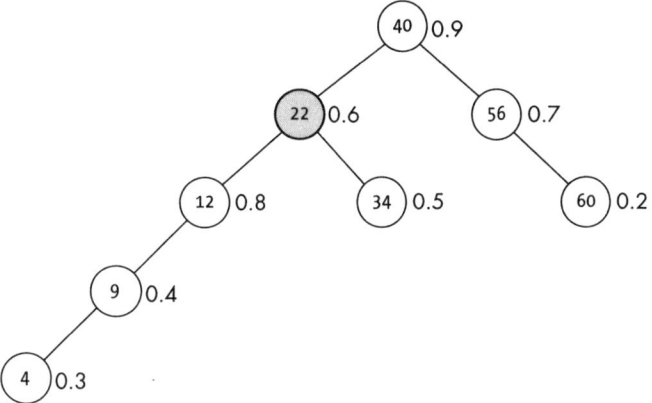

Figure 14-23: Rotations are applied until the heap property is satisfied, but this tree is still wrong.

The rotation still leaves a valid binary search tree, but the process hasn't ended, because the heap property isn't yet fully satisfied. The 12 node has a higher priority than its parent; perform a right rotation to solve this, which leads to the tree shown in Figure 14-24.

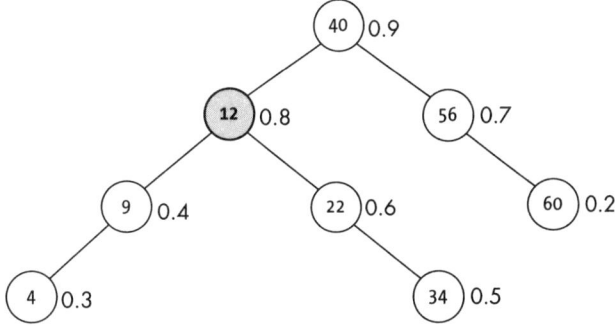

Figure 14-24: Now the heap property is satisfied.

After this second rotation, you can check that the heap property is satisfied, thus the addition to the treap was done correctly.

The final code follows, and there's a single line that's different from common binary search trees:

```
const add = (tree, keyToAdd) => {
  if (isEmpty(tree)) {
    return newNode(keyToAdd);
  } else {
    const side = keyToAdd <= tree.key ? "left" : "right";
    tree[side] = add(tree[side], keyToAdd);
    return tree[side].priority <= tree.priority ? tree : _rotate(tree, side);
  }
};
```

The line in bold makes sure that the heap property is satisfied. After adding the new key on tree[side], if the priority of that subtree is not greater than the priority of the root, you're done; otherwise, apply a rotation on that side to bring up the greater priority.

The last method needed is to remove a key from a treap.

Removing a Key from a Treap

The algorithm for removing a key from a binary search tree involves finding it first, possibly finding its successor, and putting it in the removed node's place. With treaps, given that you must maintain the heap property, it's a tad more complex, but just as with insertions, you can use rotations to make things come out right. To remove a node, use a different logic from earlier with binary search trees:

- If you search for a key to remove in an empty tree, there's nothing to be done.
- If the key to remove is lower than the key at the root, delete the key from the tree at the root's left child.
- Otherwise, if the key to remove is greater than the key at the root, delete the key from the tree at the root's right child.
- Otherwise, if the key has neither a left child nor a right child, just delete it.
- Otherwise, if it has a right child but no left child, set it to the right child.
- Otherwise, if it has a left child but no right child, set it to the left child.
- Finally, if it has both left and right children, apply a rotation to move the key lower in the tree and attempt to delete it again.

The last step is the one that may be surprising, and it's certainly different from binary search trees. With the rotations shown earlier when inserting in a treap, it's possible to rotate a node with one of its children, and the rotated node will be lower in the treap. If you carefully choose what rotation to use, you can keep satisfying the heap property; so if you had a valid treap before the rotation, you'll still have a valid one after it. And finally, as the key to delete moves lower and lower, it can't always keep having two children. At some point, it will have one or none, and then you can finish the job quickly.

Consider a more complex case. Start with the treap shown in Figure 14-25 and remove the 9 node (as in the section "Adding a Key to a Treap" on page 333, the priorities are shown to the right of the nodes).

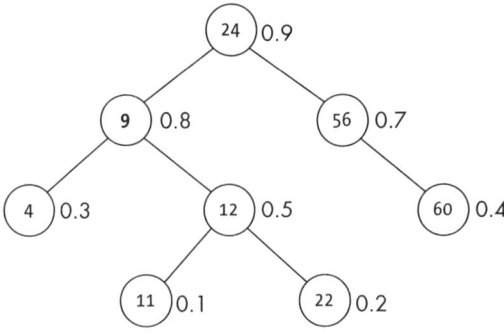

Figure 14-25: An initial treap, with a node to be removed

After finding the node, it happens that it has two children, so you need to do a rotation. The right child of 9 has greater priority, so apply a left rotation, leaving the intermediate situation shown in Figure 14-26 (notice that the heap property is not satisfied, but it will be after you remove the 9 node).

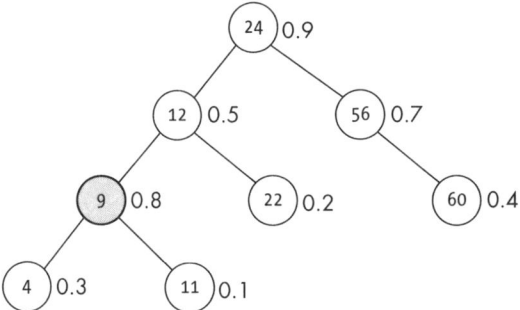

Figure 14-26: A left rotation brings down the node to be removed.

The 9 node again has two children, so do a new rotation. This time, the child with the greater priority is the left one, so you can do a right rotation, resulting in a new situation (Figure 14-27).

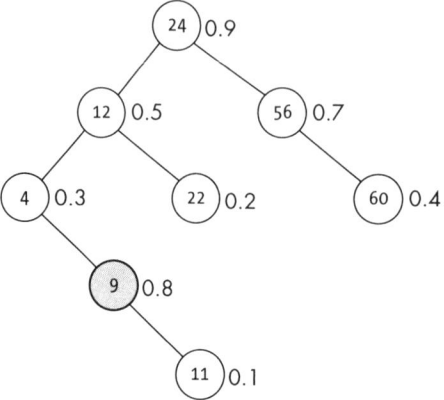

Figure 14-27: A new rotation moves the node
to be deleted further down the treap.

Now you've reached a simple case, since the 9 node has only one child,
which allows you to remove it, resulting in the final treap (Figure 14-28).

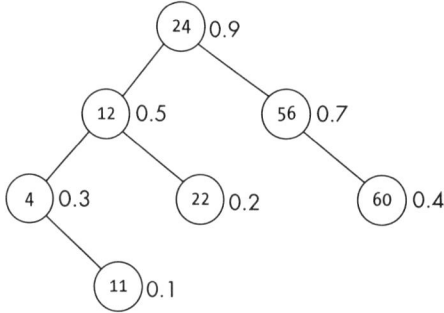

Figure 14-28: The final treap after removing
the node you want to delete

The code for removing a key is as follows (notice that the implementation closely follows the previous bulleted list):

```
const remove = (tree, keyToRemove) => {
  if (isEmpty(tree)) {
    // nothing to do
  } else if (keyToRemove < tree.key) {
    tree.left = remove(tree.left, keyToRemove);
  } else if (keyToRemove > tree.key) {
    tree.right = remove(tree.right, keyToRemove);
  } else if (isEmpty(tree.left) && isEmpty(tree.right)) {
    tree = null;
  } else if (isEmpty(tree.left)) {
    tree = tree.right;
  } else if (isEmpty(tree.right)) {
    tree = tree.left;
❶ } else {
```

```
❷ const [side, other] =
    tree.left.priority < tree.right.priority
      ? ["right", "left"]
      : ["left", "right"];
❸ tree = _rotate(tree, side);
❹ tree[other] = remove(tree[other], keyToRemove);
  }
  return tree;
};
```

The code is the same as for common binary search trees, except that when a key is found that has two children ❶, you decide which rotation you need ❷, apply it ❸, and go down recursively to attempt deleting the key ❹. If you had done a left rotation, the original root (with the key you wanted to delete) would be displaced to the left subtree, so the removal process continues there. With a right rotation, the process would continue at the right subtree.

You've now used heaps to extend binary search trees. Let's look at the results of this change.

Considering the Performance of Treaps

As mentioned earlier, the expected height of treaps is $O(\log n)$, which means that adding, removing, and finding keys all have that same expected order. The randomization implied by the assignment of priorities does not ensure, however, that there won't be bad cases, and in fact, the worst-case scenario is the same as with binary search trees: trees having $O(n)$ depth and an effect on the algorithms' performance.

A main difference with common binary search trees is that in practice, getting a "bad" sequence of data may not be unexpected, always leading to bad trees. However, in a treap, because of the random priorities, the probability of building a "bad" treap is very low, no matter the order of the original data. In fact, to get a badly balanced treap, priorities would have to be correlated with the key values, which is very unlikely to happen with random numbers. Thus, the average performance of algorithms will be independent of the sequence of key insertions (see Table 14-4).

Table 14-4: Performance of Operations for Treaps

Operation	Average performance	Worst case
Create	$O(1)$	$O(1)$
Add	$O(\log n)$	$O(n)$
Remove	$O(\log n)$	$O(n)$
Find	$O(\log n)$	$O(n)$
Traverse	$O(n)$	$O(n)$

The key to treaps is that randomization makes it highly probable that some balance will be achieved, thus providing high performance. (This was the same argument for randomized binary search trees.) In addition, treaps allow you to implement other methods, like partitioning a treap into two or rejoining two treaps to make one. We won't discuss those methods directly here, but see questions 14.15 and 14.16 at the end of the chapter.

Ternary and D-ary Heaps

If binary heaps are a good structure for priority queues, the logical generalization is that as with B-trees, having more children at each level makes for a shorter tree and faster algorithms, such as with *ternary* (also known as *trinary*) heaps, in which each node has three children, or *quaternary* heaps with four children, and in general *d-ary* heaps with *d* children for each node.

Basically, all the differences are in the _bubbleUp() and _sinkDown() methods, which now have to deal with more than two children, as shown:

```
const { newHeap, isEmpty, top } = require("./heap.func.js");

❶ const ORDER = 3; // with ORDER===2, we get classic heaps

const _bubbleUp = (heap, i) => {
  if (i > 0) {
  ❷ const p = Math.floor((i - 1) / ORDER);
    if (heap[i] > heap[p]) {
      [heap[p], heap[i]] = [heap[i], heap[p]];
      _bubbleUp(heap, p);
    }
  }
};

const _sinkDown = (heap, i, h) => {
❸ const first = ORDER * i + 1;
❹ const last = first + ORDER;
  let g = i;
❺ for (let j = first; j < last && j < h; j++) {
    if (heap[j] > heap[g]) {
      g = j;
    }
  }
  if (g !== i) {
    [heap[g], heap[i]] = [heap[i], heap[g]];
    _sinkDown(heap, g, h);
  }
};

const add = (heap, keyToAdd) => { ...exactly as before... }

const remove = (heap) => { ...exactly as before... }
```

Take a look at the changes in the code. We added an ORDER variable (here set to 3) to store the order of the new heap ❶. Calculating the *parent* of a node requires using a corrected formula; instead of dividing by 2 as in binary heaps, divide by the heap order ❷. Then make the same change (substituting the heap order for the 2) to find the children of an element ❸❹. Since you may have any number of children for a node, loop over them to find the greatest ❺.

If you create a new heap (ternary, in this case) and add values 22, 9, 60, 34, 24, 40, 11, 12, 56, 4, and 58, in that order, you'll get the heap in Figure 14-29 (shown as both a tree and an array).

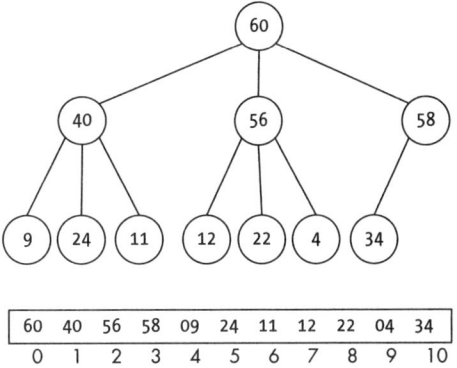

Figure 14-29: In a ternary heap, the implementation is similar to binary heaps.

What about the order of *d*-ary heaps in general? Since the height of the tree is always $O(\log n)$, the order for all operations is the same. However, some operations may have better or worse performance. For instance, bubbling up becomes faster (because the tree is flatter), but sinking down is slower (because you have to find the largest out of more values).

Summary

This chapter introduced a new data structure, a heap in several variants: binary and *d*-ary, as well as min and max heaps. We saw how heaps could be used to implement a new ADT: priority queues. Another usage of heaps was a sorting algorithm with good constant performance. Finally, we applied the heap concept to create a randomized binary search tree: a treap. In the next chapter, we'll continue exploring related concepts and consider some variants of heaps that allow for new operations.

Questions

14.1 Is It a Heap?

Given an array, write a function that returns whether the array is a max heap. You don't need to build a heap; just answer whether it already is one.

14.2 Making Do with Queues

Suppose you could work only with priority queues but not with stacks or queues. How could you emulate those two ADTs with a priority queue? (Hint: since stacks and queues do not use a priority field, you can assign them whichever values you want.)

14.3 Max to Min

Suppose you have a max heap; can you convert it into a min heap in linear $O(n)$ time?

14.4 Max or Min

What changes would you have to apply to your max heaps in order to get min heaps instead?

14.5 Merge Away!

Suppose you are given several ordered lists and want to merge them into a single list. Implement this algorithm using a min heap to decide what node to choose at each step.

14.6 Searching a Heap

Even though it makes little sense (because the heap isn't structured for it), how could you implement a find() function to search for a value in a heap?

14.7 Removing from the Middle of a Heap

In a heap, you always remove the value at the top, and if you want to remove the last value of a heap, it's trivial, but can you write an algorithm that allows you to remove any element whatsoever from a heap?

14.8 Faster Build

Floyd's enhancement builds a heap in $O(n)$ time. Modify newHeap() so if given an array of values, it will use Floyd's method to initialize the heap.

14.9 Another Way of Looping

In the heapsort_original function, you could have easily used forEach() to build up the heap; can you see how?

14.10 Extra Looping?

In the heapsort_enhanced function, what would have happened if you had done a complete loop when building up the heap? More specifically, what if that code had been written as follows:

```
for (let i = v.length - 1; i >= 0; i--) {
  sinkDown(i, v.length);
}
```

14.11 Maximum Equality

What's the order of heapsort if you use it to sort an array filled with the same value?

14.12 Unstable Heap?

Heapsort isn't stable, and trying to sort a short array is enough to verify this. Can you produce such an example and show the lack of stability? Tip: you won't need a very large array.

14.13 Trimmed Selection

You can use a heap to select the k greatest values out of n by removing the top of the heap k times. However, if $k \ll n$, you may speed up things a bit. Show that the k greatest values must be found in level 1 (the root) up to level k (but not beyond), and use this finding to prune the heap before doing the selection.

14.14 Is It a Treap?

Given a binary tree whose nodes have key and priority fields, can you write a function that will check whether that tree is actually a treap?

14.15 Splitting a Treap

Given a treap and a limit value, partition it into two separate treaps: one with all the keys smaller than the limit value and the other with all the keys larger than the limit. Assume that the limit value isn't in the treap.

14.16 Rejoining Two Treaps

Consider the inverse to question 14.15: assume that you have two separate treaps, such that all keys in the first are smaller than all keys in the second. Can you find a way to join those two treaps to form a single one?

14.17 Removing from a Treap

If in the `remove()` method for a treap you changed the line

```
tree[other] = remove(tree[other], keyToRemove);
```

to

```
tree = remove(tree, keyToRemove);
```

would it still work? For reference, this is how the code would look (the changed line is in bold):

```
const remove = (tree, keyToRemove) => {
  if (isEmpty(tree)) {
    // nothing to do
  } else if (keyToRemove < tree.key) {
    tree.left = remove(tree.left, keyToRemove);
  } else if (keyToRemove > tree.key) {
    tree.right = remove(tree.right, keyToRemove);
  } else if (isEmpty(tree.left) && isEmpty(tree.right)) {
    tree = null;
```

```
    } else if (isEmpty(tree.left)) {
      tree = tree.right;
    } else if (isEmpty(tree.right)) {
      tree = tree.left;
    } else {
      const [side, other] =
        tree.left.priority < tree.right.priority
          ? ["right", "left"]
          : ["left", "right"];
      tree = _rotate(tree, side);
      tree = remove(tree, keyToRemove);
    }

    return tree;
};
```

14.18 Trees as Heaps

What would happen if you used binary search trees to represent heaps? What would the performance be of the three basic operations: add(), remove(), and top()? Can you think of ways to make top() faster?

15

EXTENDED HEAPS

In this chapter, we'll explore some new data structures that allow extra operations on heaps. For example, we'll be able to change or alter the value of a key by increasing or decreasing it, or produce a new heap out of two or more other heaps by melding or merging. These new structures are based on several concepts we've looked at in previous chapters, such as common heaps, linked (and doubly linked) lists, forests, and more.

First, we'll consider *skew heaps*, which are heaps implemented as binary trees, and *binomial heaps*, which are a new heap implementation based on a forest that allows you to merge heaps quickly. We'll also study an enhanced version of heaps called *lazy binomial heaps* that provide better amortized performance, and finish by looking at *Fibonacci heaps*, which allow you to change (increase or decrease) a key in (amortized) constant time, and *pairing heaps*, which present an easier alternative with surprisingly similar performance.

Meldable and Addressable Priority Queues

First review the priority queue abstract data type (ADT) you implemented with binary heaps. This ADT requires three operations: add() adds a new value to the queue, remove() gets the highest priority from the queue, and top() shows the priority of the current top element of the queue. With the binary heap implementation that represents the heap as an array, the two first operations run in $O(\log n)$ time, and the third runs in $O(1)$. The extra change() operation alters the priority of an element in the queue, and that also is $O(\log n)$.

If those three (or four) operations were all that were needed, you'd have a good enough implementation. Working with arrays and no pointers is usually quite fast, providing performance that is tough to beat. However, if you need specific enhanced times (for example, adding new values in constant time) or extra operations (like being able to merge two queues into one), you'll need other solutions. Each structure considered in this chapter allows us to meld (or merge) two separate priority queues into one. This is called a *meldable priority queue (MPQ)*.

Several algorithms need to be able to meld priority queues, so allowing that operation enhances their performance. Imagine you're implementing a priority queue for the printers in your system. If one printer goes down, you want to be able to reassign all of its printing jobs to another. A meldable heap would provide the fastest performance for such a reassignment.

You also might want to include a second operation: changing a key value. This change is a specific one where the old key is replaced with a new one that should go higher in the heap—for instance, a lower value in a min heap for several graph algorithms. In fact, when working with min heaps, the operation is usually called decreaseKey(). Since you're working with both max and min heaps, you'll use the name changeKey() but will check to make sure the new value should be nearer the top of the heap.

For this, you need a reference to a value that you insert, and you'll use the add() operation for that purpose. A heap that provides such an operation is called an *addressable heap*. The main focus of this chapter is on meldable heaps, but all of the structures here will be addressable, so we'll consider both new operations together.

NOTE *We considered max heaps with their highest value at the top in Chapter 14. Since extended heaps are often used with the lowest value at the top, we'll use the function goesHigher(a,b) to determine whether a value a should be higher than a value b. For max heaps, you have a > b (so the greater value goes to the top), and for min heaps, you have a < b. Simply changing a single line in the definition of this new function provides max heaps or min heaps as needed. All the examples in this chapter use max heaps.*

Table 15-1 shows the operations in terms of a meldable priority queue ADT, starting with creating the priority queue, modifying the addition operation, and including merging and changing.

Table 15-1: Operations on Meldable Priority Queues

Operation	Signature	Description
Create	→ MPQ	Create a new MPQ.
Empty?	MPQ → boolean	Determine whether the MPQ is empty.
Add	MPQ × key → MPQ × node	Given a new key, add it to an MPQ and provide a reference to the new node.
Top	MPQ → key	Given an MPQ, produce its top value.
Remove	MPQ → MPQ × key	Given an MPQ, extract its top value and update the MPQ correspondingly.
Change	MPQ × node × key → MPQ	Given an MPQ, one of its nodes, and a new key value, change the node's key to the new value and update the MPQ.
Merge	MPQ1 × MPQ2 → MPQ	Given two distinct MPQs, merge them into a single MPQ.

Let's now move on to a different variety of heap.

Skew Heaps

In Chapter 14, we represented binary heaps with an array, which has the advantage of maximum simplicity. However, with that representation, the best performance you can achieve when merging two heaps of sizes m and n is $O(m + n)$ by using Floyd's algorithm. (As an alternative, think of choosing a heap and adding all the other heaps' values to it, which would not be optimal; see question 15.1.) *Skew heaps*, which are based on representing a heap as a self-adjusting binary tree, provide better (although amortized) performance. In later sections, we'll look at even speedier data structures, but those will include an added degree of complexity.

> **NOTE** *Skew heaps are related to a data structure (which we don't discuss in this book) called a leftist heap. Skew heaps, however, require less space, are competitive as to running time, and are easier to implement.*

Skew heaps have one characteristic in common with the binary heaps considered previously: the heap property must be satisfied, so a root must be greater than its children—in other words, a skew heap is a *heap-ordered binary tree*. (Remember, we are working with max heaps; for min heaps, a root would be smaller than its children.) However, one important difference is that skew heaps have no structural constraints, so the tree can have any shape, and its height may be not logarithmic. Figure 15-1 illustrates this fact. It shows a valid heap, but its shape is different from what we saw in Chapter 14.

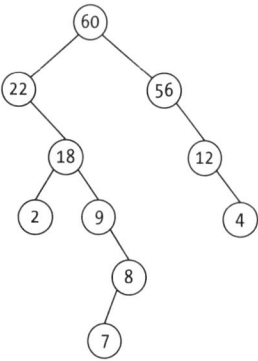

*Figure 15-1: A skew heap isn't
complete as common heaps are.*

Apart from the heap property and no structural constraints, a third particular detail is that all adding and removing operations need to use the *skew-merging* operation to assure good amortized performance.

Representing a Skew Heap

Since a skew heap is a binary tree at its core, you can represent it using the same kind of code for binary search trees and others:

```
const goesHigher = (a, b) => a > b;

const newSkewHeap = () => null;

const newNode = (key, left = null, right = null) => ({ key, left, right });

const isEmpty = (heap) => heap === null;

const top = (heap) => (isEmpty(heap) ? undefined : heap.key);
```

The goesHigher() function determines whether to have a max heap (as shown here) or a min heap (by changing the comparison to a < b). The following three functions are copies of the binary search tree code, and the final top() is simple, given that the heap's top will be the tree's root.

Merging Two Skew Heaps

The logic for merging two skew heaps together is as follows:

1. If you merge two empty heaps, the result is an empty heap.
2. If one heap is empty and the other isn't, the result is the nonempty heap.
3. If both heaps are nonempty, the heap with the greater root becomes the merged heap, but it swaps its children and then merges its left subtree with the heap that has the smaller root.

The first two cases are pretty clear, so let's consider the third, more interesting one. If you start with the two skew heaps shown in Figure 15-2, what's the result of merging them?

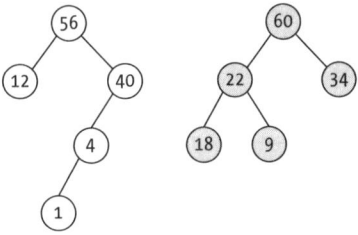

Figure 15-2: Two skew heaps to be merged

The root of the merged heap should be 60, because it's the greater root. (Again, keep in mind that you're working with max heaps.) The left subtree of the corresponding tree would switch sides with the right subtree, and the (now) left subtree would be merged with the heap with root 56. Recursively, you'd compare 56 and 34, so the new root would be 56, and so on. Can you follow all the steps to the result shown in Figure 15-3? For instance, it's easy to see that the old left child of the 60 key now is its right child; the 56 key also swapped its subtrees.

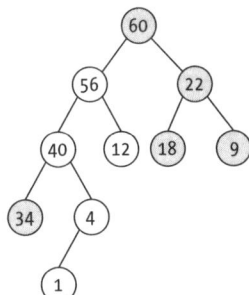

Figure 15-3: The result of merging the two skew heaps

Switching subtrees is what provides good amortized performance; if you merge without swapping, the performance is worse. (See question 15.2 for two specific sequences of operations that produce badly shaped structures.) The logic is as follows:

```
const merge = (heap1, heap2) => {
❶ if (isEmpty(heap2)) {
    return heap1;
❷ } else if (isEmpty(heap1)) {
    return heap2;
  } else if (goesHigher(heap1.key, heap2.key)) {
❸ [heap1.left, heap1.right] = [merge(heap2, heap1.right), heap1.left];
```

```
    return heap1;
  } else {
❹ return merge(heap2, heap1);
  }
};
```

If the second heap is empty, return the first one ❶. (This also covers the case when both heaps are empty. Can you see why?) Otherwise, if the first heap is empty ❷, return the second heap. If no heap is empty and the first heap has the highest key, produce a new heap as described earlier ❸. If the second heap has the highest key, just swap them around for the merge ❹. (See question 15.3 for an alternative.)

Adding a Key to a Skew Heap

How do you add a new key to a heap, if all you know is how to merge? Simply build a new heap with a single value and merge it to the existing heap. You've already looked at how to merge heaps, so let's skip to the actual code:

```
const add = (heap, keyToAdd) => {
❶ const newHeap = newNode(keyToAdd);
❷ return [merge(heap, newHeap), newHeap];
};
```

This just creates a single-node heap ❶ and merges it ❷.

Removing the Top Key from a Skew Heap

Removing the top key from a heap is straightforward. When you remove the root, you're left with two subtrees, so all you need to do is merge them to form the new heap:

```
const remove = (heap) => {
❶ if (isEmpty(heap)) {
    throw new Error("Empty heap; cannot remove");
  } else {
❷ const topKey = top(heap);
❸ return [merge(heap.left, heap.right), topKey];
  }
};
```

If the heap is empty ❶, throw an error. Otherwise, get the top key ❷ and merge the left and right subtrees ❸.

Considering Performance for Skew Heaps

When first thinking about skew heaps and realizing how their structure could become quite awful (in the same way that binary trees can degenerate into linear shapes), it's hard to believe that their performance can be good. However, the amortized time of additions, removals, and merging can be proven to be $O(\log n)$. Table 15-2 shows the results; asterisks denote amortized values.

Table 15-2: Performance of Operations for Skew Heaps

Operation	Performance
Create	$O(1)$
Empty?	$O(1)$
Add	$O(\log n)$*
Top	$O(1)$
Remove	$O(\log n)$*
Change	$O(\log n)$*
Merge	$O(\log n)$*

The changeKey() method was added, which you can do by first removing the old key and then inserting the new one; both are $O(\log n)$ methods. (See question 15.4.)

Binomial Heaps

Skew heaps have amortized logarithmic performance all around, but if you go beyond single trees, you can do better. *Binomial heaps* are based on a forest of heaps, and not only do they perform well, they also are the basis for some enhanced variations with even better performance.

Binomial Trees

Start with a definition: a *binomial tree of order k*, or *BT(k)*, is a *k*-ary tree that is either empty or consists of a root whose children are binomial trees of order 0, 1, 2, . . . up to $(k-1)$. Figure 15-4 shows the first five binomial trees, from BT(0) to BT(4), each consisting of a root and one copy of each previous binomial tree.

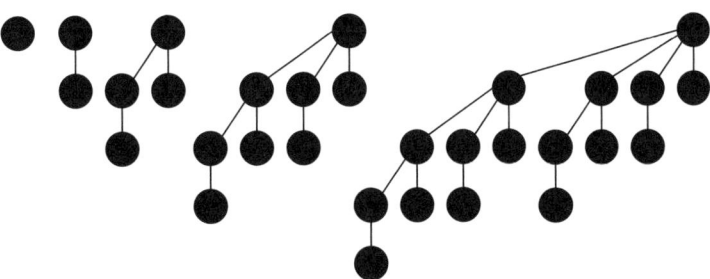

Figure 15-4: The first five binomial trees

When looking at the trees, you may notice some mathematical properties that can be properly proven, but the most important ones are that BT(k) has exactly 2^k nodes and its height is k. They share those two properties with full binary trees. (Actually, full binary trees of height k have $2^{k+1} - 1$ nodes, but it's close enough. Let it go!)

NOTE *The order of a binomial tree equals the degree of its root, so in this code, we'll use degree instead of order, because that will help when we consider Fibonacci heaps, which don't use the order concept but work with the root's degree.*

If you count how many nodes each level has, you'll notice another property that explains the name of this heap: for the trees in Figure 15-4, the results are 1; 1 and 1; 1, 2, and 1; 1, 3, 3, and 1; and 1, 4, 6, 4, and 1; and so on. These are *binomial coefficients*, as in Pascal's triangle where each value is the sum of the two above it, as shown in Figure 15-5. Another property of the triangle is that the sum of the numbers in a row is a power of 2, which also matches the size property for a BT(k).

```
            1
        1       1
      1     2     1
    1     3     3     1
  1     4     6     4     1
 ...   ...   ...   ...   ...   ...
```

Figure 15-5: Pascal's triangle provides the number of values at each level of binomial heaps.

You can look at these trees another way. Notice that each tree is actually built out of two trees of the previous order; modifying Figure 15-4 a bit makes this clearer, as shown in Figure 15-6.

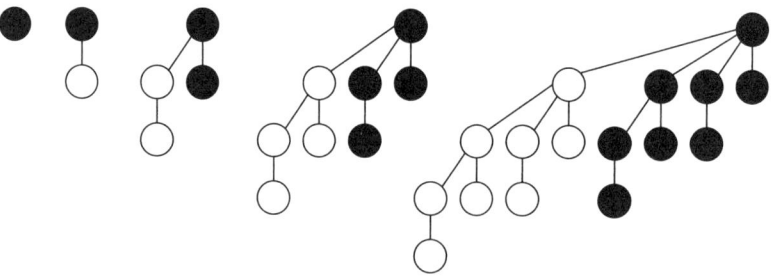

Figure 15-6: Each binomial tree actually consists of two smaller previous ones.

Figure 15-6 shows the first four binomial trees. Each one (except the first) is built out of two copies of the previous tree, joined at a new root. If these trees also satisfy the heap order property (all nodes are greater than or equal to their children), they're called *heap-ordered binomial trees*. With such trees, you can easily represent a heap whose size is a power of 2, but how do you represent other sizes of heaps?

Defining Binomial Heaps

A *binomial heap* is a forest of heap-ordered binomial trees. (Some textbooks specify that those trees must be ordered by ascending size, but we won't apply that extra rule.) How will this allow us to represent any size of heaps? Binary numbers provide an easy answer.

The number 22 in binary is 10110, which means that $22 = 16 + 4 + 2$; thus, you can represent any integer number as a sum of powers of 2. In the same way, you represent a heap of any size with a set of binomial trees whose sizes add up to whatever size you need. (You won't have two binomial trees of the same order. Can you see why?) Keep in mind another property: the representation of a number n in the binary system requires $\log n$ bits, rounded up (we'll use this later).

For simplicity, we represent the forest with an array of trees (you could also work with a list of linked trees, but that added complexity isn't needed). Since binomial trees are multiway trees, you also need a way to represent them, so we'll use a binary tree representation. Nodes have a down pointer (instead of left) to their first child and a right pointer to their next sibling. Nodes also include an up link to their parents, in case you want to implement the changeKey() operation. Finally, each node also has its degree.

This representation is necessary because these algorithms require two operations: splitting a tree into several smaller ones and fusing two trees together, and those operations are fast and easy to implement with the links. For example, fusing two BT(2)s to build a BT(3) requires changing only two pointers (we'll look at that in the next section). Start with a few basic functions and then add the rest:

```
❶ const newBinomialHeap = () => [];

❷ const newNode = (key) => ({
     key,
     right: null,
     down: null,
     up: null,
     degree: 0
   });

❸ const isEmpty = (heap) => heap.length === 0;
```

A binomial heap, then, is an array ❶, and each of its elements is a binomial tree. The nodes have the five attributes described earlier ❷. Checking whether the heap is empty just requires testing the length of the array of trees ❸.

NOTE *You might have wondered about using functions for binomial trees. However, since you aren't actually going to use the trees for searching, you'll use plain records in the code, with key and degree attributes, as well as with up, down, and right pointers. Remember that the order of a binomial tree equals the degree of its root; that's why you have degree instead of order. Having the degree for all roots will help with Fibonacci heaps.*

First, write the top() procedure, using an auxiliary _findTop() function to go through all the roots:

```
❶ const _findTop = (trees) => {
    let top;
    trees.forEach((v, i) => {
      if (top === undefined || goesHigher(v.key, trees[top].key)) {
        top = i;
      }
    });
    return top;
  };

❷ const top = (heap) => (isEmpty(heap) ? undefined : heap[_findTop(heap)].key);
```

The _findTop() function ❶ finds which tree has the greatest root; it just goes through the trees array looking for the greatest key. With this method, top() is simply a matter of checking whether the heap is empty. If so, return undefined; otherwise, use _findTop() to get the top value of the heap ❷.

Adding a Value to a Binomial Heap

When adding a value to a heap, start by creating a binomial tree of order 0 with just the key and add it to the heap; however, this may cause a problem if another binomial tree of the same order already exists. (Remember, trees of repeated orders aren't allowed.) You can solve it with merging.

Assume you have two binomial trees of the same order, as shown in Figure 15-7, and want to merge them into one. In general, one of the trees should become a subtree of the other.

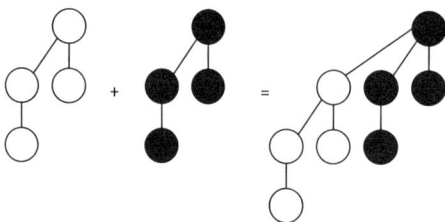

Figure 15-7: Merging two binomial heaps where one becomes a subtree of the other

This example, however, needs further consideration because you're not dealing with just any trees; you're dealing with trees that satisfy the heap condition. So, what do you do in this case? Assume the two trees are as shown in Figure 15-8.

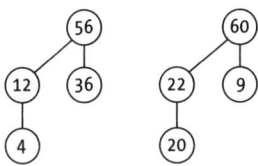

Figure 15-8: Two binomial heaps to be merged

The idea is simple. The greater root becomes the new root, and the smaller root becomes its child, as shown in Figure 15-9.

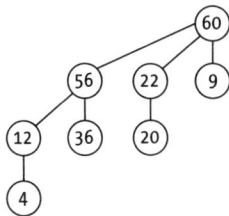

Figure 15-9: A way to merge the two trees from Figure 15-8

But Figure 15-9 doesn't show how to actually do it. With the binary representation discussed earlier, both original trees would look like the ones shown in Figure 15-10.

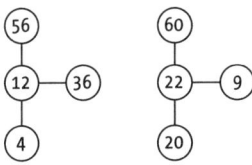

Figure 15-10: The actual binary tree representation for the two binomial heaps

Correspondingly, the result also follows the same scheme of "down to child, right to sibling," as shown in Figure 15-11.

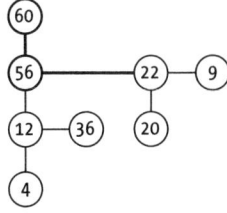

Figure 15-11: The merged binomial trees in binary tree representation

In this way, you can always merge two BT(k)s into a single BT($k + 1$), and thanks to the binary representation, that procedure requires you to change only two links. The code for this is as follows, where low is the tree with the smaller root and high is the one with the greater root:

```
const _mergeA2B = (low, high) => {
❶ low.right = high.down;
❷ low.up = high;
❸ high.down = low;
❹ high.degree++;
  return high;
};
```

It's pretty straightforward. The lower tree will have the higher tree's children as siblings ❶ and that node as the parent ❷❸, making the new root's degree go up by one ❹. (You'll use this operation repeatedly in this chapter.)

Now you have the means to merge two trees, but there could be an extra complication. In the example, what would happen if the original heap already had another BT(3)? In that case, you'd keep merging the original BT(3) with the new one to get a BT(4). And, of course, that might lead to a repeated BT(4), and so on. (You'll look at the complete algorithm in the next section.)

Assume for a moment that you already have a method to add a new tree into a heap by merging; then the add() method is short:

```
const add = (heap, keyToAdd) => {
❶ const newHeap = newNode(keyToAdd);
❷ return [merge(heap, [newHeap]), newHeap];
};
```

You just have to create a new basic BT(0) with the key to be added ❶, which is an array with a single object in it, and merge it with the heap ❷. It's basically the same technique used with skew heaps. The difference is how you go about merging the heaps.

Merging Two Binomial Heaps

Consider the problem of merging two heaps in terms of summing binary numbers. If you're familiar with the 2048 game where you try to merge boxes together to reach 2048, you'll understand the examples fairly quickly.

Start with an easy case. Say you have a binomial heap with 22 elements ($22 = 2 + 4 + 16$) and want to merge it with another binomial heap of a single element, as when adding a new value, as shown in Figure 15-12.

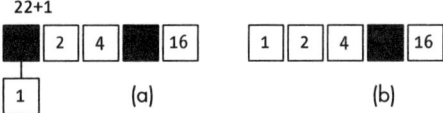

Figure 15-12: Merging two heaps has a
lot to do with binary numbers.

Start at (a)—the heap consists of three binomial trees of sizes 2, 4, and 16 (corresponding to the binary representation of 22)—and represent it with an array, using the trees' order as an index. The tree with 2 elements is at position 1; the one with 4 is at position 2, and the one with 16 is at position 4, while other positions remain empty (black). Match the tree to be added with the corresponding place in the current heap, and it's empty, so just move it into place (this is the equivalent of doing $1 + 0 = 1$), and you get (b). No more trees remain to be merged, so you're done.

Now consider the harder case of merging two separate heaps with the resulting heap of size 23 ($23 = 1 + 2 + 4 + 16$) and another heap of size 5 ($5 = 1 + 4$), as shown in Figure 15-13.

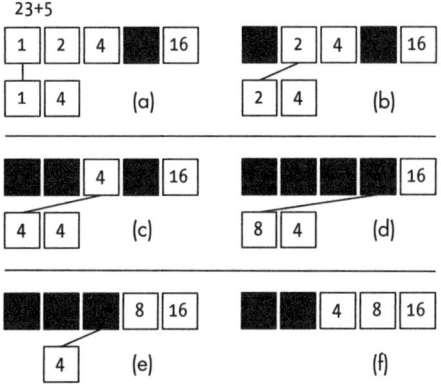

Figure 15-13: Working through the merging,
step by step

Start at (a), matching 1 and 1: remove the 1 from the top heap and merge it with the 1 at the bottom heap, which becomes a 2. Then you arrive at (b) and match 2 and 2. Again, remove the 2 from the top heap and merge it with the 2 at the bottom, which becomes a 4. In (c), the situation repeats: you again remove from the top and merge at the bottom. (The bottom values are no longer in ascending order—you have an 8 first and a 4 second—but that won't impact the final result.) At (d), you have an easy case because there is no 8 at the top, so just put the 8 there and remove it from the bottom. At (e), you now have a 4, and it has no match, so place it at the top and remove it from the bottom, getting to (f), where you are finished, since there are no more values to merge.

You have now solved the problem: merging goes on until no trees of duplicate order remain. Implementing the method isn't complex:

```
const merge = (heap1, heap2) => {
  const merged = [];
❶ heap1.forEach((v) => {
    merged[v.degree] = v;
  });

  let j = 0;
❷ while (j < heap2.length) {
    const i = heap2[j].degree;

  ❸ if (!(i in merged) || merged[i] === null) {
      merged[i] = heap2[j];
      j++;
    } else {
      if (goesHigher(heap2[j].key, merged[i].key)) {
      ❹ heap2[j] = _mergeA2B(merged[i], heap2[j]);
      } else {
      ❺ heap2[j] = _mergeA2B(heap2[j], merged[i]);
      }
    ❻ merged[i] = null;
    }
  }

❼ return merged.filter(Boolean);
};
```

First, you place all the binomial trees of the heap in the merged array ❶, setting each tree in its place according to its order. Then, start processing all the trees in the second list ❷, "adding" them as described. If you don't have a match in merged, just put the new tree in there ❸ and advance to the following tree to merge. Otherwise, when there's a match, merge both trees. You have one case if the second tree has the greater root ❹ and a different case if the merged tree has it ❺. In both cases, place the resulting tree in the second array, emptying the place in merged ❻. At the end, after going through all the trees in the second array ❼, filter out the merged array to remove empty trees.

Removing a Value from a Binomial Heap

The method to remove the top value from a heap is based on breaking a binomial tree apart by removing its root and then melding the separated trees with the original heap. Assume the heap consists of two binomial trees (sizes 2 and 8) and remove the top value (60), as shown in Figure 15-14.

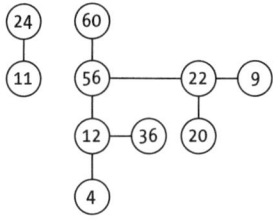

Figure 15-14: A heap with
two binomial trees

After removing the top value, separate its subtrees (of sizes 4, 2, and 1),
leaving four binomial trees, as shown in Figure 15-15.

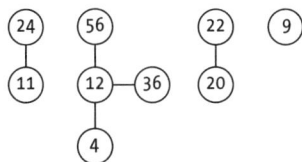

Figure 15-15: After you remove the 60 value,
the tree becomes several heaps.

Next, use the same method as before to merge these four trees with
an initially empty set of trees. The first step melds together the two 2-sized
trees, as shown in Figure 15-16.

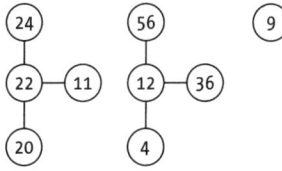

Figure 15-16: The first step of
the merge

Then, as you have two 4-sized (order 2) trees, do another merge, as
shown in Figure 15-17.

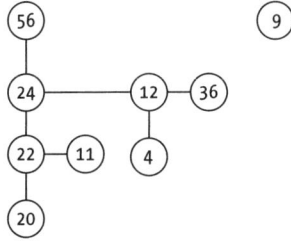

Figure 15-17: The second merge
finishes the job.

And you're done! The new binomial heap has one fewer element, and you managed that easily by splitting the tree into subtrees and using the merging code you had already written. The remove() method is as follows:

```
const remove = (heap) => {
❶ if (isEmpty(heap)) {
     throw new Error("Empty heap");
   }

❷ const top = _findTop(heap);
❸ const heapTop = heap[top].key;

❹ const newTrees = [];
❺ let bt = heap[top].down;
   while (bt) {
   ❻ newTrees.push(bt);
   ❼ const nextBt = bt.right;
     bt.right = null;
     bt.up = null;
     bt = nextBt;
   }

❽ heap.splice(top, 1);
❾ return [merge(heap, newTrees), heapTop];
};
```

First, check whether the heap is empty ❶. Then find which tree has the greatest root ❷ and get its value ❸ to return it at the end. Create a newTrees array ❹ and set up a loop ❺ to push the tree ❻ and its siblings ❼ there. After splitting the original tree, remove it from the heap ❽ and use the merge() function ❾ to merge the new trees into the rest of the original heap.

Changing a Value in a Binomial Heap

Some graph-oriented algorithms often need to change a value already in a heap. In such cases, the heap typically includes full records instead of just a priority and keeps an external reference to the heap node that includes the record. (We'll look at such graph algorithms in Chapter 17.) We won't go into all of that here, but the logic remains the same.

NOTE *The most common case is working with min heaps and decreasing a priority with a decreaseKey() method. If you want to work with a min heap, all you need to do is change the direction of the comparison in the goesHigher() function described earlier in this chapter.*

We've already discussed how to change a value in binary heaps: after effecting the change, it's simply a matter of bubbling it up or sinking it down, depending on the relationship with other values. In this chapter, we'll consider only the logic to bubble up a key, moving toward the top of the heap, since that's the case actually needed in practice. In Figure 15-18, say you want to change the 4 key to 50.

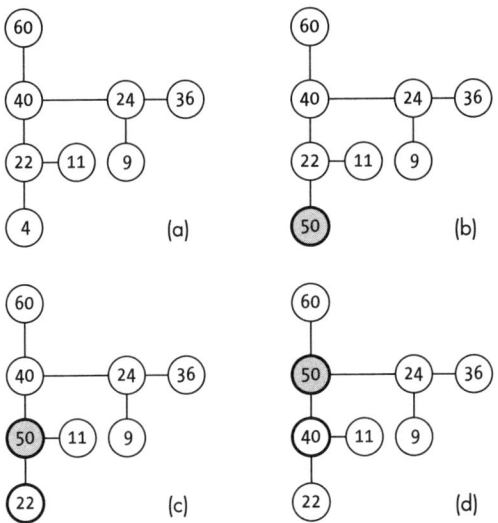

Figure 15-18: A key-changing example, step by step

Change the key and compare it to the parent, unless you already are at the root, which means you're finished. If you're not at the root, if the changed key is smaller, you're done; otherwise, you have to swap nodes and keep bubbling up. Starting at (a), change the 4 key to a 50 in (b). Now compare 50 with 22 and exchange the nodes (actually, it's the pointers that are exchanged) getting to (c). A new comparison, 50 and 40, requires yet another exchange as in (d), but now 50 is smaller than its parent (60), so you're done.

NOTE   *The need for a value to bubble up is the only reason we include an up pointer in the tree's nodes. If you don't plan to provide the changeKey() method, you can remove all instances of up in the code.*

Here's the implementation for this method:

```
const changeKey = (heap, node, newKey) => {
❶ if (isEmpty(heap)) {
       throw new Error("Heap is empty!");
❷ } else if (!goesHigher(newKey, node.key)) {
       throw new Error("New value should go higher than old value");
   } else {
     ❸ node.key = newKey;
       _bubbleUp(heap, node);
       return heap;
   }
};
```

First check whether the operation is possible: the heap shouldn't be empty ❶, and the new key should go higher in the heap ❷. If everything's okay, change the node's key to the new value and call _bubbleUp() to make it climb up the heap ❸ if that's needed.

The code for bubbling up is as follows, and it has the longest line of code in this entire book:

```
const _bubbleUp = (heap, node) => {
❶ if (node.up && goesHigher(node.key, node.up.key)) {
  ❷ const parent = node.up;
  ❸ [
      node.up,
      node.down,
      node.right,
      node.degree,
      parent.up,
      parent.down,
      parent.right,
      parent.degree
    ] = [
      parent.up,
      parent,
      parent.right,
      parent.degree,
      node,
      node.down,
      node.right,
      node.degree
    ];

  ❹ if (node.up) {
      _bubbleUp(heap, node);
  ❺ } else {
      const i = heap.findIndex((v) => v === parent);
      heap[i] = node;
    }
  }
}
};
```

First see whether any bubbling up is needed. If the node doesn't have a parent, or if it has one but the parent's key is higher than the node's key, nothing's required ❶. If you need to swap the node up, get a pointer to its parent ❷, and make all the pointer (and degree) changes we saw previously (it's quite a long line ❸ but straightforward in concept). Finally, if the node again has a parent, use recursion to check whether it still has to bubble up more ❹. If the node doesn't have a parent ❺ (meaning it got to the top of its heap), you need to fix the reference in the array of heaps, so instead of pointing to the old node's parent, now it points to the new top. (Couldn't we just have exchanged keys between the node and its parent, as with binary heaps? The answer is important; see question 15.7 for more on this.)

Considering Performance for Binomial Heaps

Table 15-3 summarizes the binomial heaps' performance; results with an asterisk are amortized.

Table 15-3: Performance of Operations for Binomial Heaps

Operation	Performance
Create	$O(1)$
Empty?	$O(1)$
Add	$O(\log n)$*
Top	$O(\log n)$ as seen; $O(1)$ with a fix
Remove	$O(\log n)$
Change	$O(\log n)$
Merge	$O(\log n)$

Given that a binomial heap may consist of up to log n heaps, getting the top is $O(\log n)$. If you only did additions but no removals, the amortized order of adding a new value would be $O(1)$—see question 15.5 for more on this—but we can prove that sequences of operations worsen this result to $O(\log n)$. Getting the top value, as implemented previously, means looking through log n heaps, but that can be enhanced to $O(1)$; see question 15.6. As for the other results, removing the top value means separating a heap into, at most, log n subtrees, doing an $O(\log n)$ procedure, followed by merging, which is another $O(\log n)$ procedure, so that's the total order for the operation.

Lazy Binomial Heaps

Binomial heaps have a potential performance problem when adding values, when you go from $O(1)$ to $O(\log n)$, which means you can think in amortized terms to find a solution that enhances this procedure at the possible risk of (not too often) costlier fixes. *Lazy binomial heaps* do exactly that.

With lazy binomial heaps, when you do additions, you don't care about merging. You just let the heap have more and more trees, so add() is a trivial operation, running in $O(1)$. Take care, however, to keep track of the greatest value, so top() is also $O(1)$. You can fix the structure when you try to remove() a value, except then you process the heap to bring it back into binomial heap shape.

Defining Lazy Binomial Heaps

Lazy binomial heaps are binomial heaps after all, although you add an extra top property to keep track of the top value in the heap. The class definition then is quite short, as most of the methods are shared with binomial heaps:

```
const goesHigher = (a, b) => a > b;

❶ const newLazyBinomialHeap = () => ({
    top: undefined,
    trees: []
});
```

```
  const newNode = (key) => ({
    key,
    right: null,
    down: null,
    up: null,
    degree: 0
  });

❷ const isEmpty = (heap) => heap.trees.length === 0;

❸ const top = (heap) => (isEmpty(heap) ? undefined : heap.top);
```

So far, there are only two differences. A heap is now a record with two
fields: top ❶, which will have the top value of the heap, and the trees array,
which are the individual heaps. An empty heap has no trees ❷, so use that
for detection. You need to update top when adding to or removing from the
heap. The top() method ❸ is quite short: if the heap is empty, return undefined; otherwise, return the value of heap.top. With this implementation,
you don't need to go through the whole trees array to find the top, which
provides enhanced performance, although later heap.top will need some
extra maintenance work.

Adding a Value to a Lazy Binomial Heap

The first important difference with lazy binomial heaps is that you won't do
any merging when adding a new key. How is this possible? First, if you want
to know the top of the heap, you can do that without any structure. The
heap.top attribute described earlier can be updated easily. As long as you
keep adding, the heap just grows a tree at a time, and you'll always know the
top of the heap. For example, assume that a binomial heap had the structure shown in Figure 15-19 at a certain moment; the triangle points to the
maximum of the heap at that time.

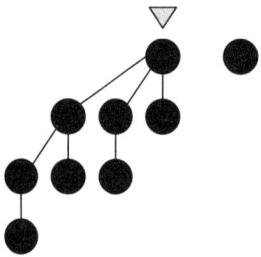

Figure 15-19: Tracking the top value
just needs a simple property.

If you add three new values, the process is quite fast, as the only
thing you do is add new trees. The diagram in Figure 15-20 with several
binomial trees of the same order, which wouldn't be allowed in binomial
heaps, shows this. Again, after each addition, update the maximum. In this

example, one of the newly added keys was greater than the previous maximum, so you now have a different top of the heap.

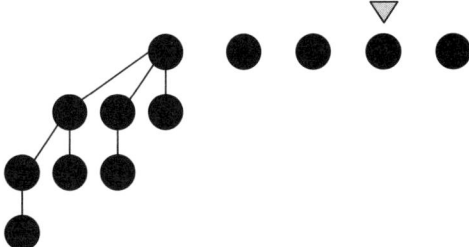

Figure 15-20: After you add values, the property needs adjustment.

But removing the heap's top is a different scenario, because then finding the new _heapTop is too slow: $O(n)$. You can restructure the heap by merging trees together only when removing a key. Some math (which we'll skip) shows that the amortized performance remains quite good.

The code for adding a value is similar to what you saw before, but instead of merging a new tree, you just add it with no further process:

```
const add = (heap, keyToAdd) => {
❶ const newHeap = newNode(keyToAdd);
❷ heap.trees.push(newHeap);

❸ if (heap.top === undefined || goesHigher(keyToAdd, heap.top)) {
    heap.top = keyToAdd;
  }

  return [heap, newHeap];
};
```

First, create a new binomial tree with the new key ❶ and push it to the end of the current trees array ❷. The only extra step is possibly updating the heap's top. If the array of trees is empty or if the current top is not greater than the newly added value ❸, reset heap.top.

Removing a Value from a Lazy Binomial Heap

As mentioned previously, the idea with lazy binomial heaps is to delay merging trees as long as possible (thus, the term *lazy*). When you remove a key, first merge together all the current binomial trees to get a binomial heap and then proceed with the removal.

You do this because although the number of trees in the heap grows slowly when adding values and eventually becomes high, it drops sharply after merging, and the balance between many fast operations and an eventual slow one ends with a nice amortized cost.

The code is as follows; notice it has a few changes and additions in comparison to the original binomial heaps:

```
const remove = (heap) => {
❶ if (isEmpty(heap)) {
     throw new Error("Empty heap");
   }

❷ const heapTop = heap.top;

❸ const top = _findTop(heap.trees);
   let bt = heap.trees[top].down;
❹ while (bt) {
     heap.trees.push(bt);
     const nextBt = bt.right;
     bt.right = null;
     bt.up = null;
     bt = nextBt;
   }

❺ heap.trees.splice(top, 1);
❻ const newHeap = merge(newLazyBinomialHeap(), {
     top: undefined,
     trees: heap.trees
   });

❼ newHeap.top =
     newHeap.trees.length === 0
       ? undefined
       : newHeap.trees[_findTop(newHeap.trees)].key;

❽ return [newHeap, heapTop];
};
```

First check whether the heap is empty ❶ and save the current top of the heap ❷ to return its value later ❽. Then find which tree had the top ❸ and do a loop to split its subtrees ❹, which you then add to the list of trees. Then remove the split tree ❺, merge all trees together ❻, and update heap.top ❼ to find the new current top. It's not very different from binomial heaps. The way you merge all trees together is pretty neat: by merging an empty heap with the list of trees, you trigger all the necessary merges that will reduce the number of trees. Can you see how it works?

Changing a Value in a Lazy Binomial Heap

There is one final method: how to change any key. Here's the code:

```
const changeKey = (heap, node, newKey) => {
  if (isEmpty(heap)) {
    throw new Error("Heap is empty!");
  } else if (!goesHigher(newKey, node.key)) {
    throw new Error("New value should go higher than old value");
```

```
    } else {
      node.key = newKey;
      _bubbleUp(heap, node);

      heap.top =
        heap.trees.length === 0
          ? undefined
          : heap.trees[_findTop(heap.trees)].key;

      return heap;
    }
};
```

This code is the same as for binomial heaps, with the addition of the single line to update `heap.top`; you did a similar calculation in `remove()`. The code for `_bubbleUp()` is unchanged, so it's not repeated here.

Considering Performance for Lazy Binomial Heaps

The performance of lazy binomial heaps is similar to that of binomial heaps, but postponing merges has a positive effect in amortized terms. In particular, adding a value is logically faster, since you don't do practically anything, as Table 15-4 shows. Remember, asterisks denote amortized results.

Table 15-4: Performance of Operations for Lazy Binomial Heaps

Operation	Performance
Create	$O(1)$
Empty?	$O(1)$
Add	$O(1)$
Top	$O(1)$
Remove	$O(\log n)$*
Change	$O(\log n)$*
Merge	$O(\log n)$*

You now have very good performance (in particular, adding a new value is faster), but you'll want a better result when changing a value. Let's look at another variant of binomial heaps that allows this enhancement.

Fibonacci Heaps

Some graph algorithms use min heaps and frequently call the `decreaseKey()` operation, which we renamed to `changeKey()` to allow for max heaps and min heaps. In that situation, being able to decrease keys in a quicker fashion than the $O(\log n)$ performance for lazy binomial heaps becomes important.

Enter *Fibonacci heaps*, which are quite similar to lazy binomial heaps but allow a faster algorithm to change a key.

NOTE *Michael Fredman and Robert Tarjan described Fibonacci heaps in their paper "Fibonacci Heaps and Their Uses in Improved Network Optimization Algorithms" (network refers to graphs in this title), but Tarjan later suggested an alternative simpler structure called pairing heaps, which we'll study later.*

What's the idea behind Fibonacci heaps? The add() and remove() methods work the same way as with lazy binomial heaps, but the difference appears when changing a key. If a key is changed and has to bubble up (in a max heap, this would happen if the new value was greater than before; in a min heap, if smaller), you obviously might need to bubble it up even to the root of its tree—that's $O(\log n)$.

Instead of doing any bubbling, just separate that node with its subtree and add it as a new tree to the heap—that's $O(1)$. However, since this process alters the expected shape of the binomial trees and doing it too often could lead to badly structured heaps, there's a compromise. You won't allow non-root nodes to lose more than one child in the described fashion. Should a node lose a second node (and you'll know it because every time a node loses a child, the node will be marked), you'll also separate it, which itself may lead to further separations. And, as with lazy binomial heaps, you'll patch things up when a removal is done; you will see all of this later, but first consider how to represent the new heaps.

Representing a Fibonacci Heap

The structure used earlier—an array of trees where each tree is represented with up, right, and down links plus a degree field with the number of children—works, but that's not efficient enough for the operation needed here. When changing a key and it bubbles up, the idea is to remove the corresponding node and its subtree, but can you quickly unlink it from its siblings? If you keep the siblings in a singly linked list, that procedure requires traversing a list with a size of up to $O(\log n)$, which will spoil the $O(1)$ goal. So, in the same way as shown in Chapter 10, use a doubly linked list here. But there's more! When merging trees, you want to merge two lists of siblings, so make the lists circular, and that will complete the solution.

Figure 15-21 shows a small BT(3) and how it would look with all the added links in a Fibonacci heap.

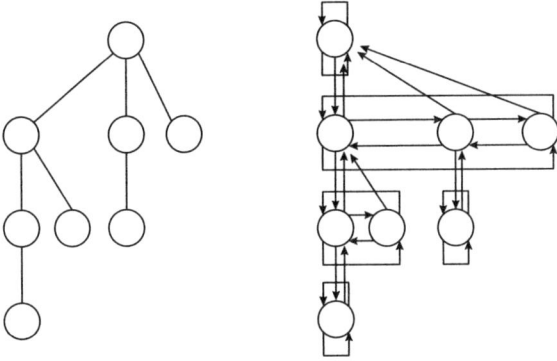

Figure 15-21: A binomial tree represented in Fibonacci heap style

Showing all the up and down and left and right links makes for a cluttered diagram, so from now on, we'll remove any unnecessary links. For instance, the up links to parents will be removed, as you can deduce those from the diagrams. We'll also omit arrowheads and circular links for single node lists for clarity, but we'll point out when we make those alterations.

To sum up the changes, we add a `left` link to the nodes (so we can build the circular doubly linked list) and a `marked` boolean field, which we'll use to mark a node that has lost a child:

```
const goesHigher = (a, b) => a > b;

const newFibonacciHeap = () => ({
  top: undefined,
  trees: []
});

const newNode = (key) => ({
  key,
  degree: 0,
  marked: false,
  left: null,
  right: null,
  down: null,
  up: null
});

const isEmpty = (heap) => heap.trees.length === 0;

const top = (heap) => (isEmpty(heap) ? undefined : heap.top);
```

The single change is the addition of the **marked** and **left** fields; the rest is the same as it was for lazy binomial heaps.

Merging Two Fibonacci Trees

When we first looked at how to merge binomial trees, the procedure was relatively simple. Now that siblings are in a circular doubly linked list, however, you'll need to make some changes. Say you want to merge the trees shown in Figure 15-22 (remember, details for subtrees that won't be affected by the changes, as well as arrowheads and up links, are hidden).

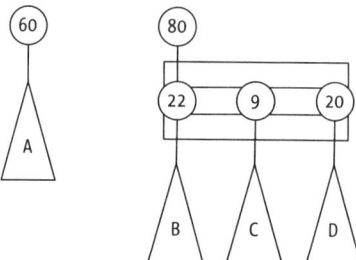

Figure 15-22: Two Fibonacci trees to be merged

After merging the trees together, you get the result shown in Figure 15-23. Pay particular attention to the changed links; arrowheads are included only for those.

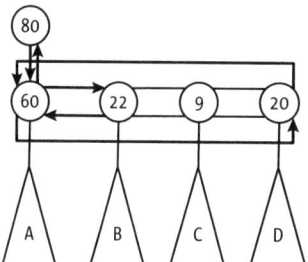

Figure 15-23: The result of merging the trees; only a few pointers needed to change.

Here's the new merging code:

```
mergeA2B(low, high) {
❶ if (high.down) {
  ❷ low.right = high.down;
    low.left = high.down.left;
    high.down.left.right = low;
    high.down.left = low;
  }

❸ high.down = low;
❹ low.up = high;
❺ high.degree++;

❻ return high;
}
```

If the tree with the higher key has no children, the logic is simple, because you just need to set the lower tree as its child ❸❹. However, if it does have children ❶, you need to add the lower tree's root as a new sibling to the higher tree's root's children ❷. (Notice the four link changes.) After that, make the higher root point to the lower root ❸ and vice versa ❹. Then increment the higher root's degree by 1 ❺ since it gained a new child and return the merged tree ❻.

Adding a Value to a Fibonacci Heap

Adding a new value to the Fibonacci heap isn't very different from lazy binomial heaps. The only change is you need to set up the circular list of siblings (initially with just the node itself) for future merging operations. The following code highlights the needed changes:

```
const add = (heap, keyToAdd) => {
  const newHeap = newNode(keyToAdd);

  newHeap.left = newHeap;
  newHeap.right = newHeap;

  heap.trees.push(newHeap);

  if (heap.top === undefined || goesHigher(keyToAdd, heap.top)) {
    heap.top = keyToAdd;
  }

  return [heap, newHeap];
};
```

Initialize the left and right pointers of the new node appropriately, so they form a single-node circular list. (Yes, you could write the two lines as a single assignment; see question 15.9.) The new tree will have marked set to false, because the node hasn't yet lost any children. To check this, see the newNode() code, in the section "Representing a Fibonacci Heap" on page 368.

Removing a Value from a Fibonacci Heap

The logic for removing a value is essentially the same as for other binomial heaps, with minor changes. Removing the top of the heap also works the same as with lazy binomial trees, except beware of infinite loops when traversing the circular list of siblings, for example. The new lines are highlighted:

```
const remove = (heap) => {
  if (isEmpty(heap)) {
    throw new Error("Empty heap");
  }

  const heapTop = heap.top;

  const top = _findTop(heap.trees);
```

```
  let bt = heap.trees[top].down;

  if (bt && bt.left) {
  ❶ bt.left.right = null;
  }

  while (bt) {
    heap.trees.push(bt);
    const nextBt = bt.right;
  ❷ bt.right = bt;
    bt.left = bt;
    bt.up = null;
    bt = nextBt;
  }

  heap.trees.splice(top, 1);
  const newHeap = merge(newFibonacciHeap(), {
    top: undefined,
    trees: heap.trees
  });

  newHeap.top =
    newHeap.trees.length === 0
      ? undefined
      : newHeap.trees[_findTop(newHeap.trees)].key;

  return [newHeap, heapTop];
};
```

To avoid infinite loops, set the rightmost link of the list to null ❶ to
ensure that the following loop will stop. (Here, bt points to an element in the
circular list. Traverse this list to the right, so bt.left points to what should be
the last element to visit. If you clear the right link of bt.left, this ensures that
the loop will stop.) The other difference is when you extract a sibling, the root
must be a circular link by itself ❷, so you have to fix its left and right links.

Changing a Value in a Fibonacci Heap

How a key change is handled is what sets Fibonacci heaps apart from the
other types of heaps. Instead of bubbling up, you directly separate that key
from the heap. The code is similar to what you saw before, but it has one
significant change (in bold):

```
const changeKey = (heap, node, newKey) => {
  if (isEmpty(heap)) {
    throw new Error("Heap is empty!");
  } else if (!goesHigher(newKey, node.key)) {
    throw new Error("New value should go higher than old value");
  } else {
    node.key = newKey;
    _separate(heap, node);
  }
};
```

When you actually change the node's key, you *separate* it from the heap instead of bubbling up. To illustrate, consider the heap shown in Figure 15-24 from earlier in the chapter.

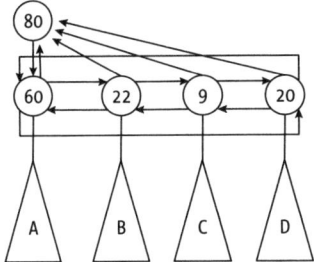

Figure 15-24: The previous
Fibonacci heap

Suppose the 9 key is changing to 99, as shown in Figure 15-25. Since it would be bubbling up, just remove it from the heap and mark its parent (80).

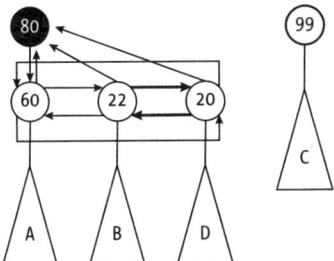

Figure 15-25: The Fibonacci heap after
changing a key from 9 to 99

The only thing you had to do is unlink the 9 (now 99) from its siblings. If you now want to change the 60 to 66, you also need to change 80's down pointer. You can make it point to 60's right sibling, as shown in Figure 15-26.

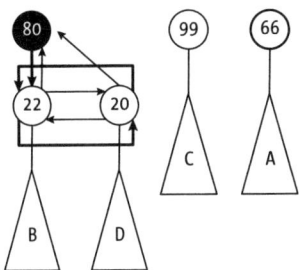

Figure 15-26: Another change
in the tree, and 60 becomes 66.

Now you need to do an extra step. After this separation, if 80 isn't the root, since it was already marked (meaning it had already lost one child), you'd also have to separate it, applying exactly the same logic as earlier. The code looks like this:

```
const _separate = (heap, node) => {
❶ node._marked = false;

❷ const parent = node.up;
❸ if (parent) {
  ❹ if (node.right === node) {
      parent.down = null;
    } else {
    ❺ if (parent.down === node) {
        parent.down = node.right;
      }
    ❻ node.left.right = node.right;
      node.right.left = node.left;
    }
  ❼ parent.degree--;

  ❽ node.up = null;
    node.left = node;
    node.right = node;
    heap.trees.push(node);

  ❾ if (parent._marked) {
      _separate(heap, parent);
  ❿ } else {
      parent._marked = true;
    }
  }

  if (goesHigher(node.key, heap.top)) {
    heap.top = node.key;
  }
};
```

Start by unmarking the node to be separated ❶. This is the *only* way a node can become unmarked again. You then get the parent of the node ❷, but if it has none (meaning the node is a root), don't do anything. Otherwise, if the node has a parent ❸, start unlinking. If the changing node has no siblings ❹, just set the parent's down link to null, and you're done. But if instead the parent is pointing down directly to the node you are changing ❺, you must change the link to a sibling, so you don't break the structure when removing the changing node. Now that you are sure the parent is pointing to a different sibling, you can easily unlink the node from the doubly linked list ❻. Then you need to reduce the parent's degree by one ❼, since it will be losing a child, and push the separated subtree after fixing its links ❽. The last check is if you are removing a child from a node that was marked (meaning it had already lost another child), recursively separate it **❾**; otherwise, just mark it **❿**, and you're finished.

Considering Performance for Fibonacci Heaps

Binomial heaps are formed of binomial trees, and each tree has a number of nodes that is a power of 2. Before any changes (which start pruning the trees), a Fibonacci heap has same-sized trees, but how low can they go? Figure 15-27 shows a Fibonacci heap with as many nodes removed as possible; the white nodes were removed, and black nodes are left.

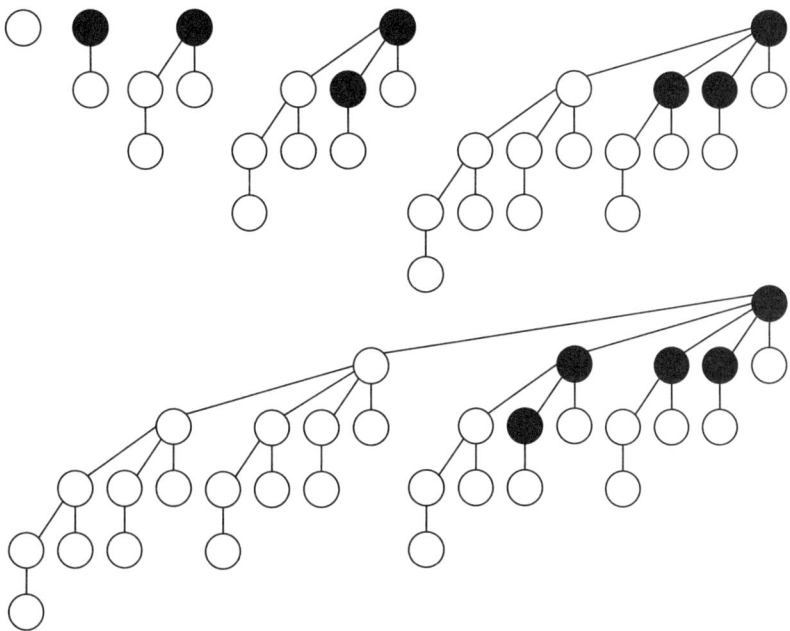

Figure 15-27: Fibonacci trees with a minimum number of nodes

How do you remove as many nodes as possible from a tree without causing any cascading? Or, how do you build the next tree, out of the previous ones, all pruned as much as possible? The worst you can do is promote the largest subtree of every node. In that case, the individual trees would at least have 1, 1, 2, 3, 5, 8, . . . nodes. Recognize the sequence? The trees in this scheme would have at least as many nodes as a Fibonacci number (instead of a power of 2). This also helps because Fibonacci numbers are exponential in nature, implying that the algorithms' performance will still be logarithmic.

Table 15-5 summarizes the performance of Fibonacci heaps; the values with an asterisk are amortized.

Table 15-5: Performance of Operations for Fibonacci Heaps

Operation	Performance
Create	$O(1)$
Empty?	$O(1)$
Add	$O(1)$
Top	$O(1)$
Remove	$O(\log n)$*
Change	$O(1)$*
Merge	$O(1)$

You can't do better than $O(1)$ for insertions, but removals possibly could be better than $O(\log n)$. However, in that case, you could sort a set of n values in $O(n)$ time by inserting all of them into the heap and then removing them in order, but you already know you can't have a sorting algorithm that depends on key-to-key comparisons run with a faster time than $O(n \log n)$, so you can't do removals with better speed.

What you can do is use a simpler structure, with less complex algorithms. The last type of extended heap we'll consider in this chapter, the pairing heap, does exactly that.

Pairing Heaps

A *pairing heap* is a multiary data structure that satisfies the heap property. It basically consists of a root that has the top value of the heap and an ordered set of heaps, so you could call it an orchard in terms of the definitions from Chapter 13. In more formal terms, you could say that a pairing heap is either an empty structure or a root element plus a (possibly empty) list of pairing heaps. Each individual heap is represented in the "left child, right sibling" style; Figure 15-28 shows an example of a pairing heap.

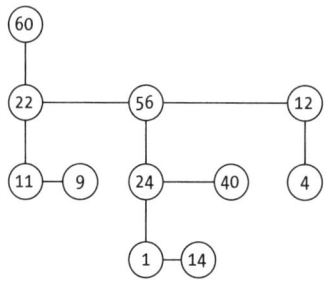

Figure 15-28: A sample pairing heap

The root is 60 and has three subheaps with keys 22, 56, and 12. The subheaps have 3, 5, and 2 elements, respectively.

Defining a Pairing Heap

We won't consider the changeKey() operation (but see question 15.4), so the representation is a tad simpler. Here's the basic starting code for a pairing heap:

```
const goesHigher = (a, b) => a > b;

const newPairingHeap = () => null;

const newNode = (key, down = null, right = null) => ({ key, down, right });

const isEmpty = (heap) => heap === null || heap.key === undefined;

const top = (heap) => (isEmpty(heap) ? undefined : heap.key);
```

It's the same code as for skew heaps, except the left pointer is named down and a small change was made in isEmpty().

Melding Two Pairing Heaps

How do we meld two heaps? If one of the two heaps is empty, just return the other one. Otherwise, if neither heap is empty, the heap with the greatest key will have the other heap added (melded) to its list of subtrees. For instance, see what happens if you want meld the first two subheaps in Figure 15-29 (this is an important example, and you'll return to it later).

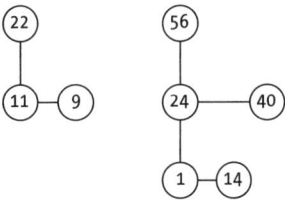

Figure 15-29: Two pairing heaps
to be merged

The new root should be 56, so the first heap (the one with root 22) becomes a child of the second heap, producing the configuration shown in Figure 15-30.

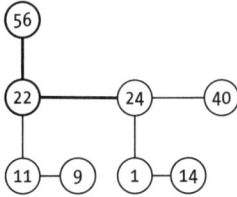

Figure 15-30: The result
of merging

You can implement this easily:

```
const merge = (heap1, heap2) => {
❶ if (isEmpty(heap2)) {
    return heap1;
❷ } else if (isEmpty(heap1)) {
    return heap2;
❸ } else if (goesHigher(heap1.key, heap2.key)) {
    [heap2.right, heap1.down] = [heap1.down, heap2];
    return heap1;
❹ } else {
    [heap1.right, heap2.down] = [heap2.down, heap1];
    return heap2;
  }
};
```

If one heap is empty ❶❷, the result of the merge is just the other tree. Otherwise, if the first heap has the highest key ❸, make the second heap its child. The last case is the same ❹ but in reverse, making the first heap a child of the second.

Adding a Value to a Pairing Heap

Adding a new value to the heap is done with the same method as for skew heaps. Create a new heap with only the new value in it and merge it with the current heap. You've already seen how merging works. The code is almost a one-liner:

```
const add = (heap, keyToAdd) => {
❶ const newHeap = newNode(keyToAdd);
❷ return [merge(heap, newHeap), newHeap];
};
```

It creates a new heap ❶ with a single value in it and merges it with the current heap ❷.

Removing the Top Value from a Pairing Heap

Removing the top value from the heap is harder than adding a new value, and it requires a lot of melding. Basically, you want to remove the heap's root and produce a list of subheaps, which you'll meld in pairs from left to right and then meld the resulting list of heaps from right to left.

NOTE *The name* pairing heaps *comes from the procedure described previously where many heaps are always merged in pairs, two by two.*

First, take a look at how to meld several heaps together. Figure 15-31 shows an example for seven heaps A through G.

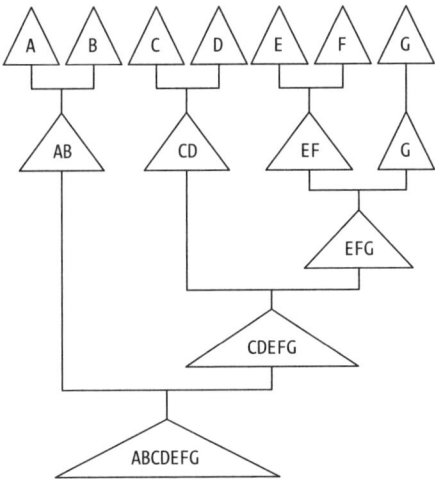

Figure 15-31: Melding seven heaps together, always working two at a time

First merge A and B, and combine that with the result of merging C through G. The second merge starts by merging C and D, and then it merges E through G. The third merge merges E and F first and then waits for the merge of only G (which is trivially just G), so you can complete merging E through G, and then C through G, and finally A through G.

Implementing this seesaw from left to right and then from right to left is actually easy. It's based on a recursive idea: given a list of heaps, merge the initial two heaps on the list to produce a first heap, apply recursion to meld all the others into a second heap, and finish by merging both results together:

```
const _mergeByPairs = (heaps) => {
❶ if (heaps.length === 0) {
     return newPairingHeap();
❷ } else if (heaps.length === 1) {
     return heaps[0];
❸ } else {
     return merge(merge(heaps[0], heaps[1]), _mergeByPairs(heaps.slice(2)));
   }
};
```

This code shows two simple, one-liner cases. If you have no heaps to merge ❶, a null heap is returned, and if there is only one heap to merge ❷, the result is that heap itself. If you have several heaps ❸, merge the two first heaps, then recursively merge all the other heaps, and finish by merging both heaps together.

This is actually an implementation of the left-to-right then right-to-left process described previously, so now redo the case with the seven heaps

A through G you looked at earlier. Writing `mbp()` for `mergeByPairs()` and `m()` for `merge()`, it would be:

```
mbp([A,B,C,D,E,F,G]) =
m(AB, mbp([C,D,E,F,G])) =
m(AB, m(CD, mbp([E,F,G]))) =
m(AB, m(CD, m(EF, mbp([G])))) =
m(AB, m(CD, m(EF, G))) =
m(AB, m(CD, EFG)) =
m(AB, CDEFG) =
ABCDEFG
```

Given this auxiliary method, you can now write the `remove()` code:

```
const remove = (heap) => {
❶ if (isEmpty(heap)) {
     throw new Error("Empty heap; cannot remove");
   } else {
❷ const top = heap.key;

❸ const children = [];
❹ let child = heap.down;
❺ while (!isEmpty(child)) {
     const next = child.right;
     child.right = null;
❻ children.push(child);
     child = next;
   }

❼ return [_mergeByPairs(children), top];
   }
};
```

If the heap is empty ❶, throw an error, because you cannot proceed with the removal. Otherwise, get the top value ❷, so you can return it later ❼, and proceed to separate the subheaps. Initialize an array for the children ❸ and set up a child variable ❹ to loop through all the root's children ❺. Then push each of the subheaps ❻, after remembering to unlink each from its siblings. After pushing all children into the array, merge them by pairs as described previously and return the removed top ❼.

For example, assume you start again with the original pairing heap (see Figure 15-32) and want to remove its root.

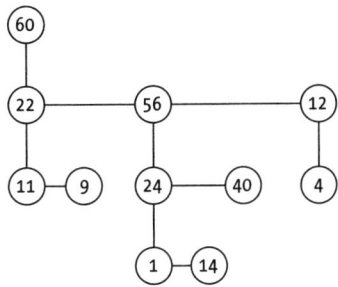

Figure 15-32: The original pairing heap revisited

Removing the root leaves three heaps. Start by melding the first two, those with roots 22 and 56 (you saw this in the previous section), resulting in Figure 15-33.

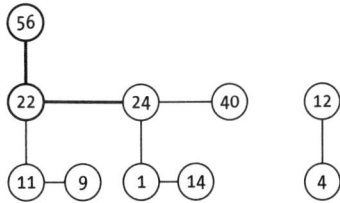

Figure 15-33: The pairing heap after removing its root

The final step is melding those heaps, resulting in the situation shown in Figure 15-34.

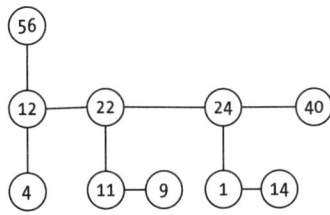

Figure 15-34: The separate heaps melded again into a single heap

Let's now consider another procedure: changing a key's value.

Changing a Value in a Pairing Heap

The procedure to change a key is based on what you've already seen. First, change the key in place, but if it needs bubbling up (as with binomial heaps), separate it from the heap (as with Fibonacci heaps). Then merge the separated heap back into the original heap. You're reusing concepts and algorithms.

For example, start with the heap shown in Figure 15-35.

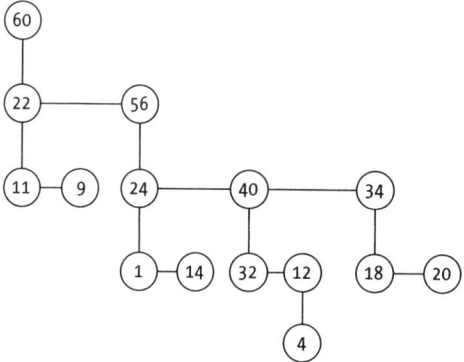

Figure 15-35: A pairing heap just before changing a key

If you want to change the 40 key, for example, and changed it to any value between 40 and 56, you'd just change the key and be done with it. However, if you change it to anything greater than its parent key (56), you have to split the heap and remerge it. This means if you want to change it to 78, after changing the key and splitting the heap away, you are left with the pair of heaps shown in Figure 15-36.

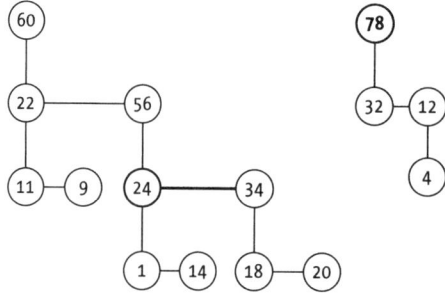

Figure 15-36: Changing the 40 to a 78 splits the heap in two.

Applying the merging function we've looked at previously, Figure 15-37 shows the final result.

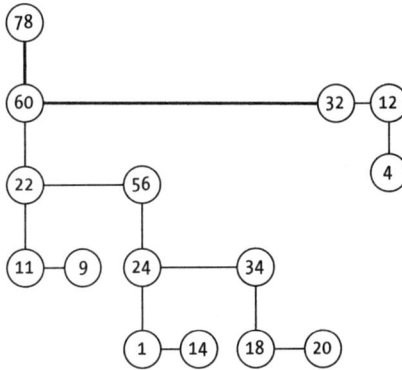

Figure 15-37: Merging results in a single heap again

If you had changed the 40 to 58 instead of 78, the result would be different (see Figure 15-38). Can you see why?

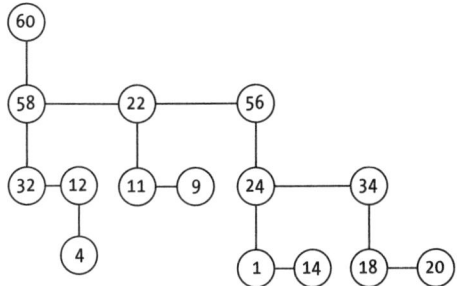

Figure 15-38: An alternative result if you had changed the 40 to 58

Here's how to code the process:

```
const changeKey = (heap, node, newKey) => {
  if (isEmpty(heap)) {
    throw new Error("Heap is empty!");
  } else if (!goesHigher(newKey, node.key)) {
    throw new Error("New value should go higher than old value");
  } else {
❶ node.key = newKey;
❷ const parent = node.up;
❸ if (parent && goesHigher(newKey, parent.key)) {
    ❹ if (parent.down === node) {
        parent.down = node.right;
      } else {
      ❺ let child = parent.down;
      ❻ while (child.right !== node) {
          ❼ child = child.right;
        }
```

```
      ❽ child.right = node.right;
        }
      ❾ node.right = null;
      ❿ heap = merge(heap, node);
        }

      return heap;
    }
};
```

The first two if( ) statements are the same ones you've seen before to check whether the change is possible. Then, actually change the node's key ❶ and get a pointer to its parent ❷. If there's a parent and the new node's key should go higher ❸, then you need to act; otherwise, nothing else is needed. If the node was the first child of its parent ❹, separate the node from the heap by changing the link down from its parent to point to the node's first sibling. Otherwise, if the node isn't the first child, loop through the siblings list ❺❻❼ until you find the node's previous sibling. Then unlink the node from the list ❽❾ and finish by merging the separated heap back into the original ❿. (Similar procedures were discussed in Chapter 7.)

Considering Performance for Pairing Heaps

The performance of pairing heaps is similar to that of Fibonacci heaps, as shown in Table 15-6; values with an asterisk are amortized.

Table 15-6: Performance of Operations for Pairing Heaps

Operation	Performance
Create	$O(1)$
Top	$O(1)$
Add	$O(1)$
Top	$O(1)$
Remove	$O(\log n)$*
Change	$O(\log n)$?
Merge	$O(1)$

Why does changing a value have a question mark in the table? There's still no consensus as to the precise amortized order of this operation. An initial estimate suggested it would be $O(1)$, but then it was proven to be at least $\Omega(\log \log n)$. Further work produced an $O(\log n)$ estimate, but no definite proof has yet appeared. In any case, this looks worse than Fibonacci heaps, but in practice, the performance of pairing heaps is reported to be excellent, despite the purported theoretical deficiency, which is most likely a result of the simpler implementation.

Summary

In the previous chapter, we studied binary heaps that were represented with an array, and in this chapter, we completed our overview of heaps, exploring several extended versions that are implemented with binary trees, multiway trees, or forests and that allow operations such as merging two heaps and changing a key in a more performant way. These changes not only maintained the functionality of the previous heaps, but they also added enhanced performance and new features, allowing us to use these structures for other types of problems that common heaps wouldn't handle as well. This chapter presents the best example of the possibilities of modified (or hybrid, if you will) structures that add speed and functionality—but at the obvious cost, clearly, of some more complex algorithms!

Questions

15.1 Intuitive but Worse

Suppose you have two common heaps of sizes m and n and you implement melding by applying the following intuitive method: successively choose all the elements from one heap and insert them into the other. What would be the order of this method?

15.2 Sequential Cases

What's the shape of a skew heap if you insert keys in ascending order? What about in descending order?

15.3 No Recursion Needed

When merging two skew heaps and the second had the greater key, no recursion was actually needed; could you do the merge directly?

15.4 Change Needed

How would you implement the changeKey() function for skew heaps? Would you need to make some structure changes, and if so, what would they be?

15.5 Just Adding

Assume you have a binomial heap with only a BT(3) in it, so it has eight values. Add a new value to that node eight times and count how many merges are needed. What can you conclude about the amortized cost of the operation?

15.6 Faster Binomial Top

You can accelerate reaching the top of a binomial heap with means used for other heaps; can you figure out how to do so?

15.7 Easier Bubbling Up?

Why can't you implement the _bubbleUp() method for binomial trees in the following way, similar to what you wrote for binary heaps? (The reason is easy to miss.)

```
const _bubbleUp = (heap, node, newKey) => {
  node.key = newKey;
  const parent = node.up;
  if (parent && goesHigher(newKey, parent.key)) {
    node.key = parent.key;
    _bubbleUp(parent, newKey);
  }
};
```

15.8 Searching a Heap

Even if it makes little sense, can you implement a find() function for heaps? Be careful with Fibonacci heaps, so you don't create an infinite loop because of the circular lists.

15.9 Two in One

In a Fibonacci heap, you can make the add(), remove(), and mergeA2B() methods a few lines shorter by joining assignments with the same right value; can you see how?

16

DIGITAL SEARCH TREES

In the previous four chapters, we explored different types of trees, including binary trees, general trees, heaps, and more. All of those trees are based on storing and comparing keys. With the digital search trees we'll consider in this chapter, we won't associate keys with nodes. Instead, the node's position in the tree will define the key with which it is associated. In other words, you won't store keys in a single place in the tree; they'll be distributed across the whole structure, starting at the tree's root. The leaves will mark where the keys end.

This might look like merely changing the way we work, in the same way that the radix sort changed the way we sorted (Chapter 6). With radix sort, instead of sorting by comparing keys, we worked with the keys character by character (or digit by digit, for numbers). With digital search trees, instead of storing and comparing keys, we'll work with paths in trees.

We'll focus on three different data structures: *tries*, which can do searches in time proportional to a key's length; *radix trees*, which are optimized versions of tries; and *ternary search tries*, an extension of binary search trees. These structures are particularly effective when we're searching for strings, just as we search in a dictionary for words.

The Classic Version of Tries

Tries often are used simply to store words, allowing for easy, fast searches where users can enter a few letters, and words starting with those letters appear. Tries also are used as generic search trees where keys and data are stored and then a search is conducted for a key to provide the associated data.

NOTE *Trie originally was pronounced like the word* tree, *but it's also pronounced like the word* try *to distinguish it from* tree. *You can pronounce it either way.*

Think of tries as a bit like old telephone indexes that had a set of buttons, one for each letter. If you wanted to find a name starting with *F*, you'd press that key, and the index would open on the F page. The analogy doesn't stop there. Assume the index had another set of buttons for the name's second letter, and pressing that button led you to yet another page with a new set of buttons. If you clicked all the buttons in order, you'd arrive at the name you were looking for or an empty page, meaning the name wasn't in the index. This analogy may be hard to understand (I wonder how many readers have ever seen such a phone index! Maybe think of how autocomplete works instead?), so let's consider the actual definition of a trie.

A trie has a link for each possible character (the same way the previous phone index example has a button for each letter in a name), but for simplicity, we'll work with only the letters from A to E, plus an end of word (EOW) character to indicate where each word ends. We'll use ■ for this. (Other languages, such as C, use the NULL \0 character for EOW, but the square symbol is more visible.) Suppose the words are ACE, AD, BADE, BE, BED, and BEE. In the trie, you'll actually see ACE■, AD■, and so on. This trie looks like the diagram in Figure 16-1, and for clarity, I've placed the root on the left instead of at the top, so you can read the words horizontally.

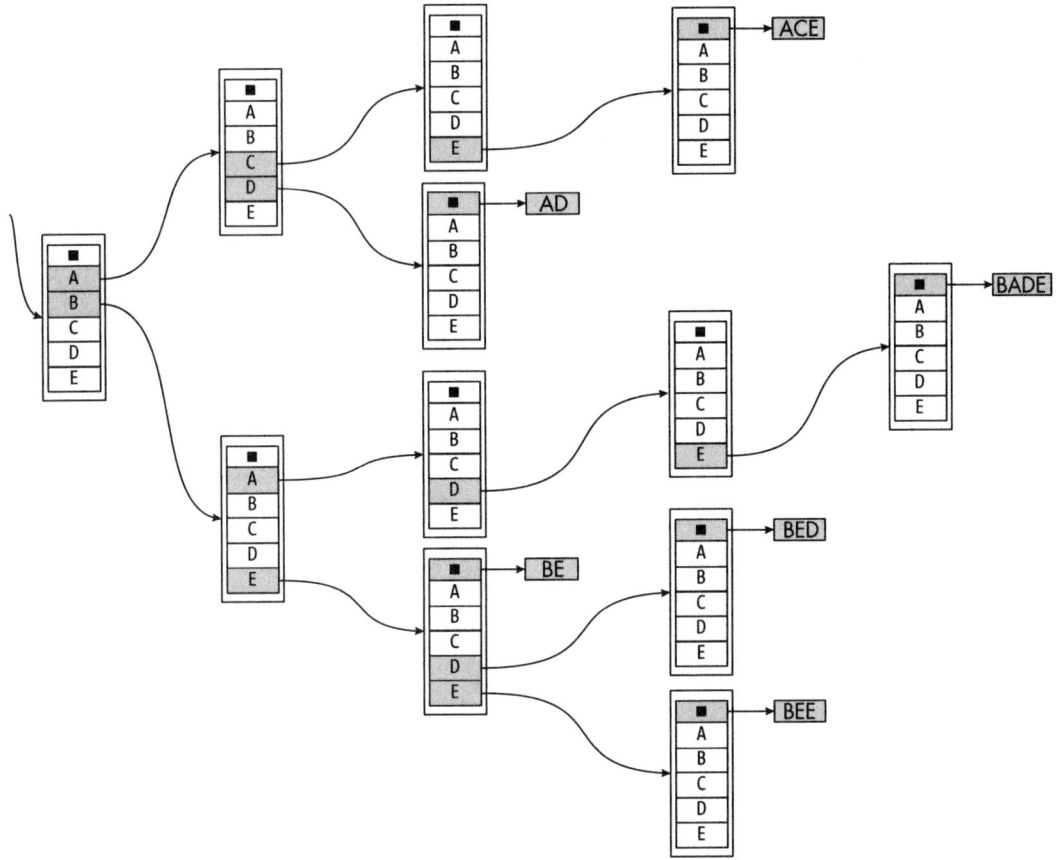

Figure 16-1: A sample trie for words using only letters from A to E

Each node in the trie consists of an array of links, one for each possible letter (A–E) plus the EOW character. (You could say it's a six-ary tree; see Chapter 13.) The empty links have a white background in the diagram, and actual links have a gray background. Each word in a little box represents some extra data or value associated with the corresponding key (we'll get to that in the next section).

You can define the basic functions to create a trie as follows:

```
const EOW = "■";
const ALPHABET = `${EOW}ABCDE`;
const newTrie = () => null;
const newNode = () => ({ links: new Array(ALPHABET.length).fill(null) });
const isEmpty = (trie) => !trie; // null or undefined
```

First, define the EOW character; you'll use the same definition through-out this chapter. The ALPHABET constant includes all the characters we'll accept; for this example, you're using only five letters, but for a real application, you'd most likely include at least the whole alphabet, A to Z. A new trie is just a null value, and a new node is an object with a links property, which is an array with one null link for each character in ALPHABET. Finally, to recognize an empty trie, simply check for "falsey" values on the last line.

You can also associate some value with every key, as shown next.

Storing Extra Data in a Trie

When we studied trees in previous chapters, we were concerned only with storing keys and searching for keys, because adding extra data was simple. Instead of a single key field, we could have a record with a key field and an extra data field. If we wanted to modify the algorithms to include extra fields, the changes were minor: searches would return the extra data instead of just a boolean value, and adding a key would also add the extra fields in the same object.

But in a trie, keys aren't stored in a single place and instead are distributed throughout the trie's branches. There is a solution to this, but because the changes needed in the algorithms aren't so minor, we'll work with keys plus data.

The dictionary abstract data type (ADT) will change slightly, specifically the add and find operations, as shown in Table 16-1.

Table 16-1: Operations on Tries

Operation	Signature	Description
Create	→ D	Create a new dictionary.
Empty?	D → boolean	Determine whether the dictionary is empty.
Add	D × key × data → D	Given a new key and data, add them to the dictionary.
Remove	D × key → D	Given a key, remove it from the dictionary.
Find	D × key → data | null	Given a key, return its data or return null if not found.

As in other chapters, we'll study the performance of the structures for this ADT. Now that we've defined the full structure for a trie and looked at how to create one, let's consider the rest of the needed functions.

Searching a Trie

How do you look for a word? For example, if you want to know whether BED is a valid word, Figure 16-2 shows the path you'd take.

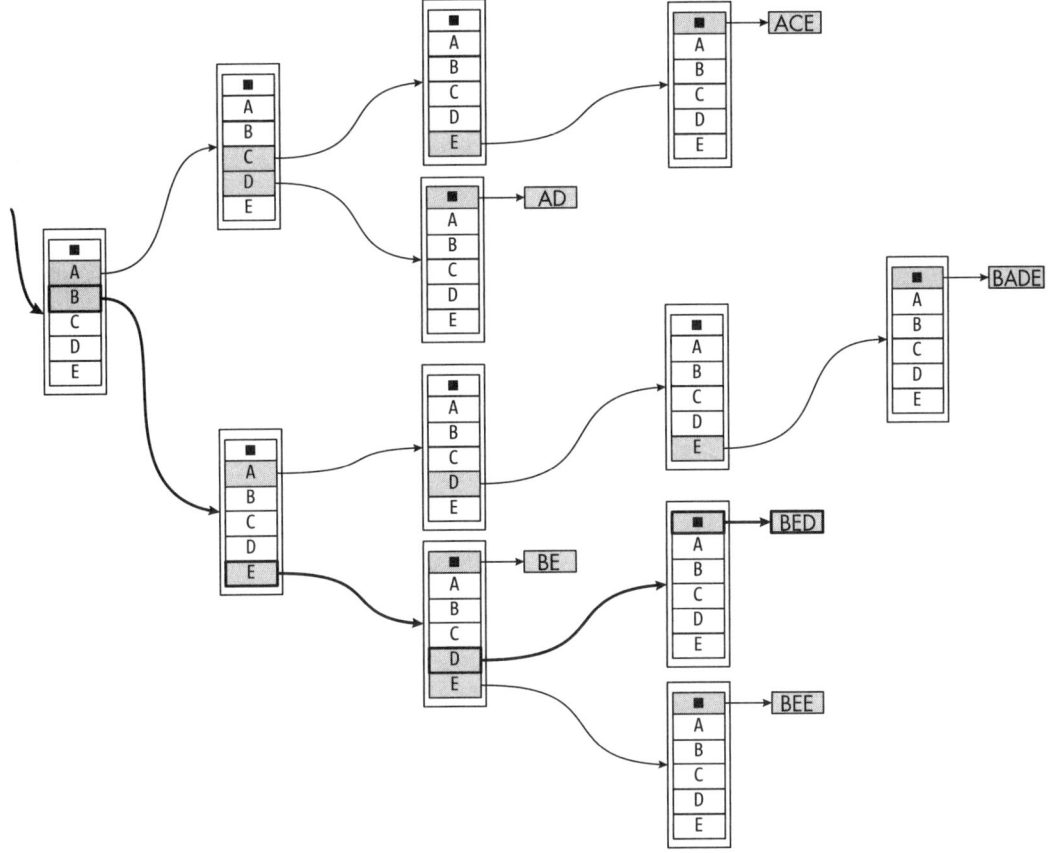

Figure 16-2: Searching successfully for BED in a trie

The search for BED starts at the root. You look at the B link and find it's not null, so you follow it to the next level. There, you look at the E link and again follow it. The next step is similar: look at the D link and follow it. Finally, you arrive at the EOW, and finding a link to some data in the corresponding link, call it a successful search and return the found data.

A failed search looks different. For example, if you want to find the word DAB, the search would fail at the beginning, since no word starts with D. What about ACED? This time, you'd start the search at the A link, then the C link, and finally the E link, but you'd arrive at a node with no D link, meaning ACED isn't in the trie. Figure 16-3 shows this failed search.

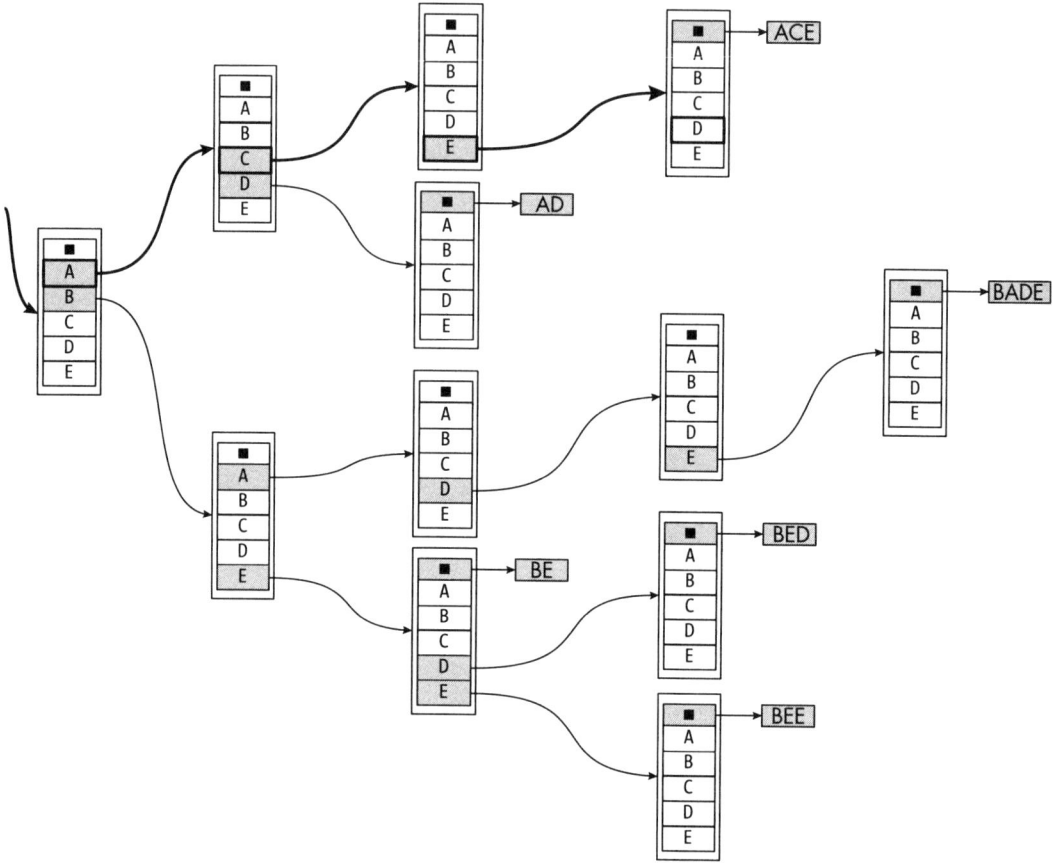

Figure 16-3: Searching (unsuccessfully) for ACED in a trie

Consider one last case and search for BAD. Remember, you're adding an EOW character, so in reality, you're trying to find BAD■. Figure 16-4 shows what happens.

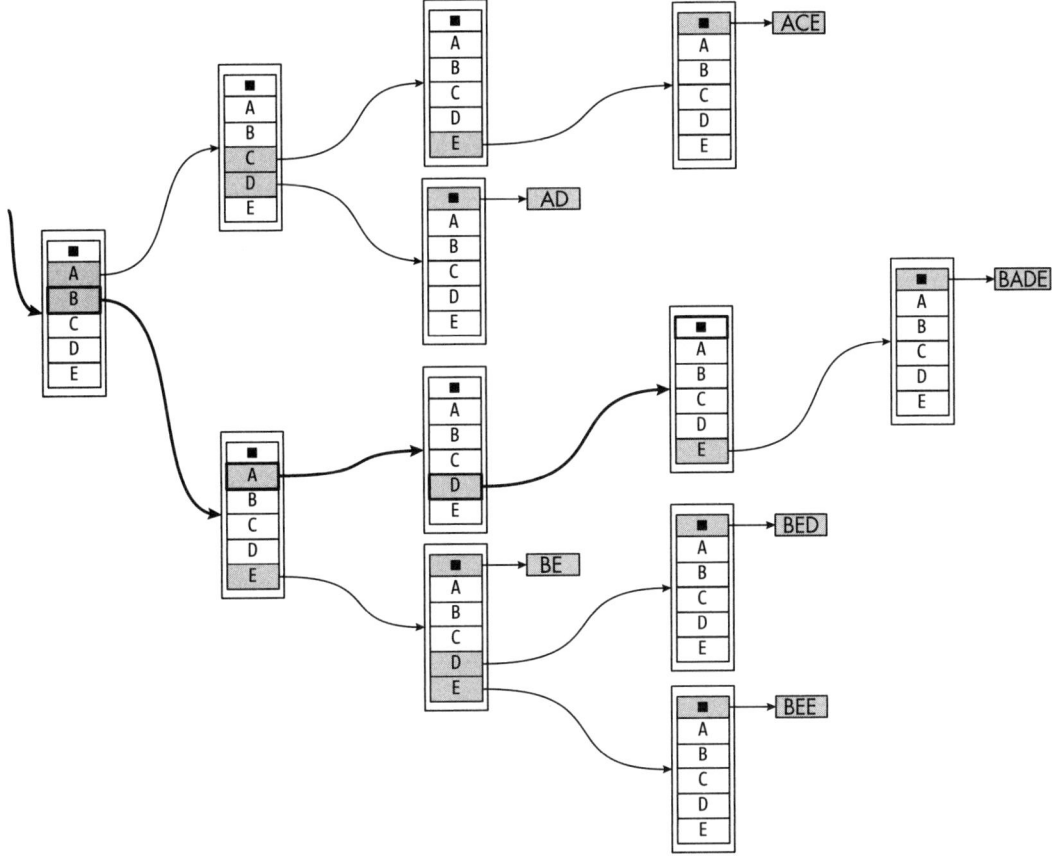

Figure 16-4: Searching unsuccessfully for BAD: BADE is in the trie, but BAD isn't.

This search starts at the root and follows the B link first, then the A link, and then the D link. BAD is a prefix of at least one word, but the EOW link is missing, so BAD isn't considered to be in the trie.

To implement this logic, first create an auxiliary _find() function that actually does the searches:

```
const _find = (trie, [first, ...rest]) => {
❶ const i = ALPHABET.indexOf(first);
❷ if (isEmpty(trie)) {
    return null;
❸ } else if (first === EOW) {
    return isEmpty(trie.links[i]) ? null : trie.links[i].data;
❹ } else {
    return _find(trie.links[i], rest);
  }
};
```

The i variable ❶ selects the proper value from the links property, and now you're ready to start searching. If the trie is empty (maybe it was empty from the beginning, or maybe you traveled down a null link), the search has obviously failed ❷. If you arrive at the end of the word (marked by the EOW character), do a simple test: if the corresponding link is null, the search fails as before; if not, the link points to an object with a data property, which you return ❸. Otherwise, if you haven't yet reached a null link or the EOW character ❹, go down the right link and keep searching recursively.

The find() function just calls the earlier _find(), but it adds the needed EOW character at the end of the string you're trying to find:

```
const find = (trie, wordToFind) =>
❶ !!wordToFind ? _find(trie, wordToFind + EOW) : null;
```

Test for an empty word to find ❶, and if it's empty, return null without any further ado.

Searching tries isn't complex, and it's similar to searches in other types of trees we looked at in previous chapters.

Adding a Key to a Trie

Adding a new key (plus data) to an existing trie follows the same strategy used for searches. Go down the links, letter by letter, and if the link exists, follow it, and if it doesn't, add a new empty node. For instance, in the trie diagram shown in Figure 16-5, if you want to add a BAD key, you just need to add a link at the last node with the data.

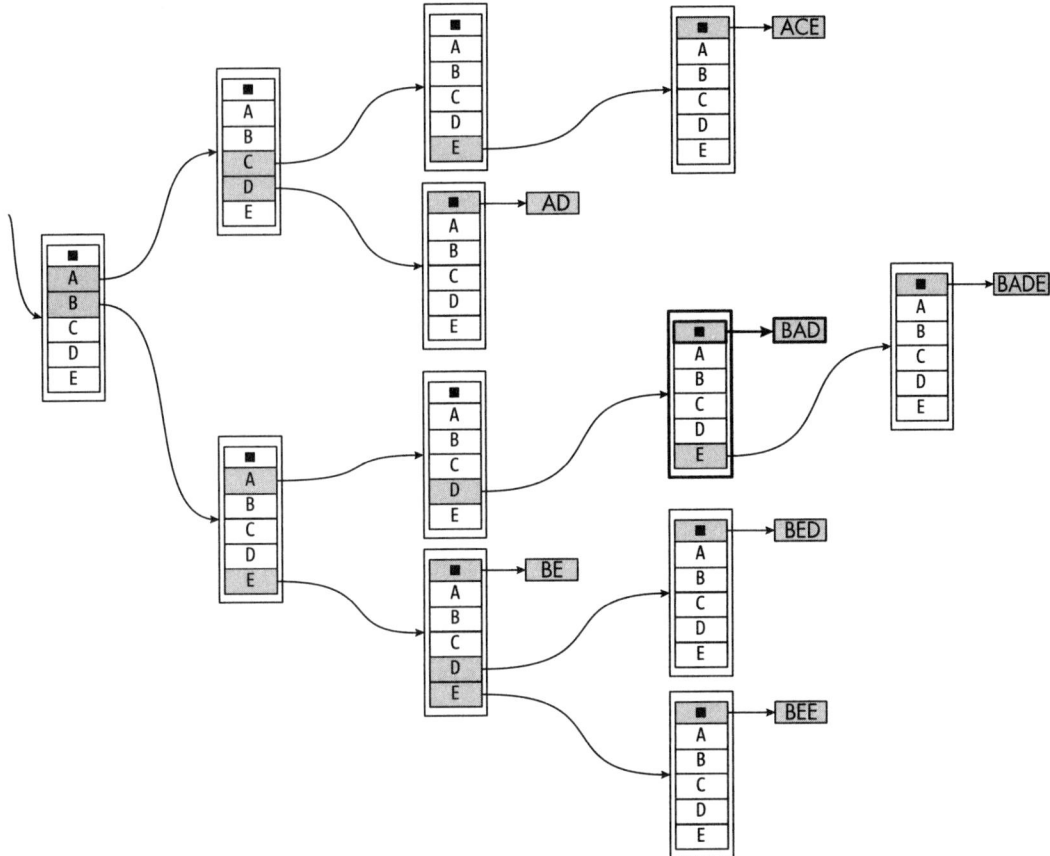

Figure 16-5: Adding BAD to a trie

The only change in the trie is that the (so far) empty EOW link now points to the data associated with the BAD key.

As another example, to add ABE, you would start by following the A link at the root, but then you'd need to add two new nodes, as shown in Figure 16-6.

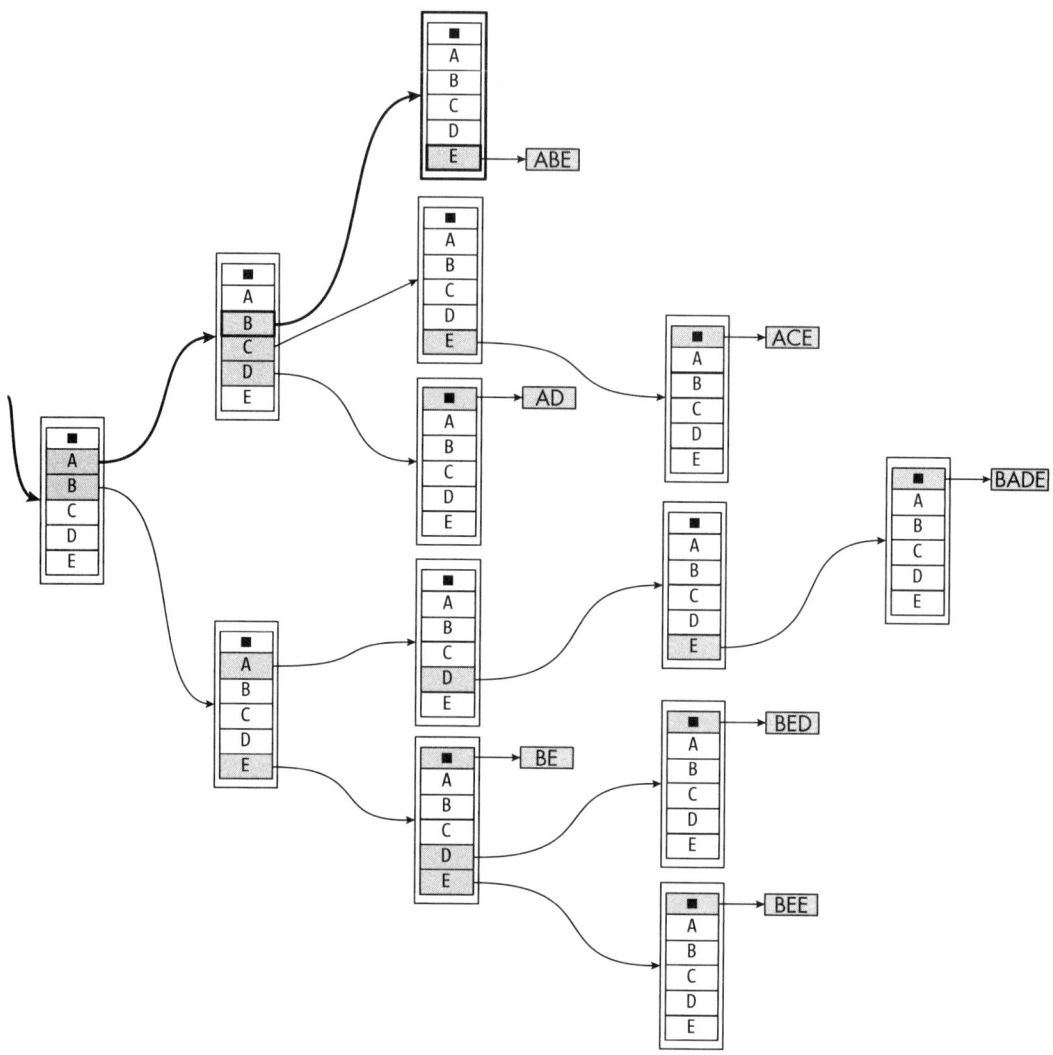

Figure 16-6: Adding ABE to the trie requires new nodes.

The code for this addition is similar to other insertion functions we've discussed in previous chapters:

```
❶ const _add = (trie, [first, ...rest], data) => {
❷ if (first) {
    ❸ if (isEmpty(trie)) {
        trie = newNode();
      }
```

```
❹ const i = ALPHABET.indexOf(first);
   if (first === EOW) {
   ❺ trie.links[i] = { data };
   } else {
   ❻ trie.links[i] = _add(trie.links[i], rest, data);
   }
 }
 return trie;
};
```

An _add() auxiliary function ❶ actually does the insertion. While you haven't reached the end of the string (including the added EOW) ❷, advance, following the links corresponding to successive letters. If you find an empty link ❸, create a new node and keep advancing until you reach the EOW character. At each step, decide what link to follow ❹, and when you reach the EOW, add a link to the extra data ❺; otherwise, recursively keep adding the rest of the string ❻.

Finally, the add() function just invokes _add():

```
const add = (trie, wordToAdd, dataToAdd = wordToAdd) =>
  _add(trie, wordToAdd + EOW, dataToAdd);
```

This function also ensures that the EOW character is added.

Removing a Key from a Trie

Now we'll look at deleting a key, which is a bit more complex than adding keys to a trie. First, try to find the key; if you can't find it, you're finished, since there's nothing else to do. If you find the key (and arrived at an EOW character), just delete the associated data, making the pointer null. If doing that left the current node empty of pointers, then delete the node and fix the parent's pointer, which may again leave an all-empty node, so keep going up, back to the root, possibly deleting more nodes on the way.

Let's consider a few cases. Returning to the example trie, if you want to delete BE, you'd clear its EOW link as shown in Figure 16-7, and you'd be done.

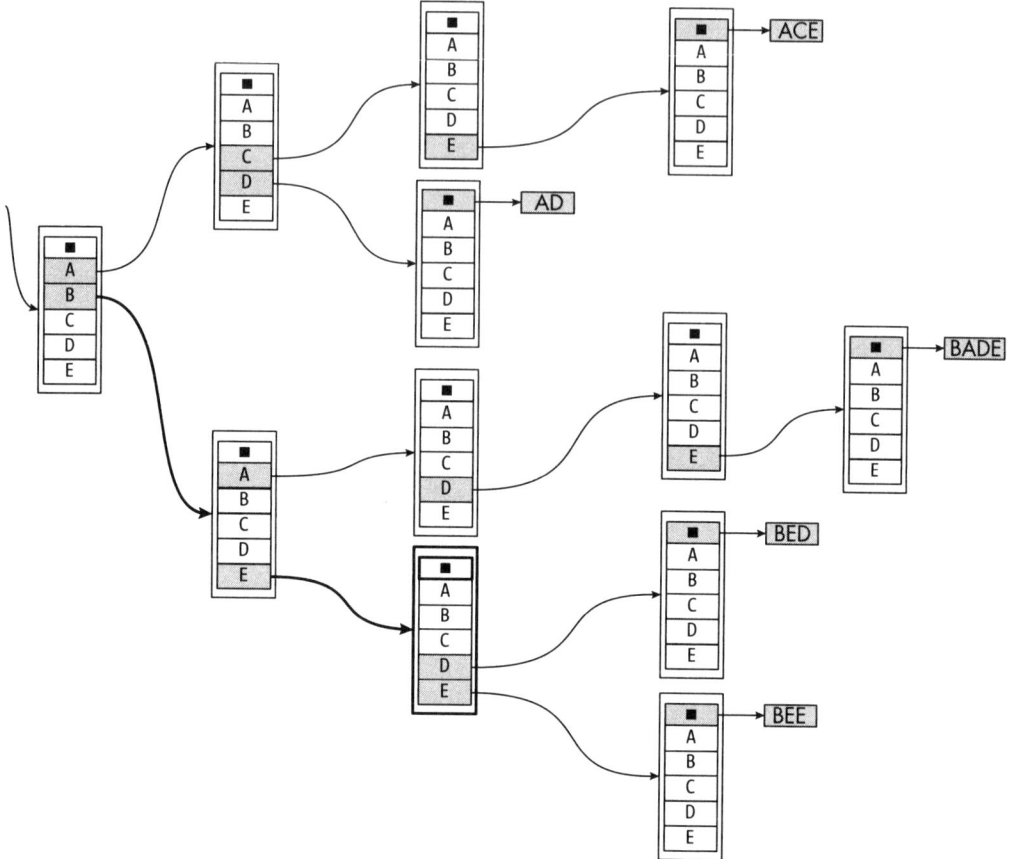

Figure 16-7: Deleting BE from the trie

To make it more difficult, after deleting BE, what if you want to delete BADE? First you clear the EOW link, but then the whole node has only null links, as shown in Figure 16-8.

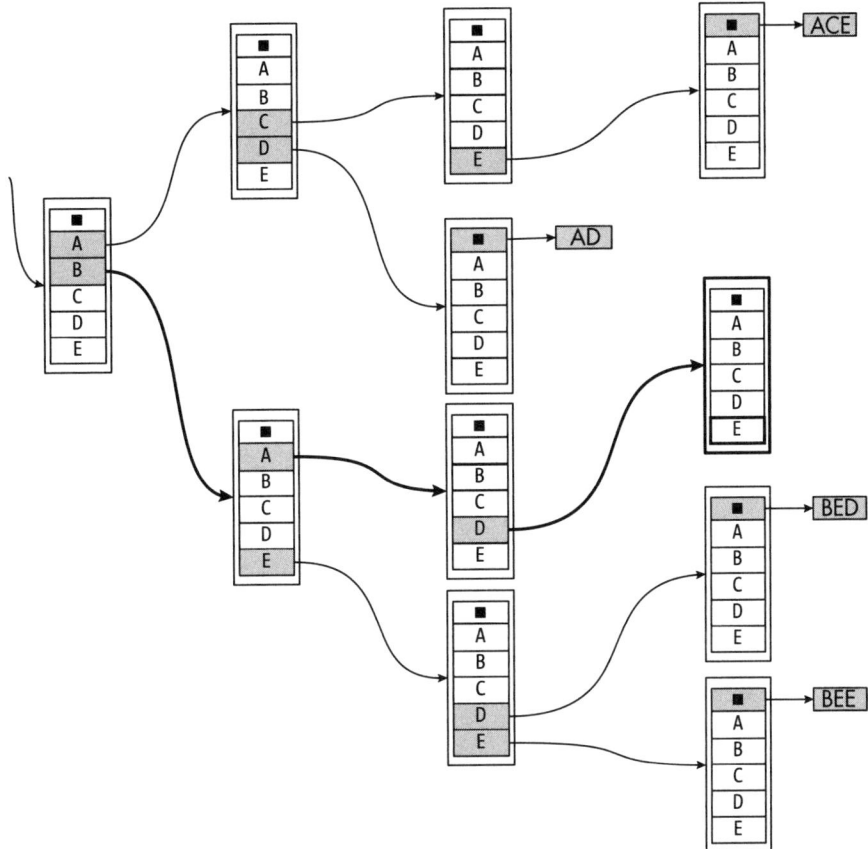

Figure 16-8: Deleting BADE implies removing several (now empty) nodes from the trie.

However, removing that node (and clearing the E link at its parent) repeats the situation, so you also delete the parent, and then the parent's parent, until you arrive at a not all-empty node. The final situation looks like Figure 16-9.

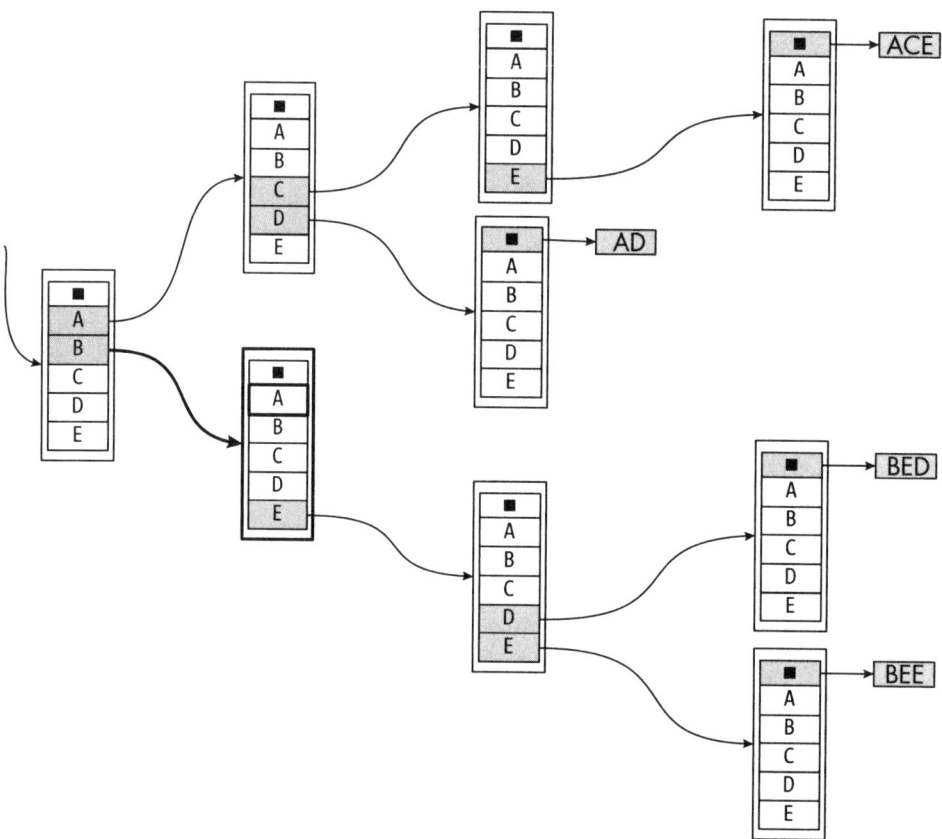

Figure 16-9: The final trie after removing BADE

As with searching and adding, you need an auxiliary function to implement this:

```
❶ const _remove = (trie, [first, ...rest]) => {
❷   if (isEmpty(trie)) {
      // nothing to do
    } else if (!first) {
❸     trie = null;
    } else {
❹     const i = ALPHABET.indexOf(first);
❺     trie.links[i] = _remove(trie.links[i], rest);
❻     if (trie.links.every((t) => isEmpty(t))) {
        trie = null;
      }
    }
❼   return trie;
};
```

The _remove() function does the actual removal ❶. You search, link by link, and when you reach a null link, you know you're done ❷. If you reach the end of the word (past the EOW character), set the current link to null ❸.

If you're still in the middle of the word, decide what link to follow ❹.
Then recursively remove the rest of the word ❺ and do a final check to see
whether the node is totally empty (all null), in which case, you also set the
node to null ❻. At the end, just return the modified trie ❼.

The final code you need is the following:

```
const remove = (trie, wordToRemove) => _remove(trie, wordToRemove + EOW);
```

The remove() function just calls _remove(), and it adds the needed EOW
character to the string you want to remove.

Considering Performance for Tries

How do the different operations perform? Table 16-2 is almost the simplest
table in this book; can you see why?

Table 16-2: Performance of Operations for Tries

Operation	Performance
Create	$O(1)$
Empty?	$O(1)$
Add	$O(k)$
Remove	$O(k)$
Find	$O(k)$

Creating a trie and checking whether it's empty are $O(1)$ constant
operations. And, given the structure, if keys are k characters long, all the
other operations take (at most) k steps, which is the maximum height of a
trie with such keys—excellent constant results! Note that the $O(k)$ performance, with constant k, should be written as $O(1)$. I wrote it as $O(k)$ here
just as a reminder that k steps will be taken, but in performance terms, any
constant implies $O(1)$.

However, despite having a very good performance, notice that this
structure is quite wasteful. All the nodes have many links (as many characters as possible in the keys), making for a lot of unused space. Even worse
with this structure, if you want to distinguish between uppercase and lower-
case characters, or deal with the entire alphabet, or store words in foreign
languages, the space requirements for nodes will grow wildly, because of
the many links that each node will require.

This situation isn't quite favorable, so let's consider another approach to
tries and aim for a more modern (and less wasteful) implementation, which
JavaScript fortunately lets you do easily.

An Enhanced Version of Tries

Consider again how to represent a trie. At each node, you need a link for
each character in a string, but if only a few of those characters are actually

necessary, you don't need to waste space for the rest. In fact, it sounds like you need some kind of set here, and that's indeed the solution.

If you have a large varying number of possible links, you should use some of the structures discussed in Chapter 13. However, because alphabets are limited, you can use an object with characters for keys and links for values. (For yet another solution, see question 16.1.)

With the same six strings as in the previous trie (ACE, AD, BADE, BE, BED, and BEE), the structure looks like Figure 16-10.

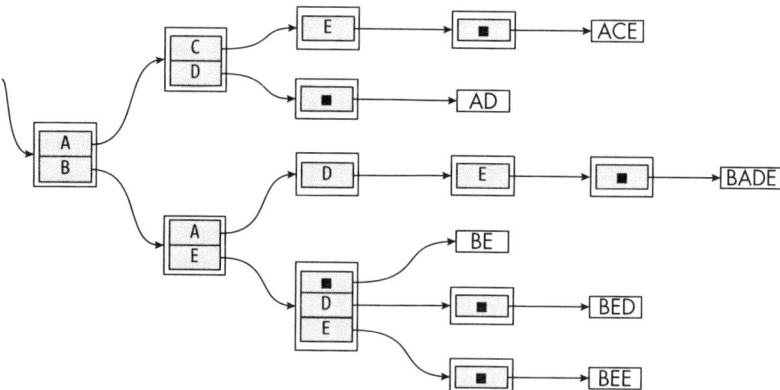

Figure 16-10: An enhanced object-based version of a trie with fewer links

The main difference is now the nodes can be much smaller, including only the strictly necessary links and nothing more.

Defining an Object-Based Trie

Instead of having an array of links, you can define this new style of space-saving tries using an object:

```
const EOW = "■";
const newTrie = () => null;
const newNode = () => ({ links: {} });
const isEmpty = (trie) => !trie; // null or undefined
```

You need to make only one simple change from the previous trie (shown in bold): a new node now has an object named `links` instead of an array with that name.

Searching an Object-Based Trie

The process for searching an object-based trie for a key is similar to searching a classic trie: start at the root and follow the links that correspond to each character in the string, one by one. Instead of an array with links to all possible characters, you have an object with only the links actually required. For instance, revisit the BED search in Figure 16-11.

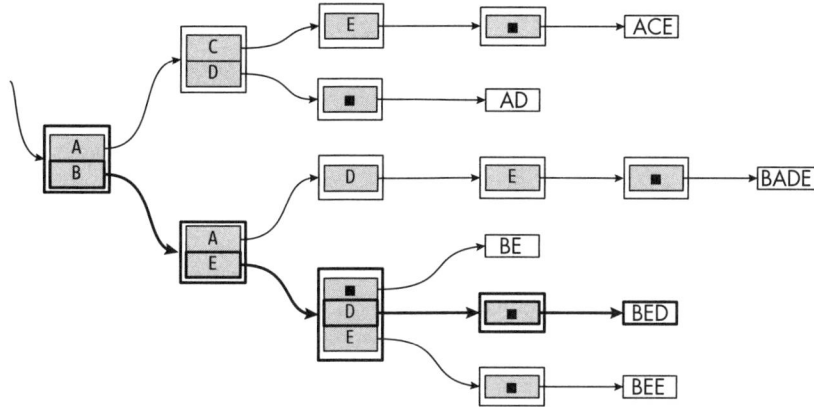

Figure 16-11: Searching successfully for BED in an object-based trie

Starting at the root in the links object, follow the B link. At the next node, follow the E link, and so on, until reaching the EOW link. Then you know the key was found and can return the associated data, whatever it is.

Try redoing a failed search as well, as shown in Figure 16-12.

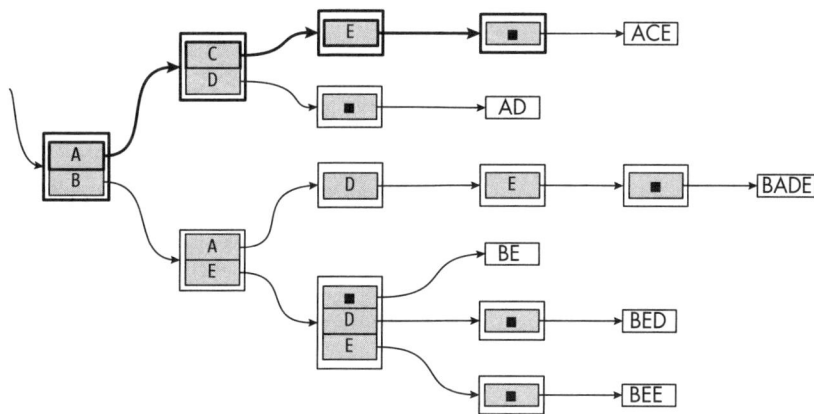

Figure 16-12: Searching unsuccessfully for ACED in an object-based trie

If you search for ACED, the process stops after following the last E link. The node you arrive at doesn't have a D link, so you can't proceed, as the key you want isn't in the trie.

Searching object-based tries isn't really different from searching the original tries; the only change is how you choose the link to follow. The following logic shows the needed changes:

```
const _find = (trie, [first, ...rest]) => {
  if (isEmpty(trie)) {
    return null;
  } else if (first === EOW) {
    return isEmpty(trie.links[first]) ? null : trie.links[first].data;
```

```
    } else {
      return _find(trie.links[first], rest);
    }
};

const find = (trie, wordToFind) =>
  !!wordToFind && _find(trie, wordToFind + EOW);
```

In comparison with the original trie search code, you have only two changes (shown in bold): you can access the needed link directly without having to check the alphabet first—for example, the link for the letter A is at links.A, (or, equivalently, links["A"]) with no further ado. The find() function itself is exactly the same as with the original tries.

Adding a Key to an Object-Based Trie

Adding a key is also similar to the algorithm for the original tries. The changes you need to make work like those you made for searching. For instance, consider adding ABE, as shown in Figure 16-13.

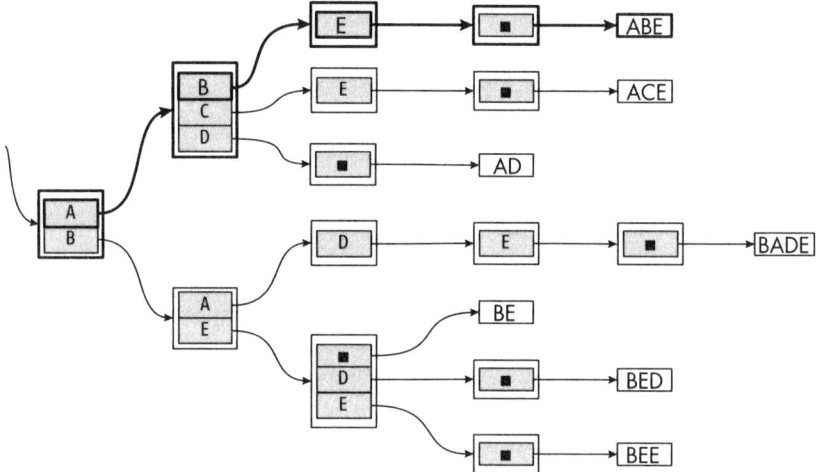

Figure 16-13: Adding ABE to an object-based trie

Start following links as you did with the search, and when links don't appear, add them. In this case, the root already has an A link, so you follow it. The next node has no B link, so add it to the existing (C and D) links. From that point onward, start adding new nodes to the trie.

It's the same process as for the original tries; here's the new logic:

```
const _add = (trie, [first, ...rest], data) => {
  if (first) {
    if (isEmpty(trie)) {
      trie = newNode();
    }
    if (first === EOW) {
```

```
      trie.links[first] = { data };
    } else {
      trie.links[first] = _add(trie.links[first], rest, data);
    }
  }
  return trie;
};

const add = (trie, wordToAdd, dataToAdd = wordToAdd) =>
  _add(trie, wordToAdd + EOW, dataToAdd);
```

Again, the only difference in the logic from the original tries is that you always know what link to use for a given character, so you don't need to search the ALPHABET array (shown in bold).

Removing a Key from an Object-Based Trie

Finally, removing a key is also similar to the previous code for original tries, but deciding whether the node is empty requires a different approach. For example, if you want to remove BADE from the original trie, you'll get the result shown in Figure 16-14.

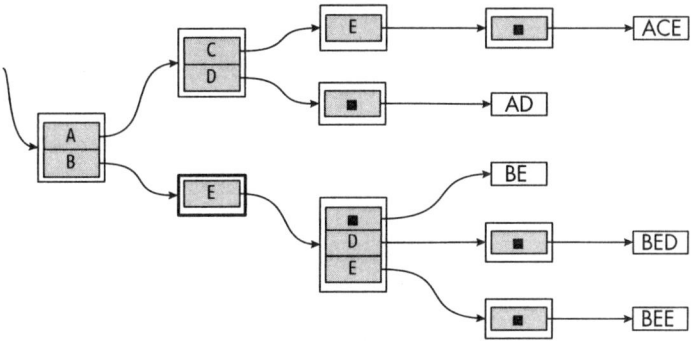

Figure 16-14: Removing BADE from an object-based trie

As with the previous array-based trie implementation, when nodes become empty, you delete them. The updated logic is as follows:

```
const _remove = (trie, [first, ...rest]) => {
  if (isEmpty(trie)) {
    // nothing to do
  } else if (!first) {
    trie = null;
  } else {
❶ trie.links[first] = _remove(trie.links[first], rest);
    if (isEmpty(trie.links[first])) {
❷ delete trie.links[first];
❸ if (Object.keys(trie.links).length === 0) {
      trie = null;
    }
```

```
    }
  }
  return trie;
};

const remove = (trie, wordToRemove) => _remove(trie, wordToRemove + EOW);
```

Most of the differences with the logic for original tries ❶ ❷ are directly related to knowing what link to use, but in order to decide whether a node became empty, you need to see how many keys the links object has ❸.

Considering Performance for Object-Based Tries

What changes with this new version of a trie? Performance is exactly the same as for the array-based tries; the only difference is the amount of memory needed. Think of it this way: if you were using a trie for a dictionary of European languages (with all the tildes, accents, and special characters, like the Danish å, or the Czech ě, or the German ß), you would need an array with more than 200 links in each node, although most of them would be empty. Using an object saves a lot of space, which could be important in some cases.

However, using objects instead of arrays isn't as efficient as it could be. After all, if the keys are long, you still need a very tall trie. As a border case, imagine having a trie with a single key that's 20 characters long. You'd have a 20-level-high tree for one single key. Radix trees provide an enhanced solution.

Radix Trees

Both the tries so far work well, but they have many levels. For instance, when looking for BADE in the example trie, you have to go down one level for each letter, while there are no other words that start with A; see Figure 16-15.

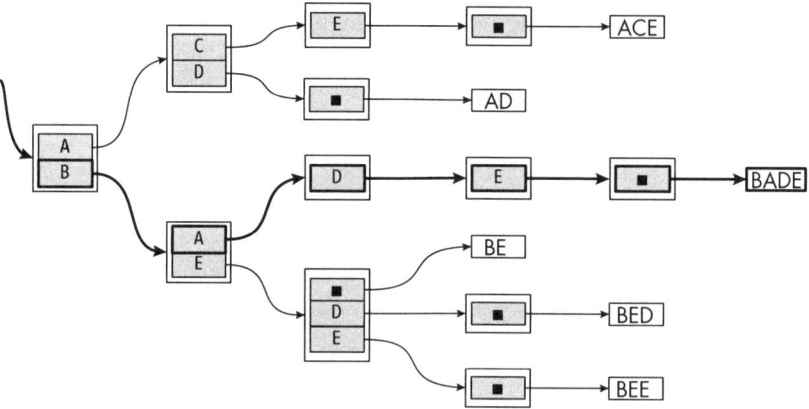

Figure 16-15: Searching successfully for BADE in a trie requires many steps.

The idea for radix trees is that if any level has a single link, we "push it up" and join it with its parent link to shorten future searches. Figure 16-16 shows what a radix tree for the ACE, AD, BADE, BE, BED, and BEE sets of words looks like.

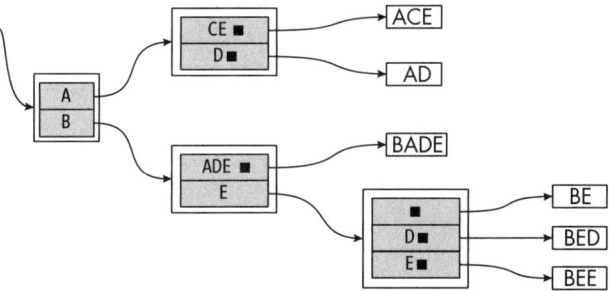

Figure 16-16: A radix tree shortens searches by joining links.

Now, you would have found BADE in just two steps. This tree is shorter than all previous tries. Some paths (for example, B to E to BE) are still the same length, but most others are shorter, because some levels have multi-character links. Let's explore how these trees would be defined and used, because they'll represent a more performant kind of digital tree.

Defining a Radix Tree

If you use an object to store links, like you did with the object-based tries, you find that the logic for radix trees is exactly the same. The difference is in *how* you use the links (they won't always be single-character links), but otherwise the structure is the same:

```
const EOW = "■";
const newRadixTree = () => null;
const newNode = () => ({ links: {} });
const isEmpty = (rt) => !rt;
```

Since there are no changes here with regard to the previous object-based tries code, let's move on to working with these new trees.

Searching a Radix Tree

Doing searches is similar to what you did for object-based tries, but now that the links may correspond to many characters, you need to work a bit differently. Start with a simple case: searching for BED—or, more precisely, BED■. Figure 16-17 shows the path to follow.

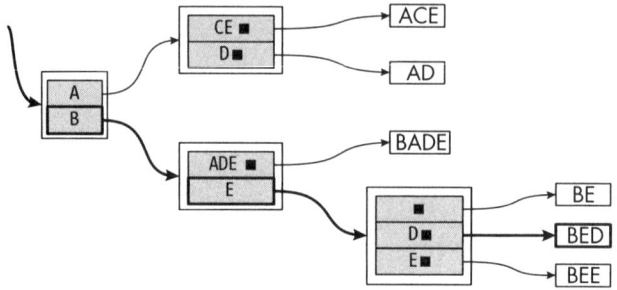

Figure 16-17: Searching successfully for BED in a radix tree

At the root, you need to find a link that is a prefix of BED■. In this case, follow a B link to the next level. (If you don't find such a link, declare the search unsuccessful right then.) Since you followed a B link, next search for ED■. Again, you find a link that matches the search and follow that E link to the third level, where you look for D■. At this point, you find a complete match, so you found the key and can return its data. You're done!

As another example, look for ACED■, which will be a failed search (see Figure 16-18).

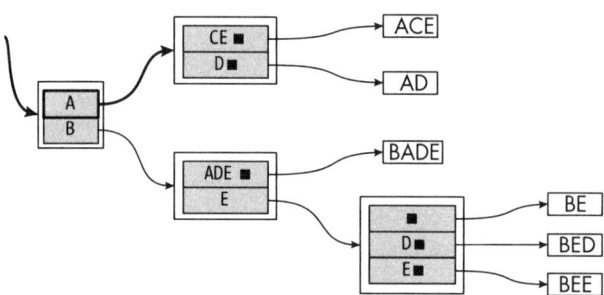

Figure 16-18: Searching unsuccessfully for ACED in a radix tree

What happens here? At the first level, you find an A link to follow, so you then look for CED■ at the second level. However, you don't find any link that has the search string's prefix, you fail. If you were searching for CAB■, you'd have the same problem, but at the root, because no link has a prefix with that string.

The idea is that you move down the tree, level by level, matching prefixes to links. Given two strings, you first need an auxiliary function to find how many characters they have in common from the start. For example, if you have BEE and BADE, you'd have only one character in common (B). If you have ACED and ACAD, you'd have two characters in common (the initial AC characters).

The auxiliary function looks like this:

```
const _commonLength = (str1, str2) => {
  let i = 0;
❶ while (str1[i] && str1[i] === str2[i]) {
```

```
  ❷ i++;
  }
  return i;
};
```

Start by comparing characters from the beginning ❶ and count ❷ until the strings end or stop matching.

With this function, you can start searching:

```
const _find = (trie, wordToFind) => {
❶ if (isEmpty(trie)) {
    return false;
  } else {
  ❷ const linkWord = Object.keys(trie.links).find(
      (v) => v[0] === wordToFind[0]
    );

  ❸ if (linkWord) {
    ❹ if (wordToFind === linkWord) {
        return trie.links[linkWord].data;
      } else {
      ❺ const common = _commonLength(linkWord, wordToFind);
      ❻ return _find(
          trie.links[linkWord.substring(0, common)],
          wordToFind.substring(common)
        );
      }
  ❼ } else {
      return false;
    }
  }
};
```

If the tree is empty, the key obviously can't be there ❶. To see whether a link is a prefix of the search key, first find the (only) object key that matches the first character ❷, because you can't have two or more keys with the same initial character in a node (it would break the structure). If you find such a link ❸, check whether it actually has the complete string you need ❹; if so, you're done. If the link doesn't match the whole string, find the common prefix ❺ and recursively search for the rest of the string ❻. On the other hand, if you don't find a link matching the initial character of the string, the search fails ❼.

The main find() function is the same as with object-based tries:

```
const find = (trie, wordToFind) =>
  !!wordToFind && _find(trie, wordToFind + EOW);
```

The logic for searching radix tries is similar to the logic used for tries, except matching links is a bit more complex.

Adding a Key to a Radix Tree

The problem with adding new keys to radix trees is if there's some previously added key that matches part of the key you want to add. Start with a simple case: adding ABE■ as with the original radix tree (see Figure 16-19).

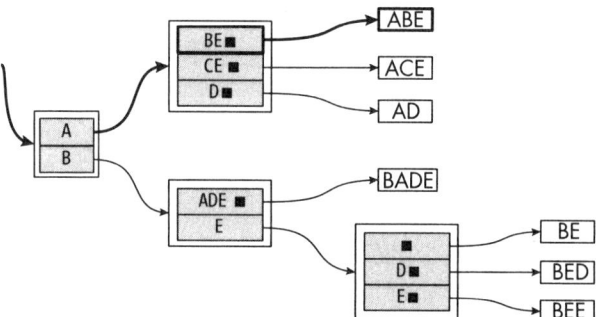

Figure 16-19: Adding ABE to a radix tree: initial structure

First, search for the string as shown in the previous section, and after following the A link, you arrive at a node with no links matching the BE■ prefix. This means you simply can stop searching right here and add the new link in that place.

Figure 16-20 shows the more complex case of adding BAD to the tree updated in Figure 16-19.

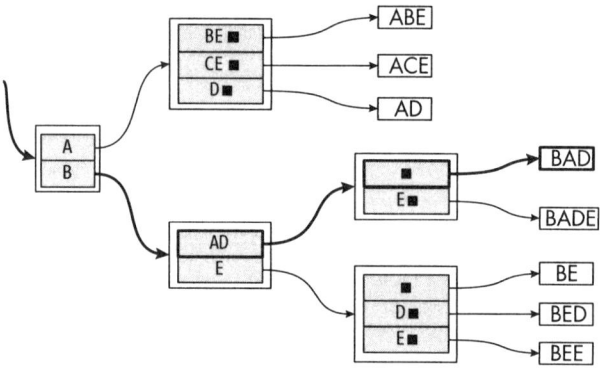

Figure 16-20: Adding BAD to a radix tree requires splitting some links.

The difference here is that after following the B link, you find there already was a partial match (ADE■), so now you need to do a split: the common prefix (AD) remains in the node, and you create a new node with the rest of the keys.

The logic thus needs to deal with two cases: not finding a prefix (as with ABE) or finding a partial match (as with BADE). Here's the code:

```
const _add = (trie, wordToAdd, data) => {
❶ if (wordToAdd) {
  ❷ if (isEmpty(trie)) {
      trie = newNode();
      trie.links[wordToAdd] = { data };
    } else {
    ❸ const linkWord = Object.keys(trie.links).find(
        (v) => v[0] === wordToAdd[0]
      );
    ❹ if (linkWord) {
        const common = _commonLength(linkWord, wordToAdd);
        const prefix = linkWord.substring(0, common);
        const oldSuffix = linkWord.substring(common);
        const newSuffix = wordToAdd.substring(common);

      ❺ if (linkWord === prefix) {
          trie.links[linkWord] = _add(trie.links[linkWord], newSuffix, data);
        } else {
        ❻ trie.links[prefix] = {
            links: {
              [oldSuffix]: trie.links[linkWord],
              [newSuffix]: { data }
            }
          };
        ❼ delete trie.links[linkWord];
        }
      } else {
      ❽ trie.links[wordToAdd] = { data };
      }
    }
  }
❾ return trie;
};
```

First check whether you are adding a nonempty string ❶; if not, return the trie unchanged ❾. If you have a string to insert, check whether the tree is empty ❷, because if it's empty, create a new node with the single key, and you are done. If the tree isn't empty, search for an existing key that matches at least the first character ❸ (as shown previously). If you don't find one, create a new node ❽; otherwise, check whether the found key matches what you're inserting in full or in part ❹ by splitting the key according to the common length you found. If the match is in full ❺, follow the link to insert the rest of the string; otherwise ❻, do a split, leaving the common prefix in the node by adding a new node and deleting the old key ❼.

The rest of the logic is the same as for the other tries:

```
const add = (trie, wordToAdd, dataToAdd = wordToAdd) =>
  _add(trie, wordToAdd + EOW, dataToAdd);
```

You have now seen how to add a new key, which had two distinct cases; now see what happens with the reverse algorithm to remove a key from a radix tree.

Removing a Key from a Radix Tree

Continuing with the same radix tree that was updated in Figure 16-20, consider two cases for removing keys: removing a link from a node, leaving it with two or more links, and removing a link from a node, leaving it with a single link. Take a look at the simpler former case first (Figure 16-21).

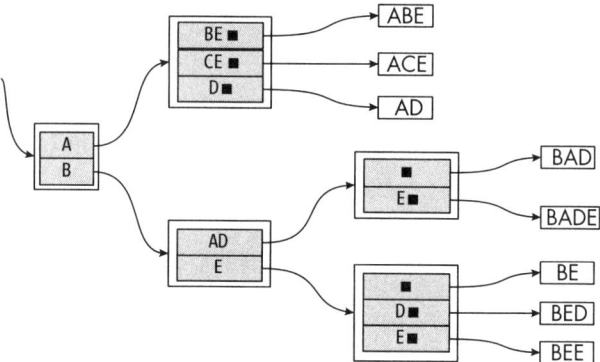

Figure 16-21: The radix tree from which you'll remove ACE

To remove ACE from the radix tree, search for it, and when you arrive at the final link, simply remove it. The tree would look like Figure 16-22.

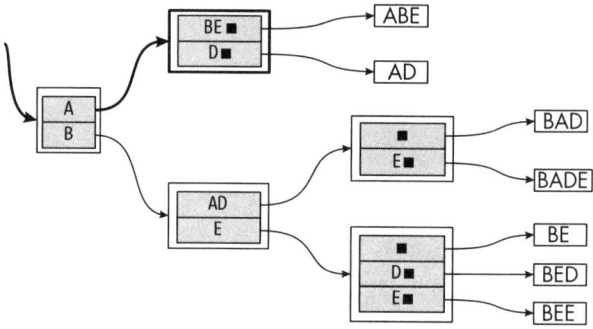

Figure 16-22: Removing ACE requires removing only a link.

However, what if you want to remove BADE? Remember, if you have a single link, you need to join it with the parent link, as shown in Figure 16-23.

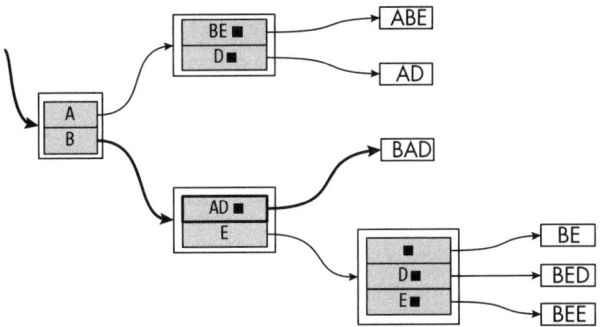

Figure 16-23: Removing BADE requires more work.

After searching for BADE and removing the last link (E■), the final node ended up with only an EOW link. You then push it up so that it joins with its parent, which is an AD link. Doing this reverses the kind of split you did when adding new keys.

Consider the actual logic. You need an auxiliary _remove() function, as in the other cases, to remove the desired key:

```
const _remove = (trie, wordToRemove) => {
❶ if (!isEmpty(trie) && wordToRemove > "") {
  ❷ const linkWord = Object.keys(trie.links).find(
      (v) => v[0] === wordToRemove[0]
    );
  ❸ if (linkWord && wordToRemove.startsWith(linkWord)) {
      const common = _commonLength(linkWord, wordToRemove);
      const prefix = linkWord.substring(0, common);

    ❹ if (wordToRemove === prefix) {
        delete trie.links[prefix];
    ❺ } else {
        trie.links[prefix] = _remove(
          trie.links[prefix],
          wordToRemove.substring(common)
        );
      ❻ if (Object.keys(trie.links[prefix].links).length === 1) {
          const single = Object.keys(trie.links[prefix].links)[0];
          trie.links[prefix + single] = trie.links[prefix].links[single];
          delete trie.links[prefix];
        }
      }
    }
  }
❼ return trie;
};
```

Start by checking whether you're done, which means reaching an empty link or the end of the word you're removing ❶. If you aren't finished, search for a link that has a common prefix with the word you're removing ❷ as described earlier. If you don't find one, you're also done, because the word you're removing isn't in the trie. If you find a suitable prefix ❸, you have two possibilities: you've found the whole word or part of the word. In the former case ❹, delete the link; no more work is needed. In the latter case ❺, search for and remove the rest of the word, discounting the prefix already found (you'll do this recursively). After that, check whether the trie ended with a single key ❻, in which case, join its only key with the prefix you found. Finally, return the updated trie ❼.

With this function out of the way, you can code the remove() function:

```
const remove = (trie, wordToRemove) => {
❶ if (!isEmpty(trie)) {
  ❷ _remove(trie, wordToRemove + EOW);
  ❸ if (Object.keys(trie.links).length === 0) {
      trie = null;
    }
  }
  return trie;
};
```

If the trie isn't empty ❶, use the auxiliary function to remove the key from the tree ❷ and then a final check remains: if the root became empty ❸, the trie is null.

Considering Performance for Radix Trees

Radix trees are like "compressed" versions of tries. They're faster when nodes include longer strings, and at worst, they behave the same way as tries if all links are "single-character" links. This means you don't need to do much analysis. Table 16-3 shows the performance.

Table 16-3: Performance of Operations for Radix Trees

Operation	Performance
Create	$O(1)$
Empty?	$O(1)$
Add	$O(1)$
Remove	$O(1)$
Find	$O(1)$

The results for radix trees are the same as for tries: all operations are $O(1)$.

Ternary Search Tries

Tries and radix trees are based on the idea of storing keys "down the links." You can also apply this idea to *ternary trees*, a new structure that saves space and provides good performance. We mentioned ternary trees in Chapter 13, as a generalization of binary trees having three links at each node instead of two, but we didn't actually work with them. The idea is that with a ternary search tree, each node has a key (a single character) and three links:

- The left link is for strings whose current character is less than the node key.
- The middle link is for strings whose current character is equal to the node key.
- The right link is for strings whose current character is greater than the node key.

Figure 16-24 shows a ternary search tree.

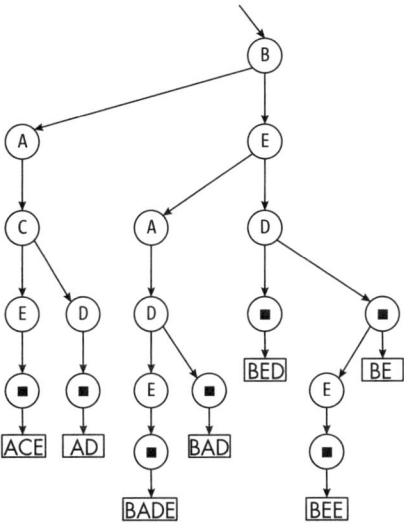

Figure 16-24: A ternary search tree with seven words

Try to follow the path to every word; you'll then understand the difference between the middle links and the right and left links. The following sections explain this in detail.

Defining Ternary Tries

We've already considered binary search trees, and defining a ternary tree is similar. All you need to do is add a middle link:

```
const EOW = "■";

const newTernary = () => null;

const newNode = (key) => ({
  key,
  left: null,
  right: null,
  middle: null
```

```
});
```

```
const isEmpty = (tree) => tree === null;
```

Creating ternary trees is quite simple, but storing extra data and doing searches require some changes.

Storing Extra Data in a Ternary Trie

With ternary tries, you have the same issue as with other common trees. You could store extra data next to the key, but again you have the problem that keys aren't stored in any single place, as they are with search trees; the keys are spread all over the trie.

To alleviate this complication, reapply the solution used for tries. When you reach the EOW character, use its middle pointer to store extra data:

```
tree.middle = { data };
```

Of course, if you're using the ternary trie just to store the keys and don't care about extra data (for example, if you were using a ternary trie to look up words to see whether they exist in a Scrabble application to check a weird-looking word your opponent entered), you can omit that line and modify the search function (which you'll look at next) to return only true or false.

Searching a Ternary Trie

Given how we designed ternary tries, it should be no surprise that searching for a key in a ternary trie is similar to the logic for binary search trees (see Chapter 12). As an example, Figure 16-25 shows how to look for the word AD (you'll actually search for AD■, with the appended EOW character).

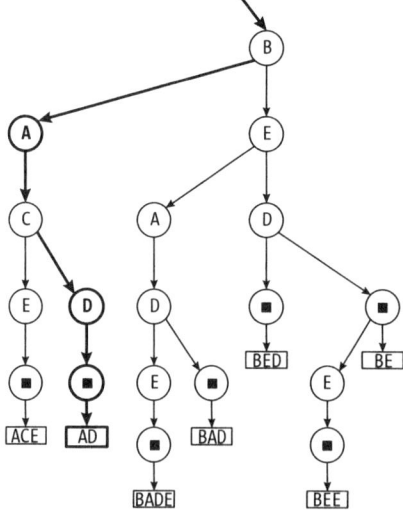

Figure 16-25: Searching successfully for AD in a ternary trie

The search starts at the root, which has a B. Since A (the first letter in AD■) is less than B, follow the left link. Then you find an A, so go down the middle link, looking for the rest of the string, which is D■. You find a C, so go down the right link. Then you find a D, so now go down the middle, looking for the EOW character. After finding it, successfully return the associated data.

If you had been looking for ADD instead, you'd have found an empty link after the first D, so the search would have been a failure.

As mentioned previously, the ternary trie search process is similar to that of binary search trees. The logic is as follows, and as with other cases, an auxiliary function comes in handy:

```
const _find = (tree, wordToFind) => {
❶ if (isEmpty(tree)) {
    return false;
❷ } else if (wordToFind.length === 0) {
    return tree.data;
❸ } else if (tree.key === wordToFind[0]) {
    return _find(tree.middle, wordToFind.substring(1));
❹ } else {
    return _find(wordToFind < tree.key ? tree.left : tree.right, wordToFind);
  }
};
```

If the tree is empty or you reach an empty subtree in the search, the search has failed ❶. Otherwise, search recursively character by character. After passing the EOW character, you've found the key and can return its data ❷; if you aren't storing any data, just return true. If the current tree key matches the first character of the word you're searching ❸, go down the middle link to match the rest of the word. If it isn't a match ❹, go down the left or right link as in a binary search tree.

To do a search, you need the same find() function as with other tries:

```
const find = (trie, wordToFind) =>
  !!wordToFind && _find(trie, wordToFind + EOW);
```

As you can see, working with the ternary trie isn't that different from working with regular tries, but things will change a bit more when adding or deleting keys from the structure.

Adding a Key to a Ternary Trie

The logic for adding a key to a ternary trie is also similar to adding a key to binary search trees. Start at the root and compare the first character of the string to be added with the character at the root and decide whether to go left or right (if there's no match) or down the middle (if the characters match). This process continues until you either find the key (and don't need to do anything) or decide you need to add it.

Try to add ABE to the ternary trie from the previous section; Figure 16-26 highlights the steps.

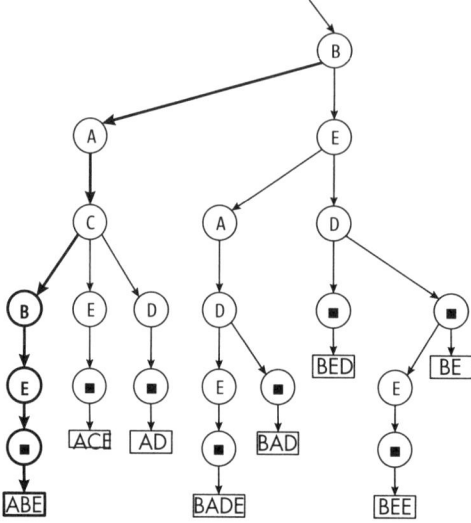

Figure 16-26: Adding ABE to a ternary trie

You want to add ABE■ (remember the EOW), so a comparison with the root's B makes you go down the left link. You find an A there, so you go down the middle link. Then, you find a C, and as you want to add BE■, again you go down the left link, but because it's null, you add a new subtree with B at its root in that location, an E down the middle link, and finally an EOW pointing to the extra data for the key.

Here's the actual logic:

```
const _add = (tree, wordToAdd, data) => {
❶ if (wordToAdd.length > 0) {
  ❷ if (isEmpty(tree)) {
      tree = newNode(wordToAdd[0]);
    }
  ❸ if (tree.key === wordToAdd[0]) {
    ❹ tree.middle =
        wordToAdd[0] === EOW
          ? { data }
          : _add(tree.middle, wordToAdd.substring(1), data);
  ❺ } else {
    ❻ const side = wordToAdd < tree.key ? "left" : "right";
    ❼ tree[side] = _add(tree[side], wordToAdd, data);
    }
  }
❽ return tree;
};
```

First, process the string you want to add character by character until the end ❶. If you reach an empty link, create a new tree ❷ and add the rest of the pending characters in the new key. If the current node's key matches the first character of the string you are adding ❸, either add the data (if you're already at the EOW character) or recursively add the rest of the string to the middle-link subtree ❹, which moves you down the tree. If there isn't a match ❺, decide whether to go down the left or right link ❻ and use recursion to add the string there ❼. At the end, return the updated trie ❽.

As before, the actual add() function makes use of the auxiliary _add() function:

```
const add = (tree, wordToAdd, data = wordToAdd) =>
  _add(tree, wordToAdd + EOW, data);
```

Addition didn't end up being that different from binary tries, if you dismiss the logic dealing with the middle links. Deletion will be similar, but with some complications.

Removing a Key from a Ternary Trie

Removing a key is harder than the other functions, but with the lessons from Chapter 12 on how to deal with maintaining the proper data structure after removing any given node, you'll get it done. (A spoiler: as with binary search trees, you'll have to deal with deleting a node differently depending on how many children a node has.) Start with a more complex ternary trie, which will provide some unique cases, as shown in Figure 16-27 (I replaced some subtrees with triangles for clarity).

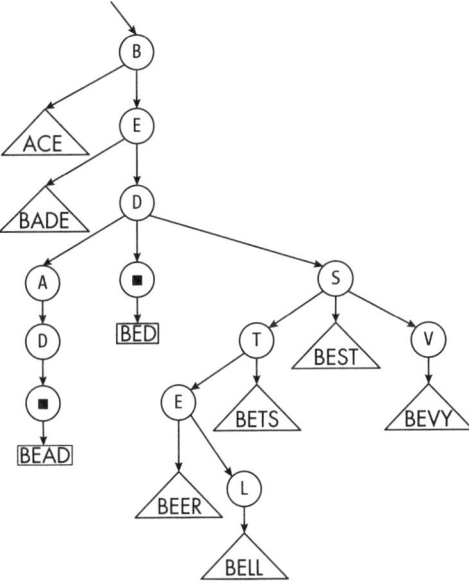

Figure 16-27: A ternary trie from which you'll delete some words

Notice that when deleting a word, you need to delete several nodes along the path from the root to the final EOW, because now the keys aren't stored in a single node. However, you can't delete the *entire* path; you stop wherever the path joins another different path.

Let's start with a simple example: What if you want to delete BEAD, BEVY, or BELL? In each of those cases, just remove all nodes from the EOW backward until you get to a node that is part of a different path, and that's it. Removing those three words results in the trie shown in Figure 16-28.

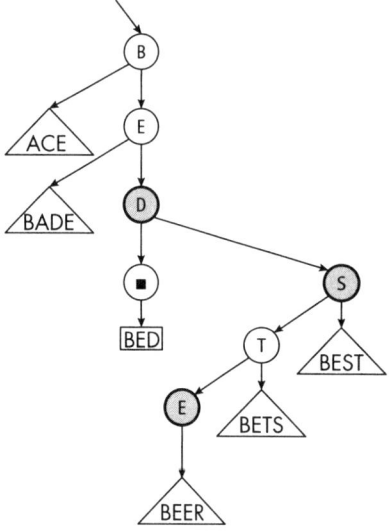

Figure 16-28: Deleting words may require removing nodes upward.

The "cut" is at the marked nodes, which still keep their middle link and (in the case of BEST) another side link. Now consider what happens if you want to remove BETS. You shouldn't leave the T node, because that's wasteful. Since that node has a single child, you can work as you did with binary search trees and just link the node's parent to the non-null child (see Figure 16-29).

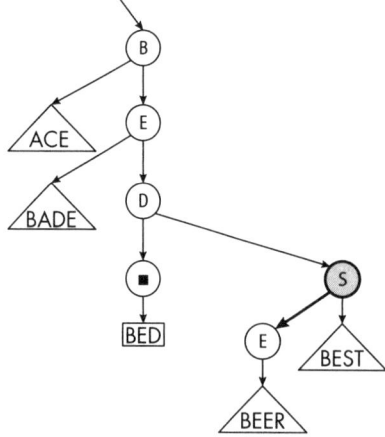

Figure 16-29: Removing BETS from the ternary trie requires changing a link from the parent.

The key problem happens when you need to delete a node that has both left and right links. Go back to the original trie and consider removing BED. To recap, Figure 16-30 shows the original trie.

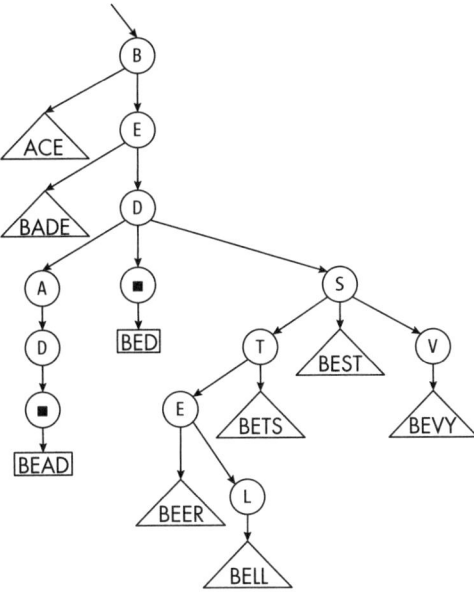

Figure 16-30: A ternary trie from which you'll remove BED

When removing BED, you can't just remove the D node, because that would break the trie, losing many other words; however, you don't want to leave it either, because it isn't a part of any word. The solution is similar to what you did with binary search trees: you can find the next word in its right subtree and use it to replace the word you're deleting. In this case, the next word is BEER, as shown in Figure 16-31.

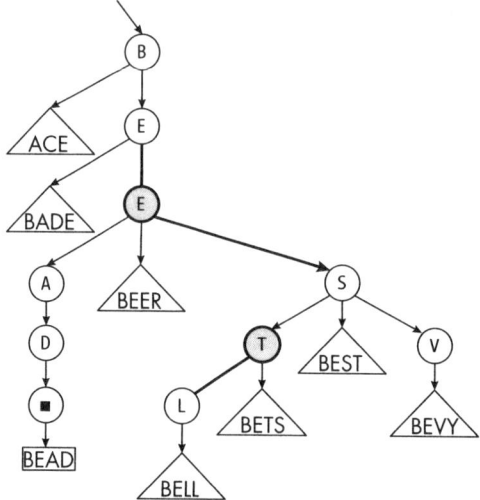

Figure 16-31: Put BEER in the place of BED to maintain the structure of the ternary trie.

After finding the word that's going to replace the deleted one (BEER replaces BED), you need to adjust several links to maintain the proper trie shape. If the right child has no left child, the changes are easier. See if you can figure out how to remove BEST and replace it with BEVY.

The logic is as follows:

```
const _remove = (tree, wordToRemove) => {
❶ if (isEmpty(tree)) {
     // nothing to do
❷ } else if (wordToRemove.length === 0) {
     tree = null;
   } else {
   ❸ if (wordToRemove[0] === tree.key) {
       tree.middle =
         tree.key === EOW
           ? null
           : _remove(tree.middle, wordToRemove.substring(1));

     ❹ if (isEmpty(tree.middle)) {
       ❺ if (isEmpty(tree.left)) {
           tree = tree.right;
         } else if (isEmpty(tree.right)) {
           tree = tree.left;
       ❻ } else {
           let treeR = tree.right;
           let prev = null;
         ❼ while (!isEmpty(treeR.left)) {
             prev = treeR;
             treeR = treeR.left;
           }
         ❽ if (prev) {
```

```
                    prev.left = treeR.right;
                    treeR.right = tree.right;
                }

            ➒ treeR.left = tree.left;
                tree = treeR;
            }
        }
    ➓ } else {
            const side = wordToRemove < tree.key ? "left" : "right";
            tree[side] = _remove(tree[side], wordToRemove);
        }
    }

    return tree;
};
```

If the tree is empty, do nothing else ➊. If you reach the end of the word (past the EOW character, as in the search), set the tree to null, and you are done ➋. While the character you are looking for matches the current node's key, follow its middle link recursively ➌, but if you reach the EOW, set that link to null. After doing the removal, see if you're at a node with an empty middle link ➍, which is a situation you don't want. If one of the two (left or right) links ➎ is empty, choose the other link, but if both are non-null ➏, you need to find the leftmost node at the right ➐ and fix the links as described earlier ➒➒. (If prev === null, there was just one single node to the right; otherwise, you need to go down the left links.) If you don't find the right character earlier, you need to go left or right ➓.

With this function out of the way, removing a key is the same function as with other tries:

```
const remove = (tree, wordToRemove) => _remove(tree, wordToRemove + EOW);
```

You have now seen all the functions that work with ternary tries, so now let's consider their performance.

Considering Performance for Ternary Tries

Ternary tries are different from the other tries we saw in this chapter in one way: the shape of the tree depends on the order of additions and removals, and that means you can have a worst case that behaves differently from the normal, average case. You may remember that this also happened with plain binary search trees. The worst-case performance differed from the average performance, and that depended on how the tree had been created.

If you insert all keys in ascending order, adding, searching, and removing will all be $O(sk)$ if you store keys that are k characters long, with an alphabet of s symbols, which is a far better result than $O(n)$! To be proper, $O(sk)$ is actually $O(1)$ because sk is constant, so the ternary tries have the same performance as other tries.

Summary

In this chapter, we looked at several variants of search trees called tries that apply a different concept: instead of directly storing and comparing keys, tries work on a character-per-character basis, providing an enhanced performance in comparison to the search trees we looked at previously, without the need for complex operations, such as rotating or balancing nodes.

The structures we studied here are frequently used for dictionaries (enabling quick look-up of words) or straightforward searches: given a key, find its related data. The assured performance of tries makes them a useful data structure, particularly if you are doing some kind of work where you need to be as fast as possible and don't want to deal with unexpected worst cases that may take too long to process.

Questions

16.1 Maps for Tries

Can you implement a trie using a map instead of an array or an object?

16.2 Ever Empty?

From the section "Adding a Key to a Radix Tree" on page 410, in the _add() function, can wordToAdd ever be empty, and if so, when?

16.3 Rotate Your Tries

Can you apply rotations to a ternary trie?

16.4 Empty Middle?

True or false: middle links can never be empty in a ternary trie.

16.5 Four-Letter Ternary Trie?

Suppose you start with an empty ternary trie and add in order keys AAAA, AAAB, . . . AAAZ, BAAA, . . . BAAZ . . . all the way to ZZZZ— with all possible combinations of four letters. How tall would that trie be?

16.6 How Do They Look?

As a recap, how would all the structures in this chapter look (array-based tries, object-based tries, radix trees, and ternary tries) if you just added a single word, ALGORITHM, to them?

17

GRAPHS

In the previous chapters we discussed several data structures, and in this chapter we'll consider a new topic, how to represent graphs. We'll also look at several algorithms related to graphs, such as finding the shortest paths, calculating distances, checking software dependencies, and more.

What Are Graphs?

An abstract definition might be that a graph is a set of objects in which pairs of those objects are somehow related. The objects are called *vertices* (plural of *vertex*), but they're also called *points* and *nodes*. The relationships between pairs of vertices are graphically represented with lines joining the pairs. These lines are called *edges*, *arcs*, *arrows*, or just plain *links*. The number of arcs connected to a point is called its *degree*. Points linked in this fashion are sometimes called *neighbors* or are considered to be *adjacent* to each other. The same word is used in a similar sense: edges are considered to be adjacent if they share a common vertex.

These definitions may sound vague or rather "mathematical" (in fact, a branch of mathematics that specifically studies graphs and their properties is called *graph theory*), so this chapter will explore some practical examples. (We've actually already studied graphs. Trees are graphs; indeed, the definition fits them.) Some use cases for graphs include the following:

- Relationships among people, where you can have people (nodes) and friendships (arcs), so that if two people are friends, they are linked
- Dependencies in code, with modules (nodes) that import components (arcs) exported from other modules
- Projects with tasks (nodes) that can't be started until some other tasks have been finished (arcs)
- Maps, as in GPS-based applications, with cities (nodes) and roads (arcs)

Figure 17-1 shows an example of the latter, using a graph to represent a part of a city or a country.

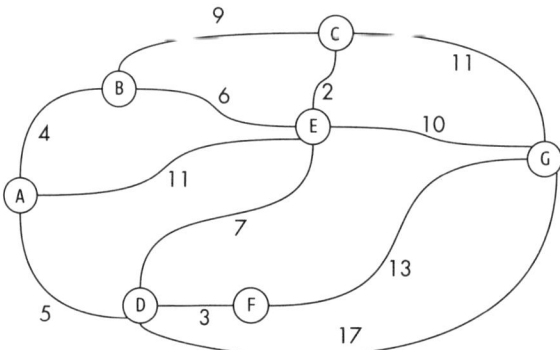

Figure 17-1: A graph representing some cities and roads linking them

In this graphical map, vertices represent cities (or street corners, or countries), and the edges represent roads (or street blocks, or flights). In Figure 17-1, the edges are *undirected*, meaning that one may travel any direction—for example, from A to E or from E to A.

In a city map, where streets may be one-way only, we'd need a *directed* graph instead, as shown in Figure 17-2. In these graphs, we can speak of the *outdegree* of a node (how many arcs lead out from it) or the *indegree* of a node (how many arcs lead into it).

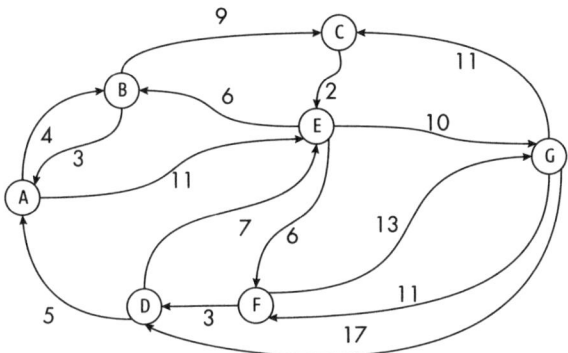

Figure 17-2: A directed graph where you can travel the roads
in only one direction

Edges usually have values associated with them, like time or cost, making them *weighted* graphs. *Unweighted* graphs with no values associated with edges are also possible; the association itself is all that matters. Don't assume that symmetry or any other rules apply to directed graphs. For instance, in the graph in Figure 17-2, you can go from B to C directly, but you can't go back from C to B in one step. Also, going from A to B doesn't cost the same as going from B to A. Finally, going from G to D via F could be cheaper than going directly from G to F. In some cases weights could be negative, but we're always working here with non-negative values.

NOTE *Graphs can also have multiple edges between any pair of vertices, but we're not going to consider those. For all the algorithms in the chapter, it'll be enough to just use the shortest edge and simply ignore the others. Another possibility we'll ignore here is edges from a vertex to itself, which are called* loops.

We'll consider the following kinds of processes:

- Given two vertices, you may want to know whether there's a path from the first to the second. As an extension to this, you might want to find the path with the minimum cost (the *shortest path*) from one vertex to another one or to all other vertices.

- A directed graph may represent a project with tasks and dependencies between them: you might want to find an ordering so that no task can start until all previous tasks have been finished; this is called a *topological sort*.

- Building on this example of tasks and dependencies, you may worry that some kind of cycle (A before B before C before A) would make sorting impossible. A related problem to topological sorting is *cycle detection*.

- An undirected graph may represent geographic points with the edges showing the cost of joining them with, say, electrical lines or communication cables. A *minimum spanning tree* shows how to choose edges so all points are connected to each other at the lowest total cost.

- Along the same lines, given the previous graph, you could ask whether it's connected (meaning that it's possible to reach every point from any other point) or unconnected. In that case, you want to implement *connectivity detection*.

This list of procedures isn't complete, but it covers the most important algorithms. Let's start by considering how to represent graphs and then move on to the necessary algorithms.

One final note: when discussing the performance of algorithms, we'll use v to stand for the number of vertices and e for the number of edges. Take care not to confuse this with the mathematical constant e, the basis for natural logarithms, $2.718281728 \ldots$!

Representing Graphs

There are several ways to represent graphs, and we'll consider the three most used methods: adjacency matrix, adjacency list, and adjacency set.

Adjacency Matrix Representation for Graphs

The *adjacency matrix* representation is the simplest and basically shows which nodes are adjacent to each other. This type of graph is represented by a square matrix, with a row and column for each vertex. If there's a link from vertex i to vertex j, the matrix has a value at position [i][j]: this is just a true value for unweighted graphs or the associated edge's cost for weighted graphs. If there's no link between those vertices, the matrix has false or a special value (zero or +infinity) at the corresponding position. For undirected graphs, note that position [i][j] will always be equal to position [j][i]; the matrix will be symmetrical with regard to its main diagonal.

Consider the directed graph from the previous section again (Figure 17-2). The matrix representation for that graph could be as Figure 17-3 shows.

	A	B	C	D	E	F	G
A	0	4	∞	5	11	∞	∞
B	4	0	9	∞	6	∞	∞
C	∞	9	0	∞	2	∞	11
D	5	∞	∞	0	7	3	17
E	11	6	2	7	0	∞	10
F	∞	∞	∞	3	∞	0	13
G	∞	∞	11	17	10	13	0

Figure 17-3: The adjacency matrix representation for the graph in Figure 17-2

We've used the option of representing missing edges with zero; an equally valid solution is to represent those with +infinity. No matter how you represent missing edges, in both cases, the diagonal of the matrix is zero. There's no cost involved in going from a point to itself; you're already there.

This representation is quite easy to work with, but it requires a lot of space for large graphs. As to performance, operations like checking whether two vertices are adjacent or adding or removing edges are $O(1)$, which is as fast as possible. On the other hand, processing the list of edges of a vertex is $O(v)$ no matter how many neighbors a node actually has.

If nodes have only a few neighbors, most of the matrix will be marked as empty (making it a *sparse* matrix), which means you're wasting space. For those cases, you can choose adjacency list representations.

Adjacency List Representation for Graphs

The matrix shown in Figure 17-3 wastes too much space and causes all procedures that need the list of arcs out of a point to be $O(v)$. You can use lists so that for each vertex, you'll have all the points to which it connects and also all of the points that connect to it.

For the same directed graph shown in Figure 17-2, the *adjacency lists* representation is as shown in Figure 17-4 (compare this with the matrix representation in Figure 17-3).

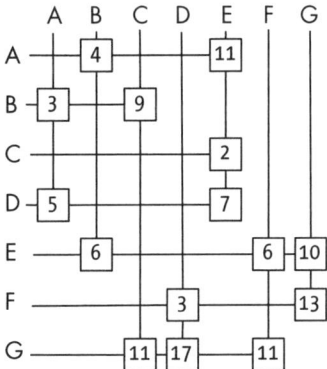

Figure 17-4: The adjacency list representation is an alternative to the adjacency matrix.

For each vertex, you have two lists: one with outgoing edges (shown horizontally, as rows) and one with incoming edges (shown vertically, as columns). For example, in the first row, you see that from A, one may reach B (at a cost of 4) or E (at a cost of 11). Looking at the first column, you see that you can reach A from B (at a cost of 3) or D (at a cost of 5). Each element in the structure would have a pointer to the next in the same row and another to the next in the same column. You could also work with doubly linked lists for easier updates. Nodes also have to carry the identities of both endpoints.

With this structure, it's easy to process all edges that start or end at a given vertex quickly, and that will speed up several algorithms. However, things become slower if you just want to know whether two given points are directly connected; with this structure, it would be an $O(e)$ operation. You may opt for using sets instead of lists, as shown later.

Adjacency Set Representation for Graphs

The more complex solution we can propose involves using sets (as in Chapter 11) or trees (as in Chapter 12) instead of lists. For instance, when working with balanced search trees, checking whether two points are connected is a $O(\log e)$ operation. (You could consider an average degree of e/v edges per node, and then it would be $O(\log e/v)$ instead.) With this structure, each vertex is associated with two maps: one for outgoing edges and one for incoming edges. The key for both maps would be the "other" point in the edge. Adding and removing edges are both $O(\log e)$ operations, so performance is better. Processing all arcs out of a node is as fast as with lists.

Finding the Shortest Paths

A common problem is as follows: given two points in a graph, find a path from the first to the second. A path is a sequence of adjacent edges (each starting where the previous ends) that begins at the first point and ends at the second one.

We already considered this kind of problem when you found a path through a maze in Chapter 5, so instead solve a more complex problem: finding the *shortest* path from a node to another node, or even more generally, finding the shortest path for a node to all other nodes. These algorithms will not only find whether a path exists, but they'll also find the best one (cheapest, shortest) among all possibilities. If you just want to find a path, any path, you can simply stop searching as soon as you reach the destination point.

Floyd-Warshall's "All Paths" Algorithm

This first algorithm is interesting because it applies dynamic programming, which you explored in Chapter 5. The *Floyd-Warshall algorithm* doesn't have the best possible performance (you'll see other options for that), but it's definitely the simplest. There's another difference: here, you'll find the shortest distance between all pairs of nodes, while in other cases you may just want to find the distance between a specific pair. This stipulation will have an impact on the performance, but in some cases, having the whole table of distances may be exactly what's required.

Assume the existence of a function $distance(i,j,k)$ that returns the length of the shortest path from point i to point j using, at most, the first k nodes in the graph for the path. (In other words, you're not considering any path through the rest of the nodes.) You want to calculate $distance(i,j,n)$ for all values of i and j.

The value of *distance(i,j,k)* must be either a path that doesn't include the *k*th node or else a path that goes from *i* to *k* and then from *k* to *j*, whichever is shortest. In other words, *distance(i,j,k)* is the minimum of *distance(i,j,k − 1)* and *distance(i,k,k − 1) + distance(k,j,k − 1)*. This formula is key; be sure you totally agree with it. The definition is recursive, but the base case is simple: *distance(i,i,0)* is 0 for all points, and for *i ≠ j*, *distance(i,j,0)* is the edge from *i* to *j*, if it exists, or +infinity instead. (What about finding the actual paths and not just the distances? See question 17.1.)

As an example, work with the graph in Figure 17-5. It's undirected for simplicity, but the algorithm works with directed graphs as well.

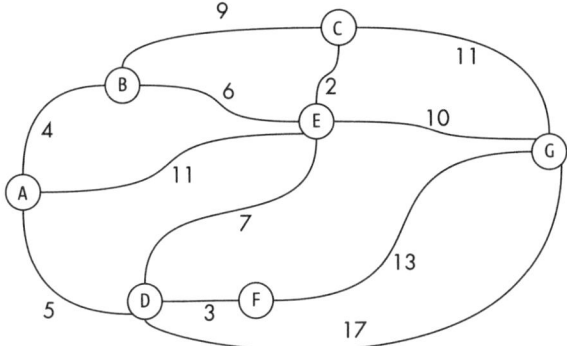

Figure 17-5: A graph for which you want to find minimum distances between any pair of points

Figure 17-6 shows the initial array of distances in Figure 17-5.

	A	B	C	D	E	F	G
A	0	4	∞	5	11	∞	∞
B	4	0	9	∞	6	∞	∞
C	∞	9	0	∞	2	∞	11
D	5	∞	∞	0	7	3	17
E	11	6	2	7	0	∞	10
F	∞	∞	∞	3	∞	0	13
G	∞	∞	11	17	10	13	0

Figure 17-6: The adjacency matrix for the graph in Figure 17-5, using infinity values for missing edges

The diagonal in Figure 17-6 is all zeros, and everything is +infinity except for the existing edges. (This is just the adjacency matrix using +infinity instead of zero for missing edges, as mentioned earlier in the chapter.) In the first iteration, check whether adding the first point (A)

as an intermediate shortens some distances. In effect, you find that now you can go from B to D at a cost of 9, as shown in Figure 17-7.

	A	B	C	D	E	F	G
A	0	4	∞	5	11	∞	∞
B	4	0	9	9	6	∞	∞
C	∞	9	0	∞	2	∞	11
D	5	9	∞	0	7	3	17
E	11	6	2	7	0	∞	10
F	∞	∞	∞	3	∞	0	13
G	∞	∞	11	17	10	13	0

Figure 17-7: When you try adding A as an intermediate point, you find a shorter distance between B and D.

The second iteration checks whether adding B as an intermediate makes for shorter routes. You find that you can now go from A to C at a cost of 13 and from C to D (C to B and then B to D via A) at a cost of 18, as shown in Figure 17-8.

	A	B	C	D	E	F	G
A	0	4	13	5	11	∞	∞
B	4	0	9	9	6	∞	∞
C	13	9	0	18	2	∞	11
D	5	9	18	0	7	3	17
E	11	6	2	7	0	∞	10
F	∞	∞	∞	3	∞	0	13
G	∞	∞	11	17	10	13	0

Figure 17-8: When you try adding B as an intermediate, you find two other shorter routes.

In the next iteration you'll check whether adding C as an intermediate helps; then, you'll try D, E, and so on. You keep iterating until all the nodes have been considered and the final result (check it out) provides all the distances between nodes (see Figure 17-9).

	A	B	C	D	E	F	G
A	0	4	12	5	10	8	20
B	4	0	8	9	6	12	16
C	12	8	0	9	2	12	11
D	5	9	9	0	7	3	16
E	10	6	2	7	0	10	10
F	8	12	12	3	10	0	13
G	20	16	11	16	10	13	0

Figure 17-9: After trying out all possible intermediate points, you compute the final distances matrix.

The code is short:

```
const distances = (graph) => {
❶ const n = graph.length;

❷ const distance = [];
❸ for (let i = 0; i < n; i++) {
     distance[i] = Array(n).fill(+Infinity);
   }

❹ graph.forEach((r, i) => {
   ❺ distance[i][i] = 0;
   ❻ r.forEach((c, j) => {
       if (c > 0) {
         distance[i][j] = graph[i][j];
       }
     });
   });

❼ for (let k = 0; k < n; k++) {
     for (let i = 0; i < n; i++) {
       for (let j = 0; j < n; j++) {
       ❽ if (distance[i][j] > distance[i][k] + distance[k][j]) {
           distance[i][j] = distance[i][k] + distance[k][j];
         }
       }
     }
   }

❾ return distance;
};
```

The n variable ❶ helps shorten the code; it's just the number of nodes in the graph. The distance array of arrays ❷ includes the distances from every node to every other node. Initially, set all distances to +infinity ❸ and then correct ❹. The distance from a point to itself is zero ❺, and the distance from a point to another, if connected, is the edge's length ❻. The distance array now has all distances with no intermediate points as described previously. The three nested loops systematically apply the dynamic programming calculation described earlier ❼, and if you find a better (smaller) distance ❽, update the table. In this case, you're keeping only the last table; the values for the *k*th iteration replace those of the previous one, because you won't need them any more. The final result ❾ is the table of distances between all pairs of points, as described.

What's the runtime order of this algorithm? The three nested loops, each $O(v)$, provide the answer: $O(v^3)$. This is steep, as mentioned, but it produces all distances among all pairs. For just the distance from one node to all the others, or even more specifically from one given node to another, you'll see better algorithms.

Bellman-Ford Algorithm

Now consider a different problem: find the distance from a given node to all the others. The idea of the *Bellman-Ford algorithm* is to see whether you can find a better path between two nodes by following a given edge, and repeat this process until no more alternatives are possible. Start by considering paths that are one edge long, and then check whether a shorter path is available when using two edges, then three edges, then four, and so on. If a graph has *n* nodes, the longest path can have *n* − 1 edges, so that's a limit for iterating.

Let's work with the same graph and calculate minimum distances from F to all the other nodes. (The algorithm works equally well with directed graphs, but you're using an undirected one for simplicity.) The initial situation is shown in Figure 17-10. Without processing any edges (not taking any routes), only F can (trivially!) be reached from F, and you cannot reach other nodes.

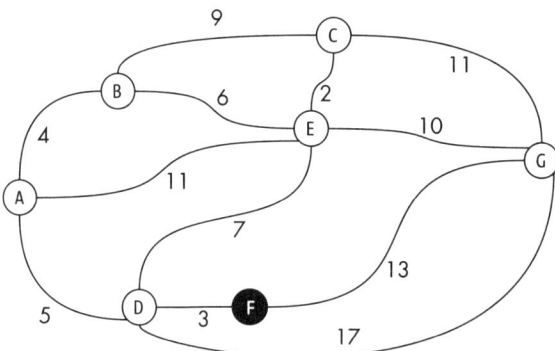

Figure 17-10: Apply the Bellman-Ford algorithm to find the minimum distances from F to all other points.

Assume that nodes are processed in alphabetical order. When you process the edges from A, B, or C, you cannot calculate distances because you don't know how to reach those nodes. When you process the (F,D) edge, you are now able to reach node D, and you have the first step in the paths, as shown in Figure 17-11.

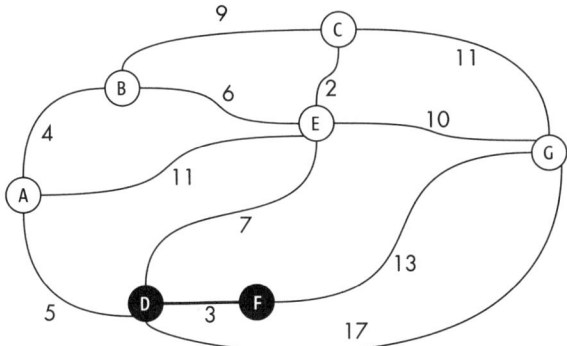

Figure 17-11: Processing nodes in order, the first other node you can reach from F is D.

You now know that you can reach D with a cost of 3. Now process the next edge, (F,G), and you can reach another node, as shown in Figure 17-12.

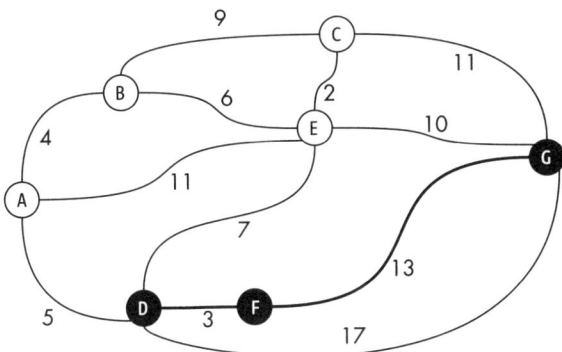

Figure 17-12: The other node you can reach from F is G; you'll reprocess D at a later iteration.

You can then process the edges out of G, because now you know that you can reach G. Processing edges (G,C) and (G,E) marks two other nodes as reachable. (Don't forget: any paths or distances you find may change later if better alternatives appear.) Figure 17-13 shows the new situation.

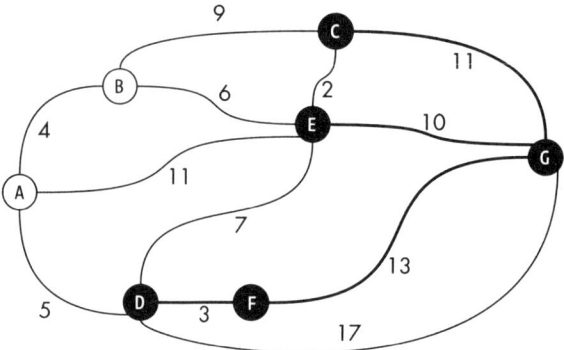

Figure 17-13: When processing G, which was reached from F, you can reach C and E as well.

The first pass is done. So far, you know you can reach five of the seven nodes, and you have possible (but maybe not optimal) paths for each.

Now start a new pass. You still can't do anything with edges out of A or B, because you still haven't gotten to those nodes, but you can process edge (C,B) and add a new path, as shown in Figure 17-14.

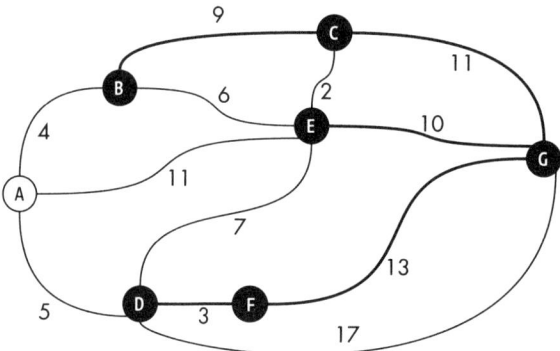

Figure 17-14: A second iteration now finds that B is reachable from the previously reached C.

The (C,E) and (C,G) edges don't represent shorter distances, so you do nothing. (For example, you can go from F to E at a cost of 23; going through C would cost 26, so it's no good.) When considering the (D,A) and (D,E) edges, things get interesting. You can now reach A, and you find a better path to E through D (see Figure 17-15).

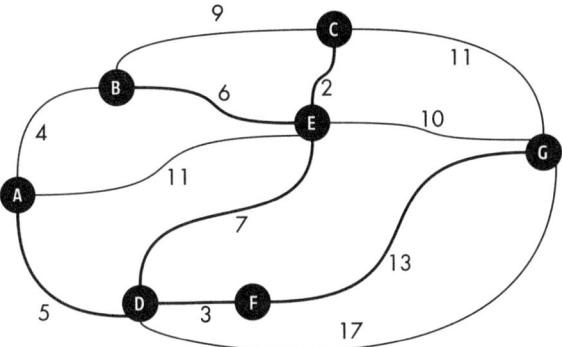

Figure 17-15: A second look at D finds a shorter path to E. The previous way was from F to G to E, which was longer than from F to D to E.

The previous best path from F to E was through G, at a cost of 23, but now you find you can go through D at a cost of 10. No more changes are possible, so start a third pass. One further enhancement is that it's a shorter route to B by going through A than through E (see Figure 17-16).

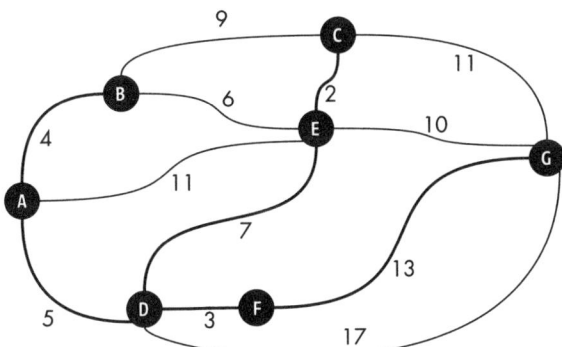

Figure 17-16: A third pass finds a better way from A to B, but no further changes.

The third pass doesn't add any more changes, and further passes won't either, so you're done. You know the shortest paths from the starting point F to all other points, and you know what edges to follow to achieve that cost.

You can code this algorithm as follows:

```
const distances = (graph, from) => {
  const n = graph.length;

❶ const previous = Array(n).fill(null);
❷ const distance = Array(n).fill(+Infinity);
❸ distance[from] = 0;

❹ const edges = [];
  for (let i = 0; i < n; i++) {
```

```
      for (let j = 0; j < n; j++) {
        if (graph[i][j]) {
          edges.push({ from: i, to: j, dist: graph[i][j] });
        }
      }
    }
  ❺ for (let i = 0; i < n - 1; i++) {
    ❻ edges.forEach((v) => {
        const w = v.dist;
      ❼ if (distance[v.from] + w < distance[v.to]) {
          ❽ distance[v.to] = distance[v.from] + w;
          ❾ previous[v.to] = v.from;
          }
      });
    }

  ❿ return [distance, previous];
};
```

Use a previous array ❶ to learn from which node you arrived at the corresponding node; if previous[j] is i, the shortest path from the start to j passed through i right before going to j. The distance array ❷ keeps track of the distances from the start to every node; initialize all distances to +infinity, except the distance from the start to itself ❸, which is obviously zero. In order to process all edges without having to go through the whole matrix, create an edges array ❹; iterating with this array will be faster. Now iterate n – 1 times ❺: for each edge ❻, see whether using it provides a shorter way between its two endpoints ❼; if so, update the distance to the second node ❽ and record from which node you came ❾. (Can you do better with fewer passes? See question 17.2.) The results of this algorithm are an array with distances and an array showing indirectly how to reach the start ❿ from any node.

What's the order of this algorithm? Given that the loop runs $O(v)$ times and each time it goes through all edges, the result is $O(ve)$. This is better than Floyd-Warshall's algorithm, but it finds all distances from a single origin to all the rest. You can do even better, as you'll see in a final algorithm for that problem.

Dijkstra's Algorithm

If you want to find the shortest path from one point to all others (or to a specific point in particular), Dijkstra's algorithm is quite efficient. It proceeds by starting at the first point (considered to be *visited*, at distance zero from itself), which becomes the initial *current* point. All other points are considered to be *unvisited* and at distance +infinity. From then on, it does the following:

1. Studies all as yet unvisited neighbors of the current node, and if there's a shorter path to the neighbor, it chooses that path and updates the distance to the unvisited node.

2. After having processed all the unvisited neighbors of the current node, mark that node as visited, and you're done with it.

3. If any unvisited points remain, choose the one with the shortest distance, make it the current node, and repeat the process.

4. The algorithm ends when no unvisited points remain (if you want the distances from the origin to all other points) or when the destination point has been marked as visited (if you just want that particular distance).

Consider an example and then the implementation. For simplicity, you'll work with the same undirected graph as before, but Dijkstra's algorithm works equally well with directed graphs. Figure 17-17 shows the initial configuration, with the origin point marked as visited.

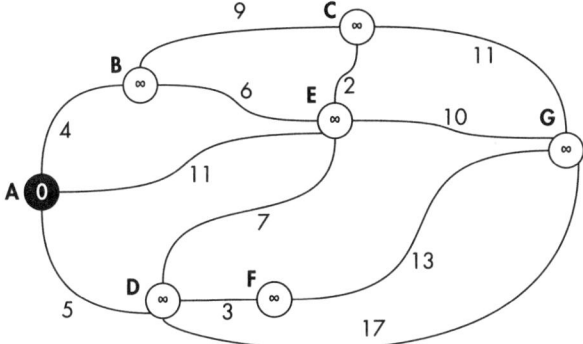

Figure 17-17: The initial setup for Dijkstra's algorithm: the distance from A to itself is zero, and distances from A to other nodes are set to +infinity.

Considering adjacencies from the current point (A), you know you can reach B, D, and E (see Figure 17-18).

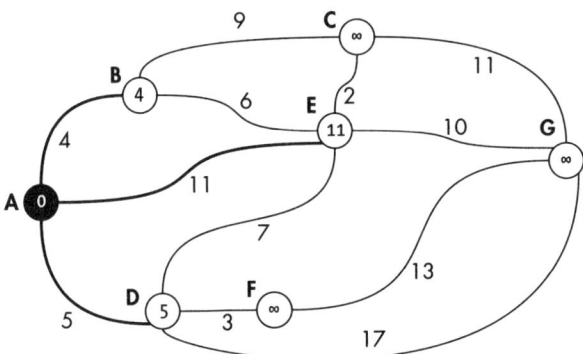

Figure 17-18: Considering the edges from A to its neighbors, you can tentatively update the distances to B, D, and E.

You've updated the distances to those three nodes, but they're strictly tentative at this point, as you may find better paths later—and you will, as there's a shorter way from A to E, for example.

The next step marks B as the current point, because it's the closest unvisited one (see Figure 17-19).

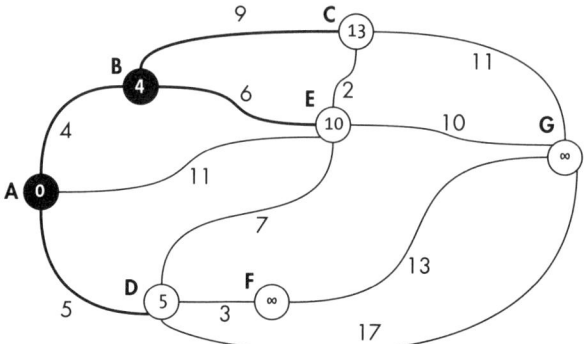

Figure 17-19: Repeat the procedure starting from B, the closest unvisited node, and find better distances to C and E.

You found a way to C with a distance of 13, because B was at a distance of 4, and the (B,C) edge costs 9. You also found a shorter way to E, so update those distances. Point B now is marked as visited, and you turn to D as the new current node (see Figure 17-20).

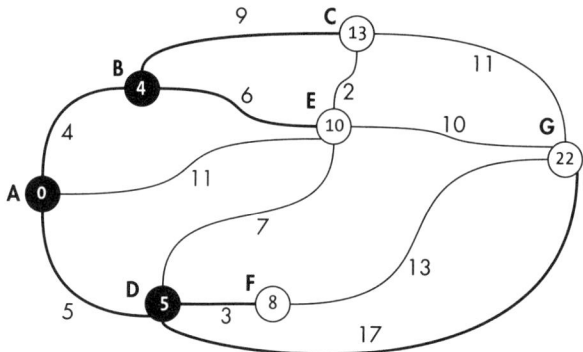

Figure 17-20: You now start from D, the next closest but not yet visited node, and find better distances to F and G.

Working from D allows you to update the distances and paths to F and G. The distance to E wasn't modified, because going through D would have required a distance of 12, and you already found a shorter path. Now, F becomes the current node, and the process will go quickly, because it allows only one path out, as shown in Figure 17-21.

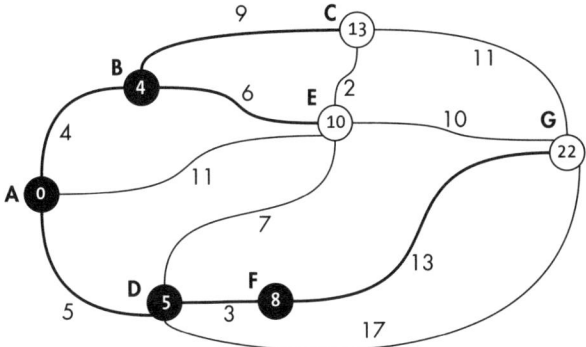

Figure 17-21: Processing F finds a better way to G.

You've updated the best (as of yet) path to G, which now is 21, going from A to D to F first. You're close to finishing, and E is the next node to process (see Figure 17-22).

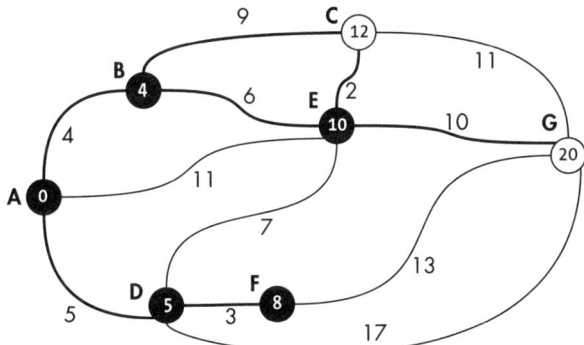

Figure 17-22: E is processed next and allows you to update distances to C and G.

Going through E makes the distances to C and G shorter, so update them. The next steps choose C and then G, and you don't need any further changes. Figure 17-23 shows the final result with the selected paths and calculated distances from A.

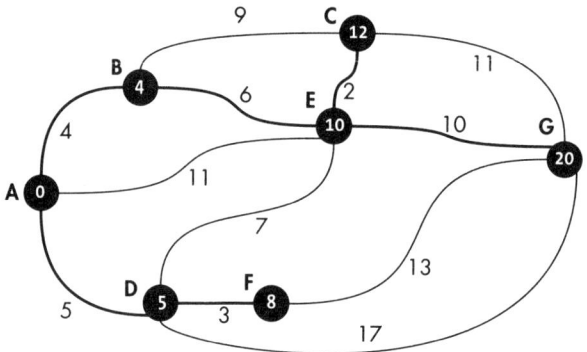

Figure 17-23: When you're done with all nodes, you get optimum distances from A to all other nodes.

To achieve good performance, it's important to be able to determine the next node (the one with shortest distance) to process quickly. A straightforward loop would be an $O(v)$ algorithm, but you've already seen an appropriate structure for that: a heap. Using that structure allows you to find the next node to process in $O(1)$ time, and updating the heap is then $O(\log v)$, which is better.

The main algorithm is as follows, but you'll look at a portion of it related to the heap later:

```
const distance = (graph, from) => {
  const n = graph.length;

❶ const points = [];
  for (let i = 0; i < n; i++) {
  ❷ points[i] = {
      i,
      done: false,
      dist: i === from ? 0 : +Infinity,
      prev: null,
      index: -1
    };
  }

❸ const heap = [from];
  for (let i = 0; i < n; i++) {
    if (i !== from) {
      heap.push(i);
    }
  }
❹ heap.forEach((v, i) => (points[v].index = i));

  // heap functions, omitted for now

❺ while (heap.length) {
    const closest = heap[0];
    points[closest].done = true;
    const dist = points[closest].dist;
```

```
❻ swap(0, heap.length - 1);
   heap.pop();
   sinkDown(0);

❼ graph[closest].forEach((v, next) => {
     if (v > 0 && !points[next].done) {
       const newDist = dist + graph[closest][next];
     ❽ if (newDist < points[next].dist) {
         points[next].dist = newDist;
         points[next].prev = closest;
         bubbleUp(points[next].index);
       }
     }
   });
  }

❾ return points;
};
```

The points array ❶ contains the distances from the starting point (from) to all others. The i attribute identifies the point; done marks whether you've finished processing it; dist is the distance, initialized to +infinity for all points except the initial; and prev shows from which point you arrived at the current one ❷. The index attribute requires explanation. As mentioned, you'll be keeping the distances in a heap and updating them, which may cause them to bubble up. However, you need to know where each point is in the heap, and that's what the index indicates. This way, whenever you update the distance for a point p you know that points[p].index is the corresponding place in the heap.

Push every point in the heap starting with from ❸ and update all the index values ❹. (Because there's a single zero distance in the points array and all others are +infinity, you've created a heap without needing any comparisons.) While the heap isn't empty ❺, you remove the top point ❻ and mark it as done and proceed to update the distances to all the nodes that it can reach ❼. If a distance gets updated with a lower value ❽, check whether it should bubble up in the heap. The final result ❾ is the updated points array with distances from the initial node to all others.

Now consider the heap code, directly based on what we saw in Chapter 14:

```
❶ const swap = (i, j) => {
   [heap[i], heap[j]] = [heap[j], heap[i]];
   points[heap[i]].index = i;
   points[heap[j]].index = j;
  };

❷ const sinkDown = (i) => {
   const l = 2 * i + 1;
   const r = l + 1;
   let g = i;
 ❸ if (l < heap.length && points[heap[l]].dist < points[heap[g]].dist) {
     g = l;
```

```
      }
      if (r < heap.length && points[heap[r]].dist < points[heap[g]].dist) {
        g = r;
      }
      if (g !== i) {
        swap(g, i);
        sinkDown(g);
      }
    };

❹ const bubbleUp = (i) => {
    if (i > 0) {
      const p = Math.floor((i - 1) / 2);
    ❺ if (points[heap[i]].dist < points[heap[p]].dist) {
        swap(p, i);
        bubbleUp(p);
      }
    }
  };
```

The swap(...) function just exchanges two values in the heap ❶ and also updates the corresponding index attributes in the points array, so you can keep track of where each node is in the heap. The sinkDown(...) function ❷ works as you saw in Chapter 14. Notice that you don't compare the heap values ❸, but rather compare the distances from the points array using the heap values as indices. (In the sorting code in Chapter 14, we directly compared the heap values.) The same change applies in the bubbleUp(...) function ❹❺.

What's the performance of this algorithm? As is, each point is processed once, and for each point you check whether you need to update the distances to all the others, so it's $O(v^2)$. You can enhance it by having a list of adjacent points, as with the adjacency list representation for graphs, and then the performance becomes $O(v \log v)$ because of the heap usage.

Sorting a Graph

At the beginning of this chapter we mentioned some real-life applications of graphs, like tracking dependencies in code (modules that import from other modules) or project management (showing tasks that rely on the completion of other tasks). In that situation, we may want to find whether a certain ordering of nodes will make everything work out smoothly. Conversely, we may want to check whether code has circular dependencies or whether completing a given task will be impossible. We want to be able detect such issues with graphs.

This type of task is called *topological sorting*, and it implies that given a graph, we sort its nodes in order so that all links "go forward" and there's never a link from a vertex to a previous node. We'll consider two algorithms for such a sort: Kahn's algorithm, which is based on a simple procedure involving counting, and Tarjan's algorithm, which applies depth-first searching to produce the order we want in a backward fashion where the last vertices are output first.

Kahn's Algorithm

Consider a basic argument: if a graph has topological ordering, there must be at least one node with no incoming edges, and *Kahn's algorithm* is based on that. (This is similar to saying that in any set of numbers, there must be one that is less than the rest.) You can select each of these nodes with no problems. If you then discard all the edges that start from those nodes, you should be left with nodes that have no incoming edges, and you can repeat the procedure. If at some point you've still got nodes to consider but all have at least one incoming edge, no topological sort is possible.

Figure 17-24 illustrates the procedure with the same directed graph we've been using for all the examples in this chapter. After first doing a count of incoming edges, the numbers in the nodes are the calculated counts.

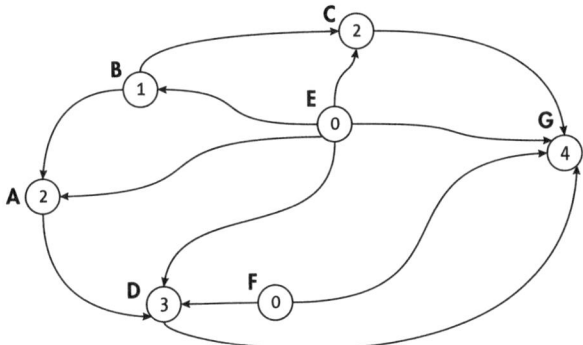

Figure 17-24: A graph is set up for topological sorting where the numbers show how many incoming edges are there at each point.

Given that you found at least one node with zero incoming edges, you can proceed. Points E and F can be output in any order, and you then reduce the counts of the nodes that you can reach from those two points (see Figure 17-25).

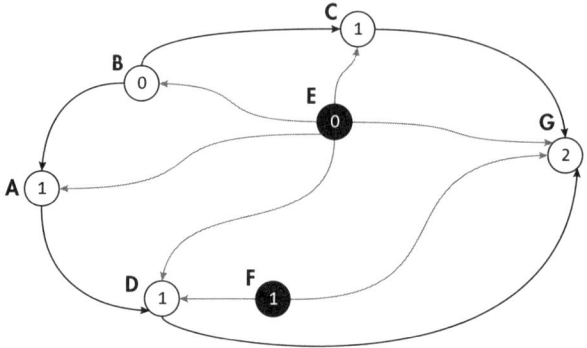

Figure 17-25: Points E and F can be output because they had no incoming edges, and you "forget" the outgoing edges from those two points.

Points E and F were output, which are in positions 0 and 1 of the output array. (The nodes in black show index values.) Again, you find at least one node with no incoming edges, so B is at position 2 of the output array, and you decrease the counts of nodes A and C (see Figure 17-26).

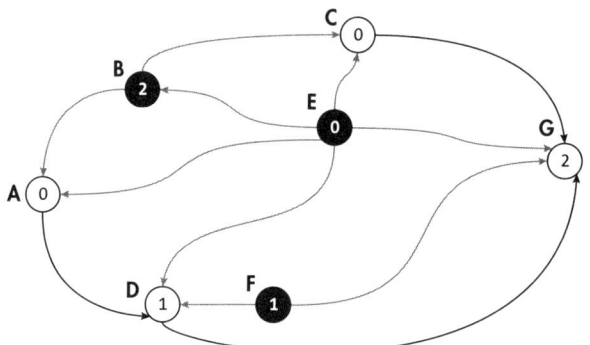

Figure 17-26: Point B now has no predecessors, so its output and edges are removed.

Now repeat the process: points A and C are output, reduce the counts, and you get to the situation shown in Figure 17-27.

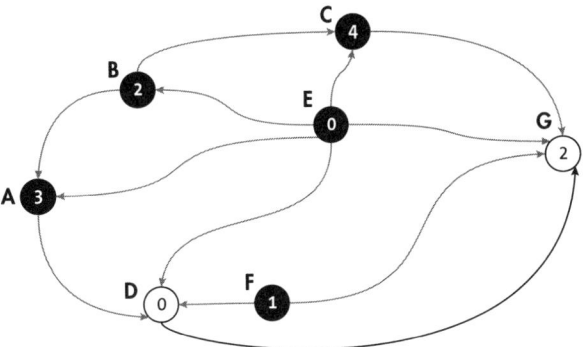

Figure 17-27: Point A is next, then D, and G will be last.

The last two steps output D first and then G. Figure 17-28 shows the final status, and the topological order you want is E, F, B, A, C, D, and G.

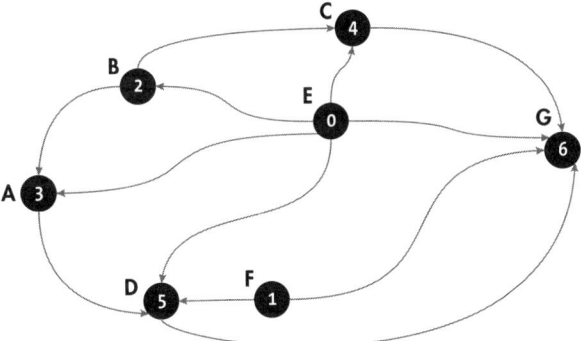

Figure 17-28: The final result, where the numbers show in which order the points were output.

You can code the algorithm as follows:

```
const topologicalSort = (graph) => {
  const n = graph.length;
❶ const queue = [];
❷ const sorted = [];

❸ const incoming = Array(n).fill(0);
  for (let i = 0; i < n; i++) {
    for (let j = 0; j < n; j++) {
      if (graph[i][j]) {
        incoming[j]++;
      }
    }
  }
```

```
❹ incoming.forEach((v, i) => {
    if (v === 0) {
      queue.push(i);
    }
  });

❺ while (queue.length > 0) {
  ❻ const i = queue.shift();
  ❼ sorted.push(i);

  ❽ graph[i].forEach((v, j) => {
      if (v) {
        incoming[j]--;
      ❾ if (incoming[j] === 0) {
          queue.push(j);
        }
      }
    });
  }

❿ return sorted.length === n ? sorted : null;
};
```

Put nodes to be processed in a `queue` ❶, and from there place them in the output `sorted` array ❷. The `incoming` array will count the number of incoming edges for every node ❸, adding 1 for every such edge. Every node with no incoming edges is pushed into the queue for processing ❹, and then you can start the sort itself. While there still are nodes to process ❺, you remove them from the queue ❻ and push them into the output list ❼. For every node you output ❽, discard its connections to other nodes, decreasing the incoming counts ❾. When there are no more nodes to process, if all were sorted, you succeeded ❿; otherwise, you failed. The remaining nodes have at least one incoming edge, which means no topological sort is possible.

Tarjan's Algorithm

We'll apply a depth-first search algorithm for another way to produce a topological sort. The idea is to start *traveling* from a node, marking the way at each node you pass (Hansel and Gretel style) and seeing how far you can get before arriving at a dead end or returning to a node that you already marked, which means you found a cycle and no topological sort is possible. You can mark the nodes from which no more movement is possible as *done* and output them and then ignore them from then on. In this fashion, you'll produce the topological sort in reverse order: you'll output the last nodes in the sort first, and the first nodes will be last.

Figure 17-29 shows an example with the same graph we've used throughout the chapter. Start at point A, mark it as *in progress* (in gray), and consider all the edges out of it. There's only one. If in future steps you get back to node A and it's gray, you'll have found a cycle.

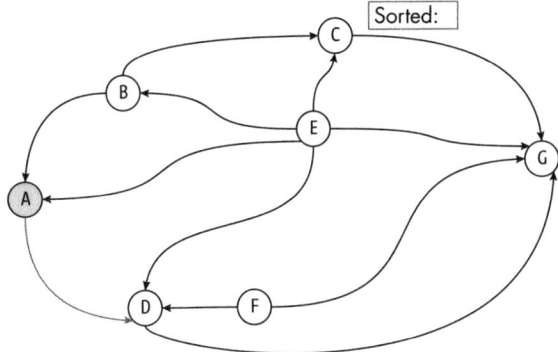

Figure 17-29: The same graph as in the previous section set up
for Tarjan's algorithm. You started at A and reached D; A is grayed out.

You're now at point D, which hasn't been visited. Mark it gray as well
and check which nodes you can reach from it. In this case, you can reach
only G (see Figure 17-30).

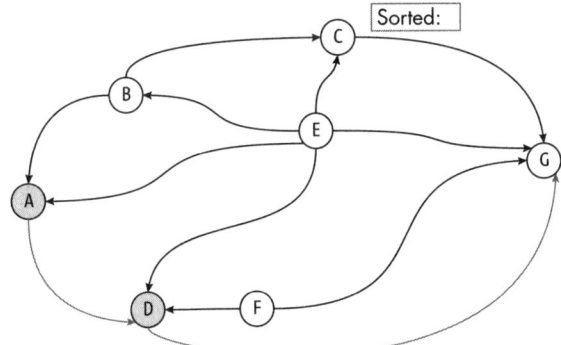

Figure 17-30: After D, you reach G; D is grayed out.

Repeat the process at G, and no edges come out of it, so mark it as *done*
and output it. The sorted list starts as shown in Figure 17-31.

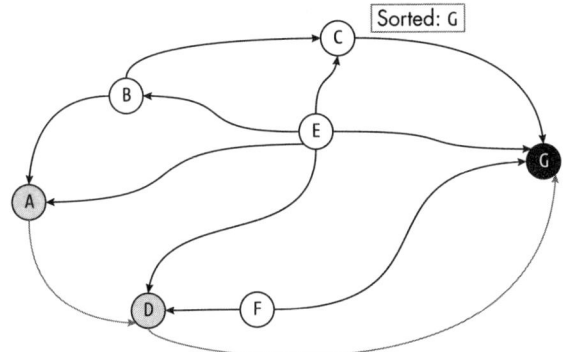

Figure 17-31: From G you can't reach any other point,
so output G to the sorted list.

Having dealt with G, you can mark D as *done* (and output it) and then do the same with A (see Figure 17-32).

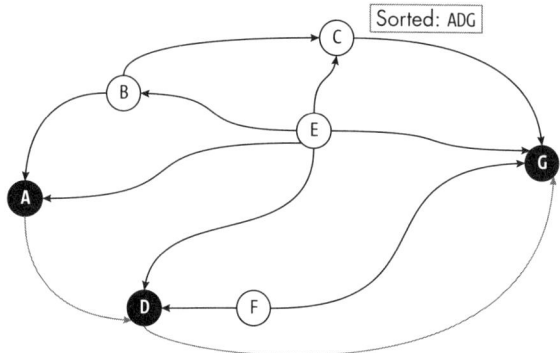

Figure 17-32: D has no further connections to other nodes, so it may be output, and then A is output too.

When you're done with A, start from B. The link from B to A goes to a node marked *done*, so you ignore it. The link from B to C needs processing, though: B is marked *in progress*, and you go to C (see Figure 17-33).

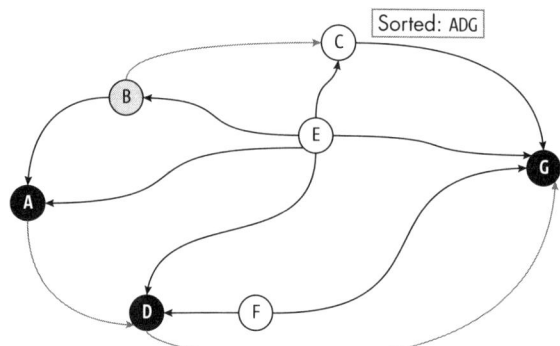

Figure 17-33: Starting at B, you can reach only one point that has not yet been output: C.

Mark C as *in progress*, but its only edge goes to a node marked *done* (G), so mark C as *done* and add it to the output. After that, you're done with B as well, which also is output (see Figure 17-34).

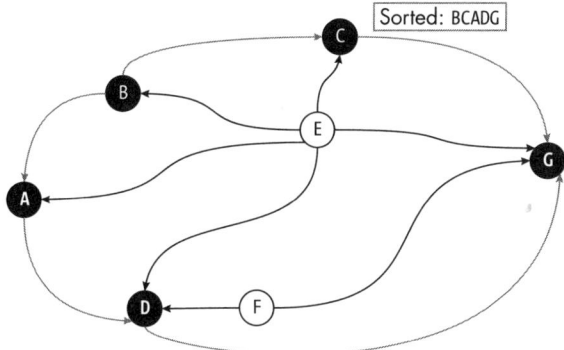

*Figure 17-34: C is linked only to an already output point,
so you can output C and, after that, output B.*

The final steps are similar. From E all the links lead to nodes marked as
done, so E becomes marked as *done* and is output; the same happens to F, and
you're finished, having visited all nodes and produced a topological sort, as
shown in Figure 17-35.

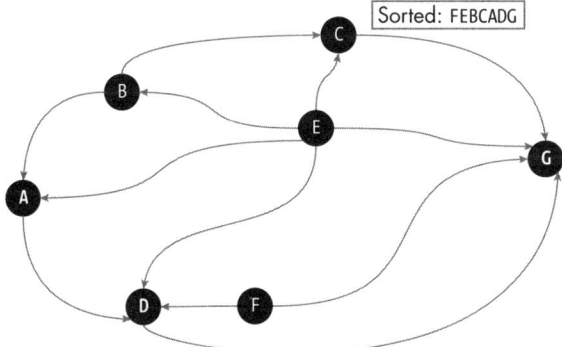

Figure 17-35: When all points have been output, the algorithm ends.

The code for this algorithm follows. The *in-progress* nodes are marked
with a 1 (a temporary mark) and the *done* nodes are marked with a 2
(a final mark):

```
const topologicalSort = (graph) => {
  const n = graph.length;
❶ const marks = Array(n).fill(0); // 1:temp, 2:final
❷ const sorted = [];

❸ const visit = (p) => {
  ❹ if (marks[p] === 1) {
      throw new Error("Not a DAG");
  ❺ } else if (marks[p] === 0) {
      marks[p] = 1;
    ❻ graph[p].forEach((v, q) => {
```

```
  ❼ if (v && marks[q] !== 2) {
        visit(q);
      }
    });

  ❽ marks[p] = 2;
  ❾ sorted.unshift(p);
    }
  };

  try {
❿ marks.forEach((v, i) => {
      visit(i);
    });
    return sorted;
  } catch (e) {
    return null;
  }
};
```

The marks array ❶ keeps track of visited and unvisited nodes. A 0 means the node hasn't been visited yet. A 1 means it was visited and you're going through all its reachable nodes, and a 2 means that the node has been dealt with and output already. The sorted array ❷ gets the output of the algorithm. You define a recursive function ❸ to visit all nodes that a p starting node can reach. If the node was marked with a 1 ❹, it means that, when starting from there, you eventually returned to it. In other words, there's a cycle, so no topological sort is possible. If the node is marked with a 0 ❺, temporarily mark it with a 1 and visit all the unvisited nodes that are reachable from it ❻; skip visiting any nodes marked with 2 ❼ because those were already analyzed. After all the visiting is finished, change the 1 to 2 ❽ and output the current node p ❾; use unshift() to get the right order. To produce the topological sort ❿, all you have to do is start from every possible node and apply the visiting logic.

What's the performance of this algorithm? Each node is visited once, and all its links are processed, but for each node, it checks the whole row for possible links to traverse, making this implementation $O(v^2)$. The algorithm would benefit from an adjacency list representation, because then you would be able to process the edges from a node directly, producing $O(ve)$.

Detecting Cycles

Another question to consider is whether a graph includes any cycles. (In other words, is the graph a tree or a forest or not?) For example, when programming, if there's a cycle in a list of dependencies among modules, something is seriously wrong! A cycle-detection algorithm just needs to check whether it can find at least one cycle in a given graph.

Fortunately, we've already seen an algorithm that does this type of detection: Tarjan's topological sort includes the logic to detect when a cycle is found, so we've already got what we need. The following code is directly extracted from that algorithm:

```
const hasCycle = (graph) => {
  const n = graph.length;
❶ const marks = Array(n).fill(0); // 1:temp, 2:final

  const visit = (p) => {
    if (marks[p] === 1) {
      throw new Error("cycle found");
    } else if (marks[p] === 0) {
      marks[p] = 1;
      graph[p].forEach((v, q) => {
        if (v && marks[q] !== 2) {
          visit(q);
        }
      });

    ❷ marks[p] = 2;
    }
  };

  try {
    marks.forEach((v, i) => {
      visit(i);
    });
  ❸ return false; // no cycles found
  } catch (e) {
    return true;
  }
};
```

All the code is the same; the only differences here are that you don't define a sorted array for the output ❶. You obviously don't add anything when marking a node as totally visited ❷, and you return a boolean value instead of an array or null ❸.

Detecting Connectivity

Now consider a different problem: How can you determine whether a given graph is fully connected? An undirected graph is connected if there's a path between every pair of points in the graph; there's no point you can't reach from any other point. (A border case is that a graph with only one vertex is also considered to be connected.) If a graph doesn't satisfy this condition, we can split it into two or more connected subgraphs.

Several algorithms can detect whether a given graph is connected; we'll consider two here. One algorithm introduces another data structure that allows merging sets, and the other uses a recursive traversal of the graph.

Detecting Connectivity with Sets

There's a rather simple way to find how many connected parts a graph has. Start by forming different sets, each with a single vertex. Then, go through all the edges in the graph, and if an edge links vertices that appear in different sets, join them so that they form a new, larger set. After going through all the edges, if you're left with a single set, the graph was connected; if you're left with several sets, the graph was unconnected.

The question is how to implement these sets. You need to be able to determine whether any two points are in the same set and be able to join two sets. There's an efficient way of doing this by working with a forest of trees, as we explored in Chapter 13.

Consider how this concept works. Each set is represented by an *upward* tree with pointers that go up from the leaves to the root (the opposite of what you did in previous chapters). The leaves of the tree are its elements, and intermediate nodes are added as needed. To see whether two values are in the same set, follow the path up to the root from each leaf. If you reach the same root, the values are in the same set. Finally, to join two sets, just add a new root and make the roots of the two sets point to it. Figure 17-36 shows an initial setup with six values, each in a separate set. Note that all pointers are implied to go up, which is a big difference from previous chapters.

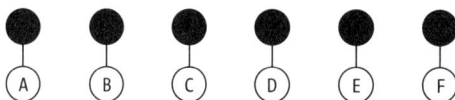

Figure 17-36: Start with each node in its individual tree.

If you want to check whether, say, D and E are in the same set, follow the pointers up, and upon arriving at different roots, you can conclude that they aren't.

To make them part of a single set, just add a new root, and you get the situation shown in Figure 17-37.

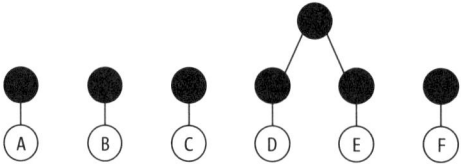

Figure 17-37: Putting D and E in the same set requires adding a new root; now D and E are in the same tree.

If you now check whether D and E are in the same set, the answer is yes, because following pointers up, you'd arrive at the same root. If you ask whether F and D (or F and E) are in the same set, the answer is no, and joining the sets produces the situation shown in Figure 17-38.

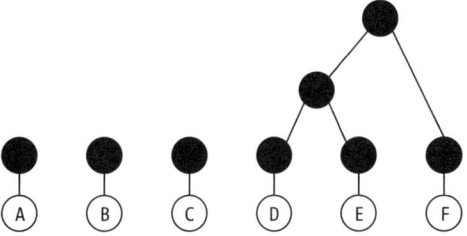

Figure 17-38: Adding F to the previous (D,E) set again adds a new root, and now the three original trees are joined.

You can handle all of this with simple pointer manipulation and end up with an upside-down forest. Joining A with C and then A with E produces a new configuration (see Figure 17-39).

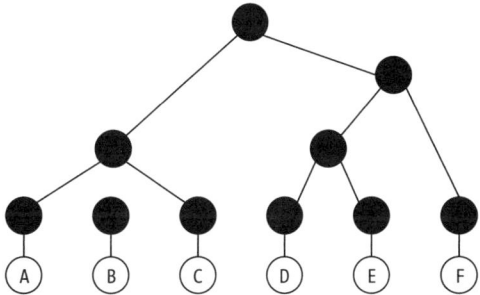

Figure 17-39: In this final scheme, you find that (A,C,D,E,F) are a set and (B) is a separate set.

In Figure 17-39, you have only two sets: a singleton (with just B) and another set that contains all the other values.

The algorithm is short:

```
const isConnected = (graph) => {
  const n = graph.length;

❶ const groups = Array(n)
    .fill(0)
    .map(() => ({ ptr: null }));
❷ let count = n;

❸ const findParent = (x) => (x.ptr !== null ? findParent(x.ptr) : x);

  for (let i = 0; i < n; i++) {
    for (let j = i + 1; j < n; j++) {
    ❹ if (graph[i][j]) {
        ❺ const pf = findParent(groups[i]);
          const pt = findParent(groups[j]);
        ❻ if (pf !== pt) {
            pf.ptr = pt.ptr = { ptr: null };
```

```
    ❼ count--;
        }
      }
    }
  }

❽ return count === 1;
};
```

Start by defining all groups with just a single element ❶, and the count variable tracks how many groups exist at any moment ❷. The findParent(...) auxiliary function goes up from each vertex to find the root of its group ❸. The rest is straightforward: go through all edges ❹ and check whether both endpoints of edges ❺ are in the same group; if not ❻, join the groups by creating a new, common root for both trees and decrease the group count by one ❼. After processing all the edges, if you're left with a single group ❽, the graph was connected. If you want to know how many subgraphs it has, you could check the count instead.

Detecting Connectivity with Searches

The second algorithm we'll consider is based on starting from any point and applying a systematic, recursive search. Check which points you can reach, and then which points you can reach from those, and so on, until you've considered all the edges. Every time you start visiting from a certain point, mark it as *visited* to avoid trying it again.

Given the same graph you've been using, shown in Figure 17-40, and arbitrarily starting at A, which would be the first node to visit?

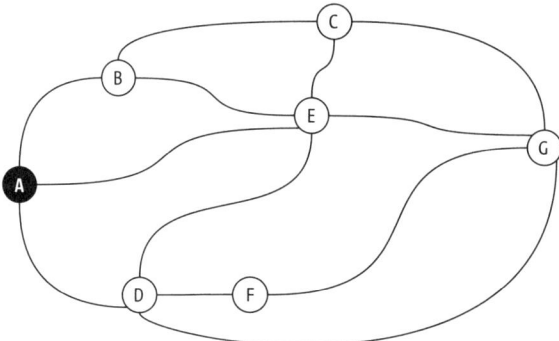

Figure 17-40: To check connectivity, start at any point, in this case A, and mark it.

From A, you reach B, D, and E, and none of them have been visited already. Mark them and start searching from B, as shown in Figure 17-41.

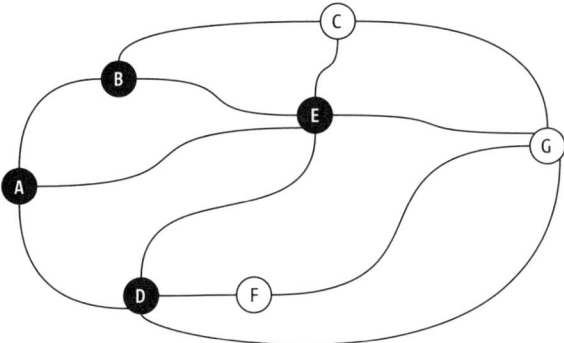

Figure 17-41: From A you can reach unmarked points B, D, and E, which you mark.

From B, you can reach A or E, but those are already marked as visited, so just add C to your search (see Figure 17-42).

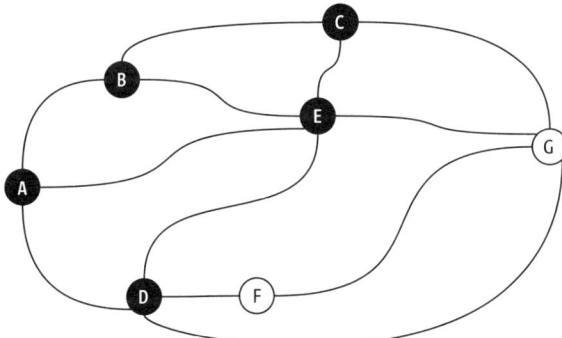

Figure 17-42: From B you can add the still-unmarked C, and from D you add the unmarked F and G.

From D, you can reach A, E, F, and G, but A and E are already marked, so just add F and G to the process (see Figure 17-43).

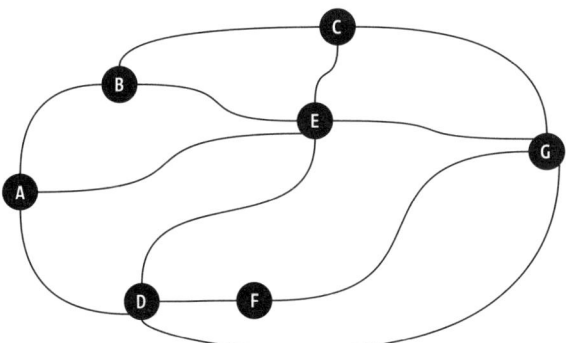

Figure 17-43: All points have been marked, so the graph is connected.

The rest of the algorithm is quick, because no unmarked nodes remain to be reached from E, C, F, or G (in the order you checked them), so you're done, and since all vertices ended up marked, the graph was connected. If there had been a separate subgraph, you wouldn't have been able to reach it, so its nodes would have been left unmarked and the algorithm would have returned false.

You can also do the search in depth-first style, which is actually simpler to code:

```
const isConnected = (graph) => {
  const n = graph.length;
❶ const visited = Array(n).fill(false);

❷ const visit = (x) => {

  ❸ graph[x].forEach((v, i) => {
    ❹ if (v > 0 && !visited[i]) {
      ❺ visited[i] = true;
      ❻ visit(i);
      }
    });
  };

❼ visited[0] = true;
❽ visit(0);

❾ return visited.every((x) => x);
};
```

Start by marking all points as unvisited ❶. The auxiliary visit(...) recursive function ❷ does the search. Given a point, it goes through all of its outgoing edges ❸. If it finds an unvisited point ❹, it marks it ❺ and visits it ❻. To run the algorithm, start by marking any point (the first in this case) as visited ❼, and call the visit function ❽ to do the search. If you finish the algorithm with every point marked as visited ❾, you've succeeded.

Finding a Minimum Spanning Tree

This problem applies to weighted undirected graphs. Imagine we want to connect people to the electrical grid or some other similar service and we know the cost of linking a given pair of points together. We don't need to build *all* possible connections between points; rather, we want to choose a set that, at minimum cost, allows all the vertices to connect to each other. Several algorithms solve this problem, and we'll consider the two best known here: *Prim's algorithm* and *Kruskal's algorithm*. If those algorithms are applied to connected graphs, the output will be a tree linking all of its nodes. If a graph is not connected, we'll find a forest of trees instead for each independent group of nodes.

Prim's Algorithm

The description for *Prim's algorithm* is simple: to build the tree, start with any node and keep adding the closest node (meaning, minimum link cost) not yet connected to the tree until no more nodes are left. It can be proven that this will produce the desired minimum tree, but we won't go into that proof here.

Start with the same undirected graph we've been using (see Figure 17-44) and arbitrarily choose A as the starting node.

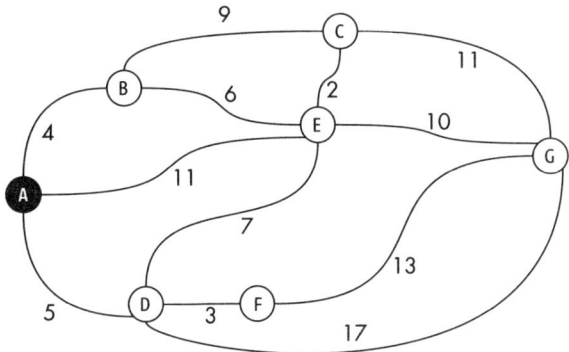

Figure 17-44: Prim's algorithm starts by choosing any node. In this case, start with A.

Now, all of the points still aren't selected, so choose the closest one, which is B in this case (see Figure 17-45).

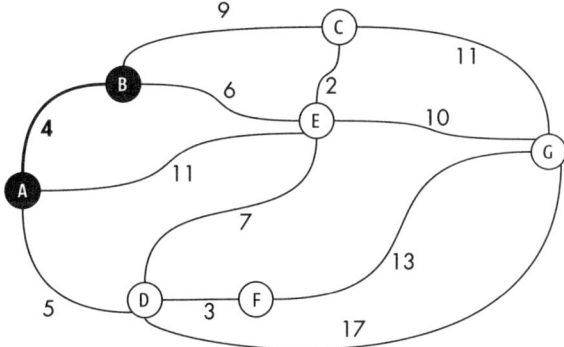

Figure 17-45: Out of all the points adjacent to A, choose the closest, which is B.

You've got two nodes in the spanning tree. Repeat the selection: the closest point to either A or B not yet selected is D, so add that one (see Figure 17-46).

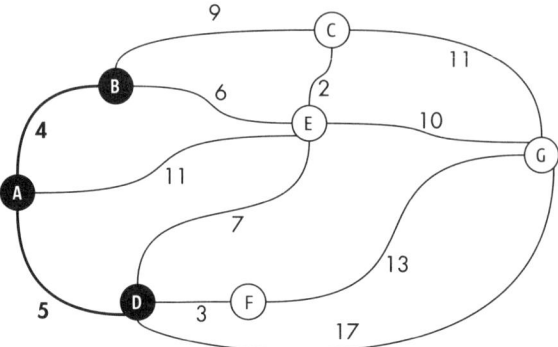

Figure 17-46: Out of the points adjacent to A or B,
you again choose the closest, which is D.

The upcoming steps are easy to predict: first, add F (which is only three units away from the selected nodes), then E, C, and finally G, for a total cost of 30 units, as shown in Figure 17-47.

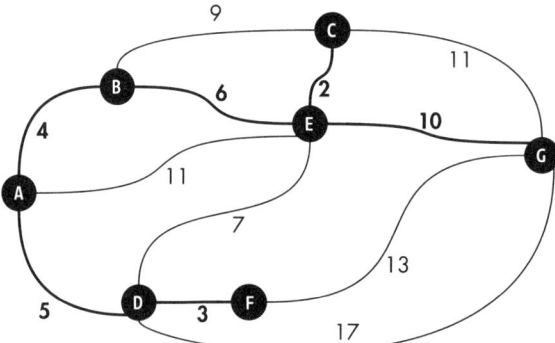

Figure 17-47: After finishing with all nodes, you get a
spanning tree.

Coding this is straightforward. In order to know which is the closest remaining unchosen point, you again use a heap. The implementation is a tad different: this time, you'll have objects in the heap with attributes `from` (a point), `to` (the closest already selected point), and `dist` (the distance to the closest point, which is the length of the edge between those two points). First explore the heap algorithms, which were coded iteratively instead of recursively just for variety:

```
const bubbleUp = (i) => {
❶ while (i > 0) {
    const p = Math.floor((i - 1) / 2);
  ❷ if (heap[i].dist > heap[p].dist) {
      return;
    }
  ❸ [heap[p], heap[i]] = [heap[i], heap[p]];
  ❹ i = p;
```

```
    }
  };

  const sinkDown = (i) => {
❺ for (;;) {
      const l = 2 * i + 1;
      const r = l + 1;
      let m = i;
      if (l < heap.length && heap[l].dist < heap[m].dist) {
        m = l;
      }
      if (r < heap.length && heap[r].dist < heap[m].dist) {
        m = r;
      }
      if (m === i) {
❻     return;
      }
❼   [heap[m], heap[i]] = [heap[i], heap[m]];
❽   i = m;
    }
  };
```

When bubbling up a value, check whether you're not already at the top ❶. If not, calculate the position of its parent and compare distances; if the parent is lower ❷, you're done. If not, exchange heap positions ❸ and repeat the procedure at the parent's position ❹. To sink down a value, set up an endless loop ❺ that exits when the value cannot sink any lower ❻ because it's smaller than its children. If the value has to sink down, do an exchange ❼ and loop again at the child's position ❽.

The code for Prim's algorithm is as follows:

```
const spanning = (graph) => {
  const n = graph.length;

❶ const newGraph = Array(n)
    .fill(0)
    .map(() => Array(n).fill(0));

❷ const heap = Array(n)
    .fill(0)
    .map((v, i) => ({ from: i, to: i, dist: +Infinity }));

  // bubbleUp and sinkDown, excluded

❸ while (heap.length) {
❹   const from = heap[0].from;
❺   const to = heap[0].to;

❻   newGraph[from][to] = graph[from][to];
    newGraph[to][from] = graph[to][from];

❼   heap[0] = heap[heap.length - 1];
    heap.pop(); // or the more unconventional heap.length--;
    sinkDown(0);
```

```
❽ for (let i = 0; i < heap.length; i++) {
    ❾ const v = heap[i];
       const dist = graph[v.from][from];
    ❿ if (dist > 0 && dist < v.dist) {
         v.to = from;
         v.dist = dist;
         bubbleUp(i);
      }
   }
}

  return newGraph;
};
```

Start by setting up the newGraph matrix for the output graph ❶ and the heap with all points in it: the from attribute is the point itself, the to attribute is the closest already selected point, and the dist attribute is the minimum distance from the point to an already selected point of the spanning tree ❷. While the heap isn't empty (implying you haven't yet considered all vertices), consider the top ❸, which is the closest point to the already chosen ones that hasn't been chosen itself yet. The path linking points from ❹ and to ❺ corresponds to the shortest pending distance, so add it to newGraph ❻. Then pop the node from the heap ❼ and proceed to adjust the distances between the from point and all the remaining heap points ❽. Then take each heap element ❾ and consider the distance from it to the from point; if there's a shorter (cheaper) edge ❿, record the edge that allows this better path and the corresponding distance and make the node bubble up if needed. (So the heap points will always have the shortest distance from them to the already chosen points.) When the heap is empty, you'll have the spanning tree in the newGraph matrix.

Kruskal's Algorithm

Kruskal's algorithm also finds the minimum spanning tree for an undirected graph. Instead of adding points one at a time as Prim's does, this algorithm works by adding edges to an initially empty graph. The idea is to sort the edges in ascending order and attempt to add each edge unless it would cause a cycle. (We won't give the proof that this algorithm is correct either, but rest assured it can be done.) How do we detect cycles? Initially, you have all the points in separate, disjointed sets, and every time you add an edge linking two nodes, join the corresponding sets (it's similar to the process in the section "Detecting Connectivity with Sets" on page 454). Never add an edge whose extremes are both in the same set.

Now explore how the algorithm works with the same example graph (see Figure 17-48).

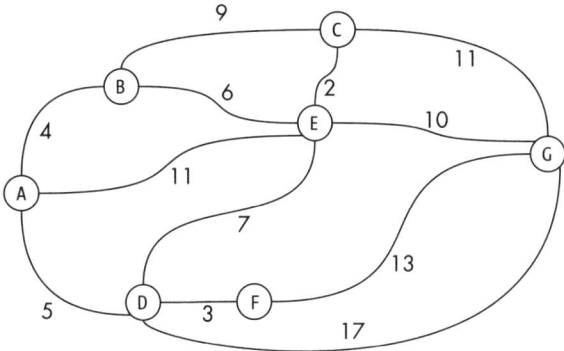

Figure 17-48: The same graph used for Prim's algorithm

Add edges one at the time, starting with the lowest, so the first step adds the (C,E) edge; now points C and E are in the same set, as shown in Figure 17-49.

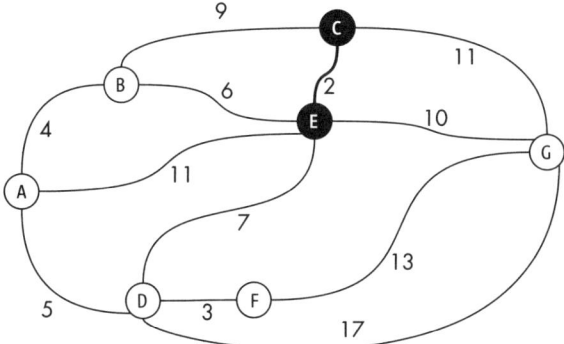

Figure 17-49: Add the smallest edge (C,E) to start.

The next two steps add edges (D,F) and (A,B); no cycles occur anywhere, as shown in Figure 17-50.

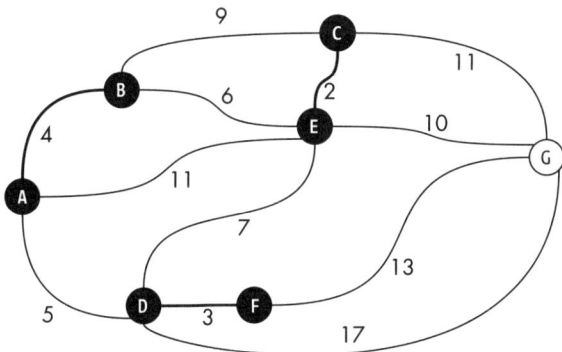

Figure 17-50: Keep adding edges, in ascending size, if they do not create cycles. First (D,F) was added and then (A,B).

The next steps add (A,D), so A, B, D, and F all end up belonging to the same set, and then add (B,E), which makes a big set with all points from A to F (see Figure 17-51).

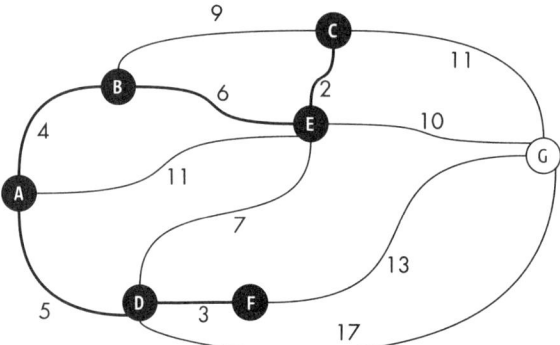

Figure 17-51: Repeating the procedure now adds (A,D).

Now things get interesting! The next edge in order is (D,E), but D and E already are in the same set, so don't add that edge. The next step adds (E,G), and you get the final tree, as shown in Figure 17-52.

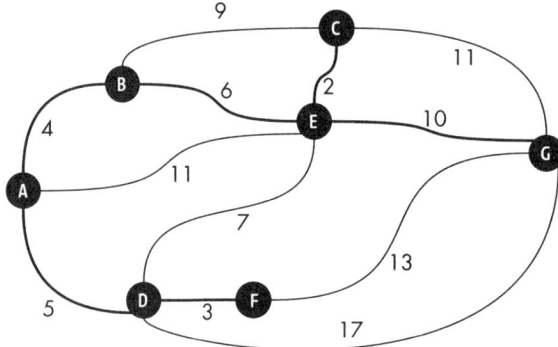

Figure 17-52: You should add (D,E), but it would create a cycle, so skip it and add (E,G) instead.

Future steps won't add anything, because the edges will always link points that are already in the same set, so you've got your spanning tree.

Kruskal's algorithm is as follows:

```
const spanning = (graph) => {
  const n = graph.length;

❶ const newGraph = Array(n)
    .fill(0)
    .map(() => Array(n).fill(0));
```

```
❷ const edges = [];
  for (let i = 0; i < n; i++) {
    for (let j = i + 1; j < n; j++) {
      if (graph[i][j]) {
        edges.push({ from: i, to: j, dist: graph[i][j] });
      }
    }
  }
❸ edges.sort((a, b) => a.dist - b.dist);

❹ const groups = Array(n)
    .fill(0)
    .map(() => ({ ptr: null }));

❺ const findParent = (x) => {
    while (x.ptr) {
      x = x.ptr;
    }
    return x;
  };

❻ edges.forEach((v) => {
  ❼ const pf = findParent(groups[v.from]);
    const pt = findParent(groups[v.to]);
  ❽ if (pf !== pt) {
    ❾ pf.ptr = pt.ptr = { ptr: null };
      newGraph[v.from][v.to] = newGraph[v.to][v.from] = graph[v.to][v.from];
    }
  });

❿ return newGraph;
};
```

Start by creating an empty matrix for the new tree ❶. Then generate
a list of all edges in the graph ❷ and sort it ❸ with the simplest method,
which is JavaScript's own. (For a better way, check out question 17.8.) You
now need to initialize all disjointed sets ❹, as you did earlier when detecting connectivity. The groups array will have a pointer to the root of each
set, all of which will start with a single element. You'll use an iterative version of the earlier recursive findParent(...) function ❺ to find to which set a
node belongs. The rest of the algorithm is as follows: go through the sorted
list of edges ❻, and for each one, find the parents of both of its extremes ❼.
If they don't match ❽, join both sets by creating a new root ❾ and add the
edge to the output graph, which you return at the end ❿.

The performance of the algorithm can be shown to be $O(e \log e)$, basically because you have to sort all the edges and then go through the list possibly joining sets, which also produces the same result. The only disadvantage
in this implementation is that getting the list of the nodes is $O(v^2)$ due to
having to go through the whole matrix, but you can enhance it if you adopt
another representation for the graph using adjacency lists, as we've seen.

Summary

This chapter introduced the concept of graphs. We considered representations for them and studied many algorithms for common requirements, such as finding paths or distances, sorting nodes, detecting cycles, and minimizing costs. These algorithms have also benefited from previous algorithms (like sorting and searching) and data structures (heaps, bitmaps, trees, forests, and lists), providing a way to apply the previous knowledge you've gained in various ways.

In the next and final chapter of the book, we'll move on to specific considerations for data structures that are meant for a fully functional programming style of work, which entails some advantages but also some challenging disadvantages.

Questions

17.1 Where's the Path?

Floyd-Warshall's algorithm finds the shortest distances between every pair of points, but what do you do if you also want to know which path to take? Modify the algorithm so that finding paths is simple. Hint: whenever you find that going from i to j is better by passing through k, make a note so that later, when trying to find the actual path, you'll know to go to k.

17.2 Stop Searching Sooner

When considering the Bellman-Ford algorithm, we mentioned that a certain number of passes ensured finding the shortest paths, but can you do better? Hint: in the example we showed in that section, fewer passes were actually needed.

17.3 Just One Will Do

How would you modify Dijkstra's algorithm if you care only about finding the shortest path to a single point?

17.4 The Wrong Way

Imagine you take a directed graph, reverse all of its edges, and then apply Kahn's topological sort algorithm to it. What will be the output of this algorithm?

17.5 Joining Sets Faster

When joining two distinct sets in the section "Detecting Connectivity with Sets" on page 454, you always add a new root, but doing so isn't necessary, because you could just have one root point at the other one. Consider adding a size attribute in each root (with the number of nodes in the corresponding subtree) and join the smallest tree as a subtree of the largest one. Can you implement these changes?

17.6 Take a Shortcut

When joining sets, doing a little work up-front can save time later. Look again at Figure 17-39 from the section "Detecting Connectivity with Sets" on page 454. Suppose you want to know whether C and D are in the same group. You would need to walk from both nodes up to the root before finding the answer. However, if you are later asked again about C or D, you'd have to redo the path, unless you modify some links, as shown in Figure 17-53.

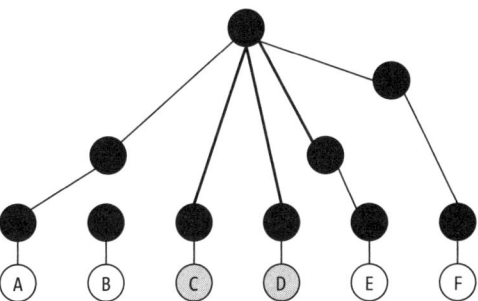

Figure 17-53: An optimized algorithm for joining sets

Three links were changed to point to the root directly, so you can get there more quickly. (From C or D, it's just one step to the root, and from E, it's one step shorter than before.) Make a change in the findParent(...) function so it creates "shortcut" paths that'll make future processes faster.

17.7 A Spanning Tree for a Tree?

What happens if you apply a spanning tree algorithm to a tree?

17.8 A Heap of Edges

Can you replace JavaScript's sort with heapsort, quicksort, or any other method discussed in the book?

18

IMMUTABILITY AND FUNCTIONAL DATA STRUCTURES

We've now considered several abstract data types (ADTs), data structures, and algorithms. Let's finish the book by considering an aspect that's not only relevant to functional programming but also to the everyday usage of libraries such as Redux for React web page developers. How do we work with data structures and not make changes to them, but instead produce new ones, in a true functional style? To do this, we'll need to consider a new concept: *persistent* (or *functional*) *data structures* that we can update without needing to clone everything for high performance and that also allow us to view the "history" of previous states and roll back changes if needed.

Functional Data Structures

Let's start with a couple of definitions. *Persistent data structures* have the interesting property that you can update them while keeping previous versions intact, without changes. This property automatically implies that these structures are ideal for purely functional programming languages, which do not allow for side effects, as mentioned in Chapter 2. It means they are also *functional data structures*: if you don't modify a data structure but produce a new one instead, you'll have both the previous and the new versions available.

Let's analyze several data structures, most of which we already covered in the book, to see whether we can make them functional.

Arrays (and Hash Tables)

We'll start with bad news first. Arrays are essentially mutable data structures, and there's no simple way to implement a functional equivalent with the same level of performance, which is namely $O(1)$ for accessing and updating array elements. Arrays support destructive updates, and such updates can't be reversed. Once you modify some position in an array, there's no way to retrieve the previous value. Arrays are the opposite of a persistent data structure, in fact.

To get around this limitation, a common technique is to use a balanced binary search tree, with indices as keys, but doing so requires $O(\log n)$ time instead. Other far more complex techniques have been explored, but the performance is still not the same as for straightforward arrays. If you'd like to learn more about it, do an online search for Melissa O'Neill and F. Warren Burton's method (it won't be an easy read).

A related conclusion from this result for arrays is that you won't have a good equivalent for hash tables, or their many variants, making them another structure that you'll have to replace with a potentially slower one.

This beginning of our study of functional data structures may seem depressing, but rest assured that we'll be able to find equivalences for many of the structures previously considered in this book.

Functional Lists

Now consider the simplest structure, linked lists, from Chapter 10. Some types of lists are quite amenable to the functional style of work. Others (like queues) require a "conversion" to make them functional, and some lists have no functional equivalent.

Common Lists

When we defined lists (see the section "Basic Lists" on page 178), given a position, you wanted to be able to either add a new value at that spot or remove whatever was there. Consider the first operation. You can achieve that by replicating the initial part of the list. Figure 18-1 shows a list you looked at in Chapter 10, and then the same list where you add an 80 value at position 3. (Remember, position 0 is first, as with arrays.) The operation involves adding a node and modifying a pointer in an already existing node.

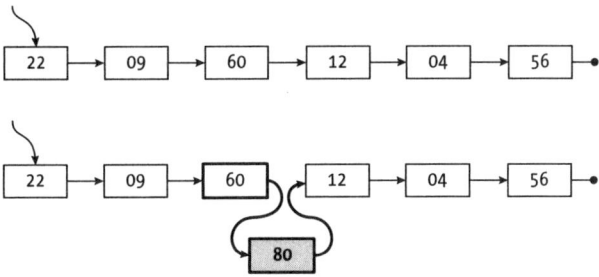

Figure 18-1: Inserting a node in a list requires that you change an original node, namely the one pointing to the new node.

With functional structures in mind, Figure 18-2 shows the way to do it.

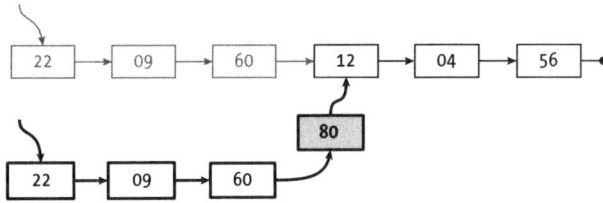

Figure 18-2: Doing an insertion functionally replicates some nodes but leaves the original ones unchanged.

In addition to the new node with the 80 value, you have some nodes that duplicate values in the previous list, but you keep part of the list unchanged. You didn't have to redo the whole list. New nodes and links have bolder lines, and discarded nodes are shown in faded gray.

The code to work in this way uses recursion:

```
const isEmpty = (list) => list === null;

const add = (list, position, value) => {
❶ if (isEmpty(list) || position === 0) {
    return { value, next: list };
❷ } else {
    return { value: list.value, next: add(list.next, position - 1, value) };
  }
};
```

If adding an element at the first position ❶, return a node that has the new value and points to the previous list. This also works if the list was originally empty. Otherwise, when the list isn't empty and you want to add the value at a position other than zero, create a new node ❷ with the same value at the head of the list and a link to the result of adding the new value to the tail of the original list.

Now, consider removing elements from a list. Going back to the initial list, say you want to remove the 60 at position 2. Figure 18-3 shows the before and after lists.

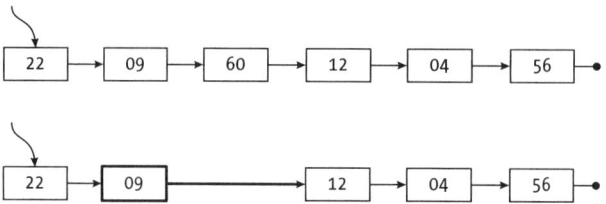

Figure 18-3: Removing a node from a list also implies modifying some original node.

To work in a functional way, replicate the initial part of the list, as shown in Figure 18-4.

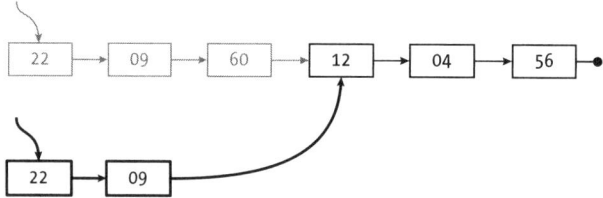

Figure 18-4: The same kind of solution used for insertions helps deal with deletions in a functional way.

As with list additions, some nodes (in faded gray) are no longer included. Part of the list is formed by new nodes (with bold lines), and part of the list remains as it was, without re-creating the whole structure. Consider the following implementation:

```
const remove = (list, position) => {
❶ if (isEmpty(list)) {
    return list;
❷ } else if (position === 0) {
    return list.next;
❸ } else {
    return { value: list.value, next: remove(list.next, position - 1) };
  }
};
```

Using recursion makes the logic clear. If you want to remove an element from an empty list ❶, you can't do anything; return the null list as is. If the list isn't empty and you want to remove its first element ❷, the new list is the one that starts with the element just after the first. Finally, if the list is not empty and you want to remove some element other than the first one, construct a new list ❸ that has the same value as the head of the list and points to the result of removing the value from the rest of the list.

As for performance, we again find that all operations are $O(n)$, although the additional creation of nodes probably implies a slower implementation. Also, given that we still have plain lists, other methods that we saw earlier (like finding the value at a position or calculating the list's size) work exactly as before. In order to implement a functional version of common linked lists, you just had to change the two methods that actually modify the list.

Let's move on to more specialized versions of lists used for other ADTs.

Stacks

The first variation of lists that we considered were stacks, which have the restriction that all additions ("pushes") and removals ("pops") occur at an extremity of the list, at its "top." It will be a nice surprise to realize that our earlier implementation already made a functional data structure. Review the diagrams from Chapter 10. When pushing a value, you had the situation shown in Figure 18-5.

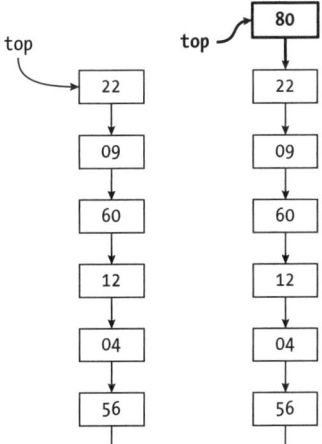

Figure 18-5: Stacks already do pushes in functional ways . . .

The updated stack shares most of the structure; the only difference is the new top element. Popping the top value has a similar behavior, as shown in Figure 18-6.

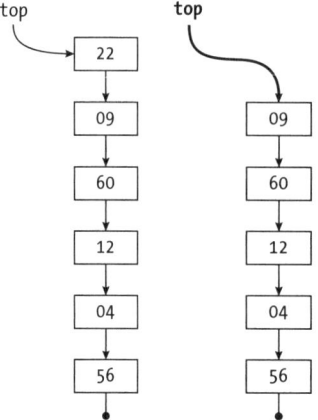

Figure 18-6: . . . and that also goes for pops.

As with pushing, you updated the stack without modifying any values or pointers in it. The original implementation already was fully functional. The performance of both operations is still $O(1)$, so it can't be enhanced. However, we won't always be so lucky.

Queues

Queues pose a challenge. They also restrict operations to the extremities of a list: you *enqueued* (entered) values at one extreme (the "back" of the queue) and *dequeued* (exited) them from the other extreme (the "front" of the queue). You also used a linked list as the basis for the queue, as shown in Figure 18-7.

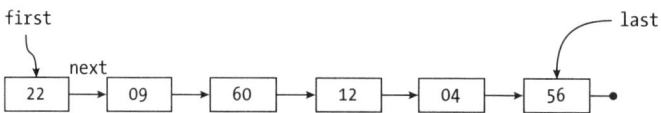

Figure 18-7: You can dequeue a node functionally, the same way you pop a stack.

Dequeueing the front element (22) is exactly the same as popping a value from a stack, so that would work. The updated queue would just have its front element removed, and its first element now becomes what originally was the second (09).

However, enqueueing a new value in this example causes a problem. You have to modify the node with the 56, and that requires modifying the node with the 04, and so on, so you end up creating a whole copy of the queue. (This would be equivalent to adding a value at the end of a simple list, as described earlier in the section "Common Lists" on page 470.) Can we do better? The answer is yes, but we'll need an ingenious trick: use a pair of stacks to represent a queue.

Consider how a queue with five values, A (first) through E (last), might look at a certain moment, as shown in Figure 18-8. (See also question 18.1.)

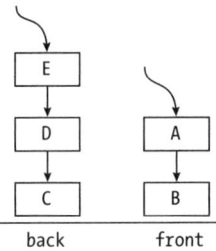

Figure 18-8: You can do functional queues, but you'll need two stacks for that: "back" and "front."

The queue is split in two stacks. Consider how you could have gotten here. You enter the queue by pushing into the "back" of the stack and exit by popping the "front" stack. For example, if F were to enter the queue, you'd get the situation shown in Figure 18-9.

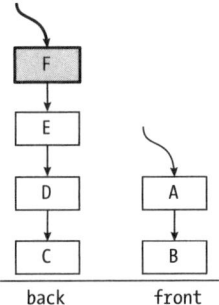

Figure 18-9: New nodes are queued by pushing them to the "back" stack.

You can push into a stack functionally, as you saw previously, so everything's fine. If a value were to leave the queue, you'd pop it and get the status shown in Figure 18-10.

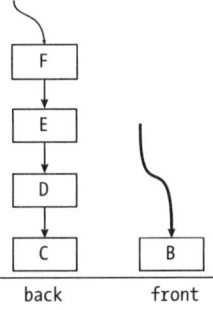

Figure 18-10: Dequeueing a node means popping from the "front" stack, so that's also functional.

Popping a stack is also done functionally, so everything is still fine. After doing another exit, you'd have the situation in Figure 18-11, and there's the problem—the front stack is empty.

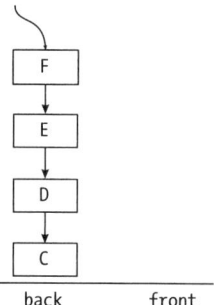

back   front

Figure 18-11: How do you dequeue a node if the "front" stack is empty?

How would we handle the next exit, now that the front stack is empty? The key to this representation for queues is as follows: if you need to exit the queue and the front queue is empty, pop all the values off the back queue and push them into the front queue (see Figure 18-12).

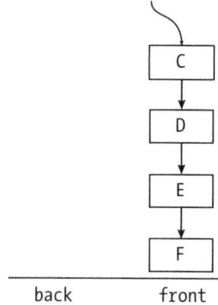

back   front

Figure 18-12: Pop everything from the "back" stack onto the "front" stack.

After this procedure (which reverses the back stack), you'll be able to keep exiting the queue, and all operations will be done in the correct order. It seems like a bit of trickery, but it works, and since all of the involved operations are done in a functional way, the result is a functional data structure for representing queues.

We're representing queues with two stacks, so the basic methods to build a queue and check whether it's empty are as follows:

```
❶ const newQueue = () => ({ backPart: null, frontPart: null });

❷ const isEmpty = (queue) =>
     queue.backPart === null && queue.frontPart === null;
```

A new queue ❶ consists of two empty stacks, and testing whether it's empty ❷ simply requires verifying whether the two tops of the stack are null.

You can enter a new value at the back end of the queue by pushing into that list, which you already know how to do:

```
const enter = (queue, value) => ({
❶ backPart: { value, next: queue.backPart },
❷ frontPart: queue.frontPart
});
```

You return a new queue, with the new value pushed onto the back end of the stack ❶ and an unchanged frontend ❷.

Things get a bit more hairy when exiting the queue. As described, if the front part isn't empty, just pop its first element, but if the list is empty, push the whole back part, element by element, into the front:

```
const exit = (queue) => {
❶ if (isEmpty(queue)) {
    return queue;
❷ } else {
    let newfrontPart = queue.frontPart;
    let oldbackPart = queue.backPart;
  ❸ if (newfrontPart === null) {
    ❹ while (oldbackPart !== null) {
        newfrontPart = { value: oldbackPart.value, next: newfrontPart };
        oldbackPart = oldbackPart.next;
      }
    }
  ❺ return { backPart: oldbackPart, frontPart: newfrontPart.next };
  }
};
```

Start by checking whether the queue is empty ❶, because in that case there's nothing to do; you'll return the very same unchanged queue. If it's not empty ❷, check whether the front stack is empty. If it is ❸, you need to do a loop ❹, popping values from the back stack and pushing them onto the front stack. At the end, now knowing that the frontend isn't empty ❺, you just return a new queue with the (possibly emptied) back part and the result of popping the top element off the front part.

How's the performance of this stack-based queue? Entering the queue is always $O(1)$, but exiting the queue may be either $O(1)$ or $O(n)$. However, in amortized terms, you can see that each item will be pushed once (at the back), popped once (from the back), pushed once again (at the front), and eventually popped once again (from the front), which are four constant-time operations. Over the history of many operations, the amortized performance is $O(1)$, because each value will pass through four $O(1)$ operations. The $O(n)$ cost of popping from the back and pushing to the front is "diluted" because after once pushing n values to an empty front, the next n exits will be $O(1)$. The final average is $O(1)$.

There was an extra operation, front(...), to access the value at the front of the queue; see question 18.2.

Other Lists

We've considered several kinds of lists, but what about the rest? We'll have no functional equivalent for dequeues (or, more generally, doubly linked lists) or circular lists because modifying a single node implies that at least one other node must be modified, and that again implies other nodes must change, and so on. Trying to update those structures in a functional way ends with creating a complete copy, which isn't very efficient.

Functional Trees

In Chapters 12 through 16 we explored varieties of trees: binary search trees, general trees, heaps of several styles, and so on. Some of these (not all, alas) allow for working in functional ways.

Binary Search Trees

How can we manage to make binary search trees behave in a functional way? In general, we apply the exact same type of solution that we used for lists and create some new nodes wherever necessary. Start by considering how to add a new value to a tree from Chapter 12 (see Figure 18-13).

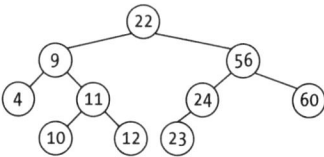

Figure 18-13: A binary search tree that you want to maintain in a functional way

In Chapter 12 you added a new 34 value, which became the right child of the 24 node. You can do the same here without modifying the existing tree. The solution lies in adding a few new nodes, as shown in Figure 18-14.

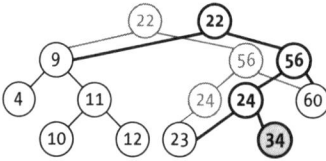

Figure 18-14: Adding a new value implies creating a few new nodes.

There's a new root, and you also have new nodes all the way to the added node itself, but otherwise, the rest of the tree remains the same. There also are some nodes (shown in gray) that are no longer part of the tree, as they were duplicated by new ones. Accessing the tree through its new root, you find the newly added 34 value, whereas accessing the tree through its old root results in the very same structure as before. You've managed to create a new tree, with the additional value, but without modifying the original structure.

Adding a new value in this functional way requires little code:

```
const add = (tree, keyToAdd) => {
❶ if (isEmpty(tree)) {
    return newNode(keyToAdd);
❷ } else if (keyToAdd <= tree.key) {
    return newNode(tree.key, add(tree.left, keyToAdd), tree.right);
❸ } else {
    return newNode(tree.key, tree.left, add(tree.right, keyToAdd));
  }
};
```

If you want to add a value to an empty tree ❶, you just need a new node with the value. (As a reminder, see the following newNode(...) function.) If you want to add a key that goes in the left subtree ❷, return a new tree that has the same value as the current root, a recursively updated left subtree, and the same right subtree as before. If the new value has to go into the right subtree ❸, the result is similar: you'll return a new tree with the same value and the same left subtree as the current node and an updated right subtree:

```
const newNode = (key, left = null, right = null) => ({ key, left, right });
```

Now, you'll turn to removing nodes and consider the most complex case: removing a node that has two children (you can work out the simpler cases). Figure 18-15 shows the original tree (the same one used to show adding a node).

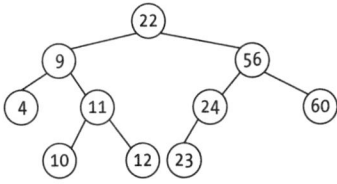

Figure 18-15: Removing a node is a bit trickier if a node (such as 9) has two children.

To remove node 9, you have to find the next greater key (10, in this case) and change its place. It needs to be removed from its current position and take the place of the 9. Working in functional terms, you again create some new nodes, as shown in Figure 18-16.

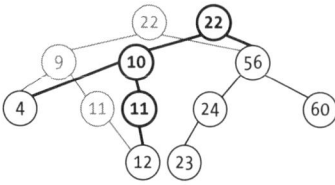

Figure 18-16: Removing a node with two children requires re-creating nodes from the root.

You have a new root, and you also re-created nodes in the path to the node that you eliminate. To remove the 10 from its tree, you also applied a functional method, so in that subtree (the one with root 11), you had to create some new nodes too. The code for removing nodes has several cases:

```
const remove = (tree, keyToRemove) => {
❶ if (isEmpty(tree)) {
     return tree;
❷ } else if (keyToRemove < tree.key) {
     return newNode(tree.key, remove(tree.left, keyToRemove), tree.right);
❸ } else if (keyToRemove > tree.key) {
     return newNode(tree.key, tree.left, remove(tree.right, keyToRemove));
❹ } else if (isEmpty(tree.left) && isEmpty(tree.right)) {
     return null;
❺ } else if (isEmpty(tree.left)) {
     return tree.right;
❻ } else if (isEmpty(tree.right)) {
     return tree.left;
❼ } else {
   ❽ const rightMin = minKey(tree.right);
   ❾ return newNode(rightMin, tree.left, remove(tree.right, rightMin));
   }
};
```

The code tightly parallels what you saw in Chapter 12, but you always return new trees instead of modifying nodes. If the tree is empty ❶, return it as is. If the value to remove is less than the current node's value ❷, return a new tree with the node's value and right child, but with its left child pointing at a new tree, which is the result of deleting the value in the left subtree. Similarly, if the value to remove is greater than the current node's value ❸, proceed symmetrically: return a new tree with the node's value and left child, but with a right child pointing at the result of deleting the value in the right subtree. When you've found the node to delete and it's a leaf ❹, just return an empty tree. If it isn't a leaf but it has only one child (if a right child ❺, and if a left one ❻), return a tree that consists of only the nonempty child, omitting the removed node. Finally, in the most difficult case, if you have to remove a node with two non-null subtrees ❼, find the minimum value in its right subtree ❽ and return a tree built with that value at the root, the original node's left subtree on the left, and the result of removing the minimum value from the original node's right subtree on the right ❾.

You have already seen an implementation of minKey(...), the function that finds the minimum value in a binary search tree, but consider a new version, just for the sake of variety (plus a one-liner is hard to resist):

```
const minKey = (tree) => (isEmpty(tree.left) ? tree.key : minKey(tree.left));
```

For all the varieties of trees that we've seen (AVL trees, red-black trees, splay trees, and so on), you can apply some variant of the methods shown in the previous removal code. After adding or removing a key, the new tree will end with a new root and several new nodes, but many parts of it remain the same, with no changes. The order of the performance of the processes will also be the same as before, so at the cost of some added complexity, we can have functional versions of all those trees.

Other Trees

We also considered other kinds of trees; in particular, Chapters 14 and 15 were devoted to heaps, which are basically binary trees or variations thereof. A simple version was built upon arrays, so we're stuck there with no easy solution. For other heaps (treaps, skew heaps, Fibonacci heaps, and so on) that are based on dynamic memory using pointers, we can apply the same kind of solution as for objects and binary trees.

What about digital trees, including tries, object-based tries, radix trees, and ternary trees? Tries had arrays in each node, with pointers to all the children of the node, but in essence, it's the same as binary trees having two pointers. The same kind of solution we saw for binary trees will work, and when updating a trie, you'll end with a new structure that has some new nodes but shares most of the previous ones. Figure 18-17 shows this solution with a trie that you saw in Chapter 17.

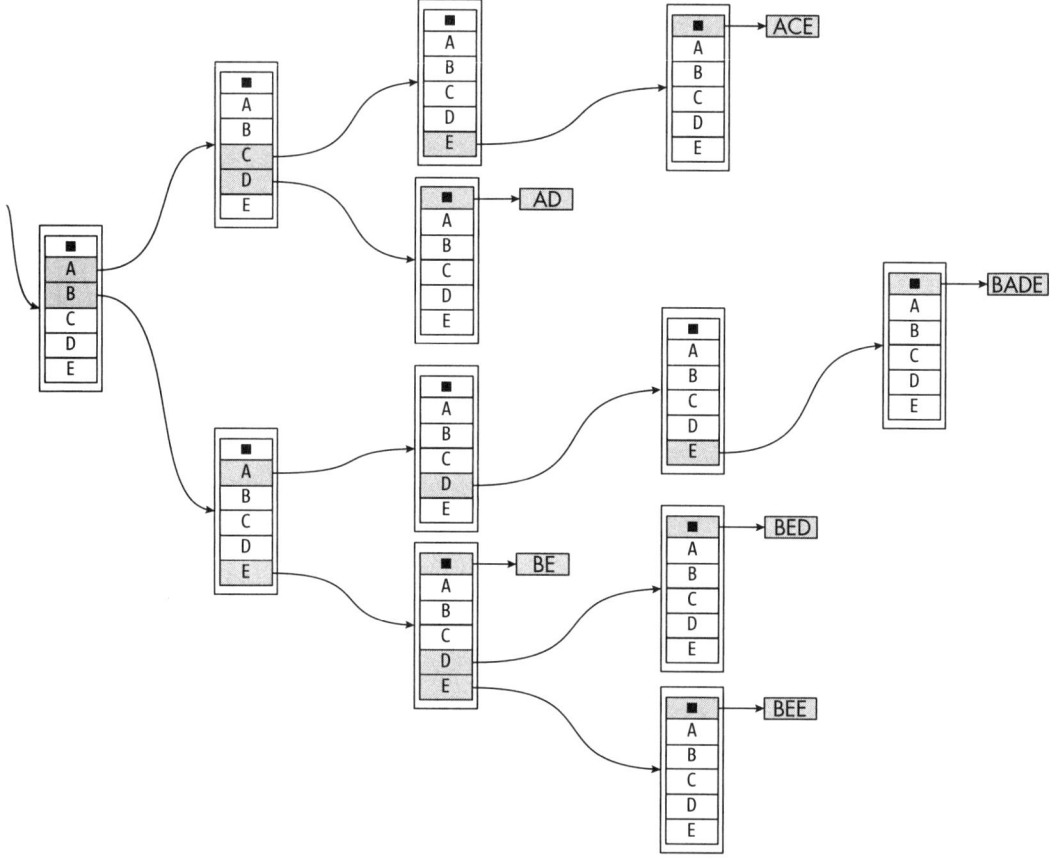

Figure 18-17: You can also maintain a trie in a functional way.

In Chapter 17, we showed how the trie would be modified if we added an ABE word. Working functionally, you'd have a new root and some new nodes elsewhere. Figure 18-18 shows the result.

You have a new root, a couple of new nodes (with a darker border), and some new links, but most of the structure is still as before. Two old nodes (in gray) are no longer part of it, in the same way as with binary trees, but that's it. The basic procedure is the same, and it also works for radix trees and ternary trees, but you won't see it here. It's always the same kind of solution.

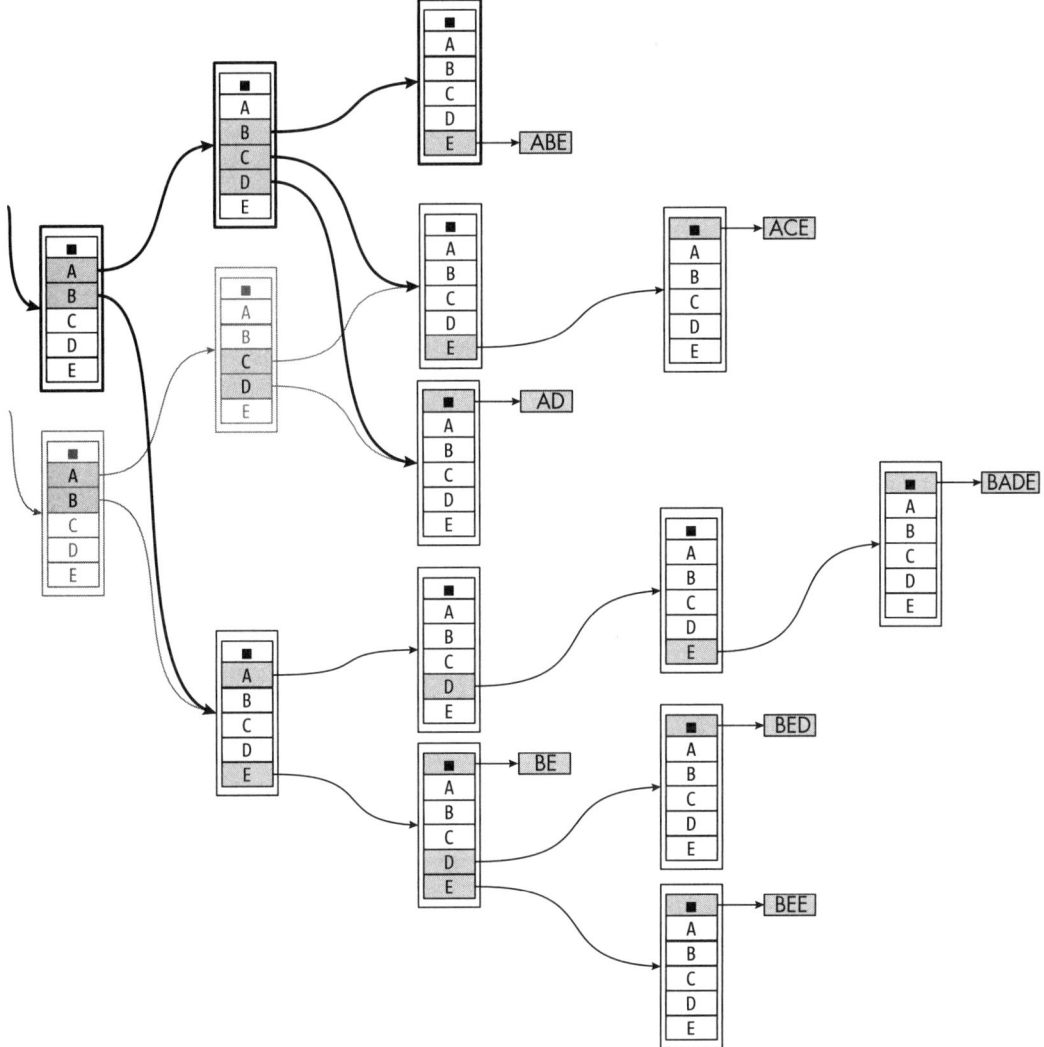

Figure 18-18: Adding a new word requires creating some new nodes, but it keeps most of the trie unchanged.

Finally, for object-based tries that were based on JavaScript objects, you can simply apply the same ideas that were described for lists and trees earlier in this chapter. In Chapter 16 you considered an example of an object-based trie; Figure 18-19 revisits it.

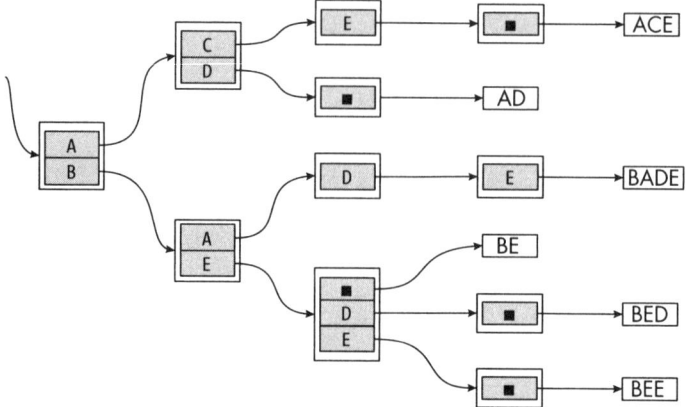

Figure 18-19: You can also update an object-based trie in a functional way.

In Chapter 16 we saw, among other examples, how to add the word ABE to the trie. You did this by modifying several objects; now you'll do it functionally by adding a few new ones and keeping most of the old structure unchanged, as shown in Figure 18-20.

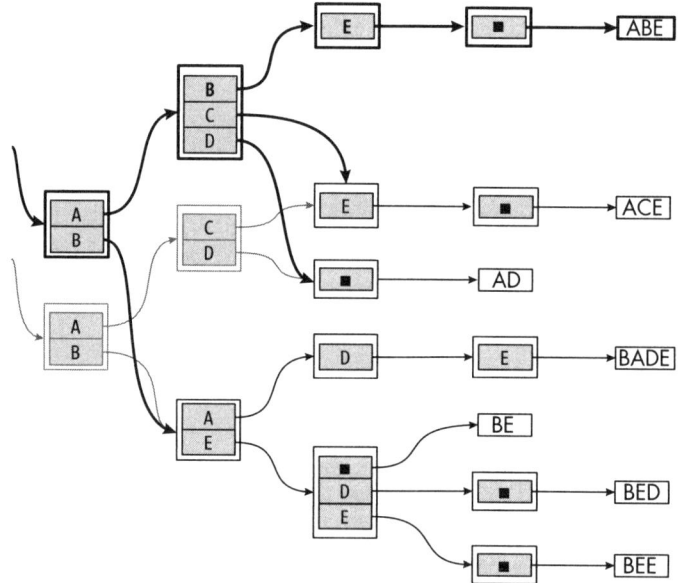

Figure 18-20: Adding a new word and re-creating some objects

As with the previous trie example, new nodes have darker borders, and borders of the dropped nodes are a lighter gray. Most of the trie didn't need to change at all. Thus, you're able to implement several varieties (although not all) of heaps and all sorts of digital trees by varying the update algorithms a bit, but always along the same lines that we already applied.

Summary

In this chapter, we've finished our overview of data structures and algorithms by considering functional data structures, which may be updated without side effects and produce new versions in an efficient way.

Working with functional data structures represents a possibly difficult balance. On one hand, there are advantages in terms of clarity and maintainability, because it's clear when and where data structures are changed. On the other hand, however, we may find some operations that originally had great performance are now replaced by slower ones.

Is there a way to avoid those problems? Unfortunately, not easily. As mentioned previously, the aim is to apply functional programming in a pragmatic way, so we'll use functional data structures when possible but accept "common" modifiable structures when needed. For instance, working with arrays is quite often necessary, and a lower-performing solution could have a great impact on performance. You must be flexible and know when and what to choose.

Questions

18.1 Getting Here

What's the minimum sequence of operations entering and exiting the queue that would produce the diagram shown in Figure 18-7?

18.2 With Apologies to Abbott and Costello, Who's on Front?

Can you implement the front(q) method that will produce the front value of a functional-style q queue?

18.3 No Change Needed

When removing a key from a binary search tree, the algorithm will produce a new tree even if the required key didn't exist. Can you modify it so that it returns the original unchanged tree in that case?

18.4 A New Minimum

Can you explain how the new version of minKey(...) in the section "Binary Search Trees" on page 478 works?

ANSWER KEY

Chapter 2

2.1 Pure or Impure?

From a purely formal point of view, this function is impure since it uses something that wasn't provided as an argument. Given that PI is a constant, you could possibly stretch things a bit and accept it; after all, nobody will be able to modify PI because of the const definition. The best solution would be to use Math.PI with no external variable, constant or not.

2.2 Prepare for Failure

You must use try and catch; the following is a possible solution:

```
❶ const addLogging = (fn) => (...args) => {
❷   console.log(`Entering ${fn.name}: ${args}`);
    try {
❸     const valueToReturn = fn(...args);
❹     console.log(`Exiting ${fn.name} returning ${valueToReturn}`);
❺     return valueToReturn;
    } catch (thrownError) {
❻     console.log(`Exiting ${fn.name} throwing ${thrownError}`);
❼     throw thrownError;
    }
};
```

The addLogging() higher-order function takes a function as a parameter ❶ and returns a new function. If the original function throws an exception, catch it and output a proper message. Upon entry ❷, log the function's name and the arguments it was called with. Then try to call the original function ❸, and if there's no problem, log the returned value ❹ and return it back to the caller ❺. If an exception was thrown, log it ❻ and throw it again for the caller to process ❼.

2.3 You Got Time?

The following function does what you need; you should note several similarities with the logging function in the previous question.

```
❶ const { performance } = require("perf_hooks");

❷ const addTiming = (fn) => (...args) => {
❸   const output = (text, name, tStart, tEnd) =>
      console.log(`${name} - ${text} - Time: ${tEnd - tStart} ms`);

❹   const tStart = performance.now();
    try {
❺     const valueToReturn = fn(...args);
❻     output("normal exit", fn.name, tStart, performance.now());
❼     return valueToReturn;
    } catch (thrownError) {
❽     output("exception thrown", fn.name, tStart, performance.now());
❾     throw thrownError;
    }
};
```

In Node.js ❶ you need to import the performance object; in a browser it's directly available. (See *https://nodejs.org/api/perf_hooks.html* and *https://developer.mozilla.org/en-US/docs/Web/API/Performance* for more information.) The addTiming() function ❷ will get a function as a parameter and return a new function. You'll use an auxiliary function ❸ to output the timing data. Before calling the original function ❹, store the start time; if the call succeeds ❺ with no problem, just output the original time and the current one ❻, returning the original returned value ❼. If any errors occur, output a different message ❽ and throw the same exception again ❾ so the timed function will behave exactly as the original one.

2.4 Parsing Problem

The problem is that .map() passes three arguments to your mapping function: the current element of the array, its index, and the whole array. (See *https://developer.mozilla.org/en-US/docs/Web/JavaScript/Reference/Global_Objects/Array/map* for more on this.) On the other hand, parseInt() receives two parameters: the value to parse and an optional radix (10 is the default value if nothing is passed). In this case, parseInt() is getting passed three arguments: it ignores the third extra one, but it uses the array index as the radix. For instance,"8" is parsed as a base 3 number, producing an error, since base 3 uses only digits 0, 1, and 2, and "32" is parsed as a base 5 number, which is equivalent to 17.

2.5 Deny Everything

A one-liner is enough:

```
const negate = (fn) => (...args) => !fn(...args);
```

2.6 Every, Some . . . None?

Just an example as a tip to get you started: if you want to check whether nobody in a group was an adult, you could equivalently check whether everybody wasn't an adult, so `people.none(isAdult)` could be tested as `people.every(negate(isAdult))` using the previous answer.

2.7 No Some, No Every

A hint for `some()`: use `findIndex()` to see whether any element satisfies the predicate; if it doesn't return `-1`, it means that at least one element fulfills your condition. For `every()`, you want to see whether `findIndex()` cannot find any element that satisfies `negate(your predicate)`, in the same manner as in the previous question; if none is found, it means that every element fulfills your condition.

2.8 What Does It Do?

Writing `Boolean(someValue)` checks whether the given argument is "truthy" or "falsy," returning true or false accordingly. In this case, `"James Bond"` and `7` are truthy values and `0` is falsy, so the result is [`true, false, false, true`]. See *https://developer.mozilla.org/en-US/docs/Glossary/* for the conversion rules involved.

Chapter 3

3.1 Chaining Calls

The `add()` and `remove()` methods should just end with `return this` to allow for chaining, and that's it.

3.2 Arrays, Not Objects

The following is a possible implementation:

❶ `const newBag = () => [];`

❷ `const isEmpty = (bag) => bag.length === 0;`

❸ `const find = (bag, value) => bag.includes(value);`

❹ `const greatest = (bag) =>`
  `isEmpty(bag) ? undefined : bag[bag.length - 1];`

❺ `const add = (bag, value) => {`
    `bag.push(value);`

```
        bag.sort();
        return bag;
    };

❻ const remove = (bag, value) => {
        const pos = bag.indexOf(value);
        if (pos !== -1) {
            bag.splice(pos);
        }
        return bag;
    };
```

Creating a bag ❶ is just a matter of creating an empty array, and checking whether the array's length is zero ❷ is a way to test whether it's empty. To find a value in the bag ❸, use JavaScript's own `.includes()` method. You'll keep the array in order, so implementing greatest ❹ is simply a matter of checking whether the bag is empty (in which case, it returns undefined, or else it returns the array's last element). To add a value to the bag ❺, push the value into the array and then sort the updated array. Finally, to see whether a value can be removed ❻, use JavaScript's `.find()`method, and if the value is in the array, use `.splice()` to remove it.

3.3 Extra Operations

Many possibilities exist, but keep in mind that you'd ask for these operations only if you actually required them for a specific problem. The following table provides a few, although this list could be extended.

Operation	Signature	Description
Count all	bag → integer	Given a bag, return how many values it contains.
Count value	bag × value → integer	Given a bag and a value, return how many times it's in the bag.
Add many	bag × value × integer → bag	Given a new value and a count, add so many copies of the value to the bag.
Remove all	bag × value → bag	Given a bag and a value, remove all the value's instances from the bag.
Find next	bag × value → value | undefined	Given a bag and value, find the closest higher value to the value in the bag.

3.4 Wrong Operations

If you want to return special values, you can do so as with the greatest() operation, where you returned undefined for an empty bag. If you throw an exception, you can also specify it as a new return value (exception), albeit one that's received in a different way (try/catch), and that would work as well.

3.5 Ready, Set . . .

We'll consider this in Chapter 13, so please skip ahead!

Chapter 4

4.1 How Fast Did You Say?

That cannot be; for large enough n, $f(n)$ becomes negative.

4.2 Weird Bound?

Yes, $n = O(n^2)$ and (more properly, since we prefer tighter bounds) also $o(n^2)$, because n^2 grows faster. The lower bound orders, big and small omega, don't apply.

4.3 Of Big Os and Omegas

In that case (and only in that case), $f(n) = \Theta(g(n))$.

4.4 Transitivity?

If $f(n) = O(g(n))$ and $g(n) = O(h(n))$, then $f(n) = O(h(n))$. We can prove this mathematically, but intuitively, the first equality means that f and g grow proportionally, and the second means that g and h also grow proportionally, and that means that f and h also grow proportionally. If you consider any other order, transitivity still applies; for example, $f(n) = \Omega(g(n))$ and $g(n) = \Omega(h(n))$ implies $f(n) = \Omega(h(n))$.

4.5 A Bit of Reflexion

We can also say $f(n) = O(f(n))$ and $f(n) = \Omega(f(n))$, but not for small omega or small o.

4.6 Going at It Backward

If $f(n) = O(g(n))$, then $g(n) = \Omega(f(n))$, and if $f(n) = o(g(n))$, then $g(n) = \omega(f(n))$. Note the symmetry: big O implies big omega, and small o implies small omega.

4.7 One After the Other

When n grows, the growth of the $O(n^2)$ part will be greater than the growth of the $O(n \log n)$ part, so that's the order of the whole process. In general, the order of a sequence will be that of the greatest order.

4.8 Loop the Loop

In this case, the result is $O(n^3)$. The order of the whole "loop within a loop" is derived from the product of both orders.

4.9 Almost a Power . . .

A formal proof would require applying induction, but consider how you'd frame it. You want to end with a single element, so 1 is a valid size. At the previous step, the array had two 1-sized parts, separated by a single element: the previous size was then 3. Before that, the array had two 3-sized parts, separated by one element: its size was 7. Going backward in this fashion, if you had an s-sized array, at the previous step, the array had to be of size $(2s + 1)$. Starting at $s = 1$, it can be formally proved that sizes are always one less than a power of 2 from the fact that $2(2^k - 1) + 1$ equals $2^{k+1} - 1$.

4.10 It Was the Best of Times; It Was the Worst of Times

In that case (and in that case only), we can deduce that the running time is $\Theta(f(n))$.

Chapter 5

5.1 Factorial in One

You can use this one-liner:

```
const factorial = (n) => (n === 0 ? 1 : n * factorial(n - 1));
```

5.2 Hanoi by Hand

On odd steps, always move the smallest disk cyclically (from A to B, from B to C, and from C to A), and on even steps, make the only possible move that doesn't involve the smallest disk. This method works for an even number of disks; for an odd number of disks, the smallest disk needs to do the cycle in the opposite direction: A to C, C to A, A to B.

5.3 Archery Backtracking

The key to this problem is not to drop an option immediately after choosing it, but rather to try it again first. The following code is basically the same as the solve() function from the section "Solving the Squarest Game on the Beach Puzzle" on page 70, with an addition:

```
❶ const solve = (goal, rings, score = 0, hit = []) => {
    if (score === goal) {
      return hit;
    } else if (score > goal || rings.length === 0) {
      return null;
    } else {
    ❷ const again = solve(goal, rings, score + rings[0], [
        ...hit,
        rings[0],
      ]);
    ❸ if (again) {
        return again;
    ❹ } else {
        const chosen = rings[0];
        const others = rings.slice(1);
        return (
          solve(goal, others, score + chosen, [...hit, chosen]) ||
          solve(goal, others, score, hit)
        );
      }
    }
};

console.log(solve(100, [40, 39, 24, 23, 17, 16]));
```

The parameters were renamed to better match the current puzzle ❶, so it's rings instead of dolls and hit instead of dropped. The additional code tries reusing the first ring ❷, and if it succeeds ❸, you're done. If trying the same ring again fails ❹, drop it and continue with the search as earlier.

For the second question, you can indeed reuse the original solve() algorithm, but you need a change, because you may hit a ring more than once, so rings must appear several times. For example, 40 and 39 should both be considered twice as an option; hitting any of those three or more times would exceed 100. Similarly, 23 and 24 could appear up to four times, 17 five times, and 16 six times. This code finds the solution:

```
console.log(
  solve(100, [
    40, 40,
    39, 39,
    24, 24, 24, 24,
    23, 23, 23, 23,
    17, 17, 17, 17, 17,
    16, 16, 16, 16, 16, 16
  ])
);
```

By the way, if you don't run the code, the answer is 16, 16, 17, 17, 17, and 17!

5.4 Counting Calls

To calculate the nth Fibonacci number, you need one call for that, plus $C(n-1)$ calls for the $(n-1)$th number, plus $C(n-2)$ calls for the $(n-2)$th one, so $C(n) = C(n-1) + C(n-2) + 1$. The solution to this is $C(n) = 2\text{Fibo}(n+1) - 1$.

5.5 Avoid More Work

Simply add a test at the beginning of the loop of the costOfFragment(...) function, as follows:

```
const costOfFragment = memoize((p, q) => {
  ...
  let optimum = Infinity;
  let split = [];
  for (let r = p; r < q; r++) {
    if (totalWidth(p, r) > MW) {
      break;
    }
    ...
  }
  return [optimum, split];
});
```

As soon as the width of blocks from p through r exceeds MW, you can stop the loop; all upcoming total widths will be even greater.

5.6 Reduce for Clarity

The following is a single-line way to calculate the partial values:

```
const partial = blocks.reduce((a, c, i) => ((a[i + 1] = a[i] + c), a),
[0]);
```

Note that the accumulator in this case is an array that you initialize with a single 0 and that will become partial[0].

5.7 Got GOUT?

Use the solve() function from the section "Solving Cryptarithmetic Puzzles" on page 83 as follows. The style of the code is exactly the same as for the SEND + MORE = MONEY puzzle:

```
const { solve } = require("../send_more_money_puzzle");

const toGoOut = (g, o, u, t) => {
  if (t === 0 || g === 0 || o === 0) {
    return false;
  } else {
    const TO = Number(`${t}${o}`);
    const GO = Number(`${g}${o}`);
    const OUT = Number(`${o}${u}${t}`);
    return TO + GO === OUT;
  }
};

solve(toGoOut);
```

The maximum sum of two 2-digit numbers is 99 + 99 = 198, so O = 1, and there was a carry-over from the center column to the leftmost one. At the rightmost column, O + O = T, so T = 2, and there's no carry-over to the center column. Finally, looking at the center column, T + G = 10 + U, but as T = 1, the only way T + G is at least 10 is if G = 9 and then U = 0; the value of GOUT is then 9102, exactly as solve() discovers!

Chapter 6

6.1 Forced Reversal

Change the signs of all the numbers. Next, sort them in ascending order, and then change the signs back. For instance, to sort [22, 60, 9], first change them to [–22, –60, –9], then sort them, and you'll get [–60, –22, –9]. Finally, change the signs again [60, 22, 9], and they'll be in the desired descending order.

6.2 Only Lower

Given `lower(a,b)`, you can implement `higher(a,b)` and `equal(a,b)` as follows:

```
❶ const higher = (a, b) => lower(b, a);
❷ const equal = (a, b) => !lower(a, b) && !higher(a, b);
```

Basically it's just like applying math: a > b if b < a ❶, and a is equal to b, if and only if neither a < b nor a > b ❷.

6.3 Testing a Sort Algorithm

Sort a copy of the data with some other method and use `JSON.stringify(...)` to get a version of the result as a string. Then, sort the data with your new method, use `JSON.stringify(...)` on its output, and compare both JSON strings; they should match.

6.4 Missing ID

This problem has two solutions. You could sort the series of numbers, and then a sequential run through the ordered series would detect missing numbers whenever the gap between consecutive numbers is greater than one. A more specific solution would be to initialize an array of size 1,000,000 with `false`, and for every number in the series, set the corresponding array entry to `true`. You can then run through the array, and entries that are still `false` represent missing IDs.

6.5 Unmatched One

Like the previous question, this problem also has two solutions. You could sort the whole series, and then a quick run through the sorted numbers would find the number that's used only once. The second, trickier solution is to apply the bitwise XOR (^) to all the numbers. If you XOR a number with itself, the result is zero, and if you XOR a number with zero, the result is the number. If you XOR the whole series, the result will be the unmatched number. However, this solution works only if there is a single unpaired number; if there were two or more, it would fail.

6.6 Sinking Sort

The logic is similar to bubble sort's; you need to change only how indices behave:

```
const sinkingSort = (arr, from = 0, to = arr.length - 1) => {
  for (let j = from; j < to; j++) {
    for (let i = to - 1; i >= j; i--) {
      if (arr[i] > arr[i + 1]) {
        [arr[i], arr[i + 1]] = [arr[i + 1], arr[i]];
      }
    }
  }
  return arr;
};
```

6.7 Bubble Swap Checking

The following logic implements the idea:

```
const bubbleSort = (arr, from = 0, to = arr.length - 1) => {
  for (let j = to; j > from; j--) {
    let swaps = false;
    for (let i = from; i < j; i++) {
      if (arr[i] > arr[i + 1]) {
        [arr[i], arr[i + 1]] = [arr[i + 1], arr[i]];
        swaps = true;
      }
    }
    if (!swaps) {
      break;
    }
  }
  return arr;
};
```

6.8 Inserting Recursively

The description is enough to write the code: to sort n numbers, first sort the initial $(n-1)$ ones and then insert the nth number in the sorted list:

```
const insertionSort = (arr, from = 0, to = arr.length - 1) => {
  if (to > from) {
    insertionSort(arr, from, to - 1);
    const temp = arr[to];
    let j;
    for (j = to; j > from && arr[j - 1] > temp; j--) {
      arr[j] = arr[j - 1];
    }
    arr[j] = temp;
  }
  return arr;
};
```

6.9 Stable Shell?

No, Shell sort isn't stable, because the first stages (for gaps larger than one) may disrupt the relative order of equal keys.

6.10 A Dutch Enhancement

The following implementation works:

```
const quickSort = (arr, left = 0, right = arr.length - 1) => {
  if (left < right) {
    const pivot = arr[right];

    let p = left;
    for (let j = left; j < right; j++) {
      if (pivot > arr[j]) {
        [arr[p], arr[j]] = [arr[j], arr[p]];
```

```
        p++;
      }
    }
❶ [arr[p], arr[right]] = [arr[right], arr[p]];

❷ let pl = p;
  for (let i = p - 1; i >= left; i--) {
    if (arr[i] === pivot) {
      pl--;
      [arr[i], arr[pl]] = [arr[pl], arr[i]];
    }
  }

❸ let pr = p;
  for (let j = p + 1; j <= right; j++) {
    if (arr[j] === pivot) {
      pr++;
      [arr[j], arr[pr]] = [arr[pr], arr[j]];
    }
  }

❹ quickSort(arr, left, pl - 1);
  quickSort(arr, pr + 1, right);
  }

  return arr;
};
```

All the code is the same up to the point where the pivot is at arr[p] ❶. You then do a loop to the left of the pivot position ❷, and if you find elements equal to the pivot, you swap; the leftmost position equal to the pivot is always at pl. After this pass, you repeat the process with a similar logic ❸ but from the pivot position to the right. The rightmost position with a value equal to the pivot is pr. After these added loops, all values from pl to pr are equal to the pivot, so you sort the rest ❹.

6.11 Simpler Merging?

If you made that change, merge sort wouldn't be stable. When you have equal values in the first and second lists, you want to choose from the former.

6.12 Try Not to Be Negative

With negative numbers, you'd get a crash when digit becomes negative. And for noninteger numbers, the algorithm disregards the fractional part, so numbers with equal integer parts may end up not sorted correctly. As an extra question, think of ways to solve these problems.

6.13 Fill It Up!

The first option would fill all the elements of bucket with a reference to the same array; instead of 10 different arrays, you'd get a single one, common to all buckets. The second option wouldn't do anything, because .map(...) skips undefined positions.

6.14 What About Letters?

In that case, you'd require more buckets, one for each possible symbol. You might have to do some fancy work if you want to have accented letters (such as á or ü) sorted together with their nonaccented versions.

Chapter 7

7.1 Tennis Sudden Death

The number of matches is easy to find: each match discards one player, and to find the champion, you must discard 110 other players, so the answer is 110, and for n players, $n - 1$. The second best could be any of the players that the first defeated—even possibly in the first round. This tournament had seven rounds, so you may need up to seven extra matches. In general terms, the number of rounds is the logarithm of n in base 2, rounded up, so the total number of comparisons to learn the two minimum values of an array is $n - 1 + \log_2 n$.

7.2 Take Five

You can do it with six comparisons, as follows:

```
const medianOf5 = (a, b, c, d, e) => {
❶ if (a > b) {
     [a, b] = [b, a];
   }

❷ if (c > d) {
     [c, d] = [d, c];
   }

❸ if (a > c) {
     [a, c] = [c, a];
     [b, d] = [d, b];
   }

❹ if (b > e) {
     [b, e] = [e, b];
   }

❺ if (c > b) {
     // b < c < d and b < e: b isn't the median, and d isn't either
     return e > c ? c : e;
❻ } else {
     return d > b ? b : d;
   }
};
```

To understand the code, follow what happens to the variables. After the first test, you know for sure that a < b ❶, and with the next test you know that c < d ❷. You can say that a < b && a < c < d ❸, so a cannot be the median. After ❹, b < e && c < d, so whichever is lowest of b and c cannot be the median either. At ❺, b < c < d && b < e, so neither b nor d can be the median, which is the smallest of c and e. Similarly, at ❻, c < b < e && c < d, so neither c nor e is the median; the smallest of b and d is it.

Just for completeness, here's an equivalent version of median5(...) that works with an array of five independent values and returns the position of the found median. The code is parallel to the previous code:

```
const swapper = (arr, i, j) => ([arr[i], arr[j]] = [arr[j], arr[i]]);

const medianOf5 = (arr5) => {
  if (arr5[0] > arr5[1]) {
    swapper(arr5, 0, 1);
  }

  if (arr5[2] > arr5[3]) {
    swapper(arr5, 2, 3);
  }

  if (arr5[0] > arr5[2]) {
    swapper(arr5, 0, 2);
    swapper(arr5, 1, 3);
  }

  if (arr5[1] > arr5[4]) {
    swapper(arr5, 1, 4);
  }

  if (arr5[2] > arr5[1]) {
    return arr5[4] > arr5[2] ? 2 : 4;
  } else {
    return arr5[3] > arr5[1] ? 1 : 3;
  }
};
```

7.3 Top to Bottom

There are two solutions. You could change the algorithm so instead of selecting minimums and going from 0 to $n - 1$, it could work by selecting maximums and going down from $n - 1$ to 0. To get an efficient algorithm, you should start by comparing k to $n/2$ and going up (as originally shown in the text) or down (as described here) depending on that comparison, whichever takes the least work.

In the particular case in which the values are numeric, you can do a trick: change the sign of all numbers to the opposite, use the algorithm to find the value at $n - k$, and change its sign; can you see why that works?

7.4 Just Iterate

Instead of recursion, you can use a loop that will exit when the code has managed to put the *k*th value of the array in place:

```
const quickSelect = (arr, k, left = 0, right = arr.length - 1) => {
❶ while (left < right) {
    const pick = left + Math.floor((right - left) * Math.random());
    if (pick !== right) {
      [arr[pick], arr[right]] = [arr[right], arr[pick]];
    }
    const pivot = arr[right];

    let p = left;
    for (let j = left; j < right; j++) {
      if (pivot > arr[j]) {
        [arr[p], arr[j]] = [arr[j], arr[p]];
        p++;
      }
    }
    [arr[p], arr[right]] = [arr[right], arr[p]];

  ❷ if (p === k) {
      left = right = k;
    } else if (p > k) {
      right = p - 1;
    } else {
      left = p + 1;
    }
  }
};
```

Set up a loop ❶ that continues as long as left and right haven't reached the same place (k). At the end, instead of recursion or returning early, just manipulate left and right properly ❷.

7.5 Select Without Changing

Simply make a copy of the input array, as follows:

```
const qSelect = (arr, k, left = 0, right = arr.length - 1) => {
❶ const copy = [...arr];
❷ quickSelect(copy, k, left, right);
❸ return copy[k];
};
```

Make a copy of the array ❶, then partition it ❷, and finally return the *k*th value from the copied and repartitioned array ❸.

7.6 The Sicilian Way

The following algorithm does the trick. You can find a different implementation (changing how elements are swapped) in "An Efficient Algorithm for the Approximate Median Selection Problem," by S. Battiato et al., available at *https://web.cs.wpi.edu/~hofri/medsel.pdf*. We'll highlight the only differences with previous code, but we've also implemented some methods in a new way just for variety, and we used iteration instead of recursion (as in question 7.4) for the same reason.

```
❶ const swapIfNeeded = (arr, i, j) => {
    if (i !== j) {
      [arr[i], arr[j]] = [arr[j], arr[i]];
    }
  };

❷ const medianOf3 = (arr, left, right) => {
    if (right - left === 2) {
      const c01 = arr[left] > arr[left + 1];
      const c12 = arr[left + 1] > arr[left + 2];
      if (c01 === c12) {
        return left + 1;
      } else {
        const c20 = arr[left + 2] > arr[left];
        return c20 === c01 ? left : left + 2;
      }
    } else {
      return left;
    }
  };

  const quickSelect = (arr, k, left = 0, right = arr.length - 1) => {
    while (left < right) {
❸    let rr = right;
❹    while (rr - left >= 3) {
        let ll = left - 1;
❺      for (let i = left; i <= rr; i += 3) {
          const m3 = medianOf3(arr, i, Math.min(i + 2, rr));
          swapIfNeeded(arr, ++ll, m3);
        }
❻      rr = ll;
      }
      const m3 = medianOf3(arr, left, rr);
      swapIfNeeded(arr, right, m3);

      const pivot = arr[right];

      let p = left;
      for (let j = left; j < right; j++) {
        if (pivot > arr[j]) {
          swapIfNeeded(arr, p, j);
          p++;
        }
      }
      swapIfNeeded(arr, p, right);

      if (p === k) {
        left = right = p;
      } else if (p > k) {
        right = p - 1;
```

```
    } else {
      left = p + 1;
    }
  }
};

const sicilianSelect = (arr, k, left = 0, right = arr.length - 1) => {
  quickSelect(arr, k, left, right);
  return arr[k];
};
```

You'll need to do lots of swapping, but the swapIfNeeded(...) function ❶ avoids some unnecessary calls by checking whether there's actually any need to swap. Since you'll always be finding the median of up to three values, it makes sense to have a specific function instead of using a generic sort ❷; medianOf3(...) returns the position of the median using up to three comparisons with no swapping. In quick-select the only part that changed are the lines in bold. You'll find the medians in ever-shortening parts of the array; the rr variable marks the right limit of the array you're processing ❸, while the left variable always points at its left limit. As long as the array has more than three elements ❹, you'll do a pass of choosing medians of 3 ❺ and packing them to the left of the array, as shown in the repeated step algorithm; the difference is that after each pass ❻ you'll shorten the array and loop again to find medians of 3. When the set of medians (of medians of medians of . . . , and so on) is short enough, you'll just choose its last element as the next pivot.

Chapter 8

8.1 Good Enough Shuffling

This is the code I used when testing the functions for the book:

```
❶ const logResults = (fn, from = 0, to = 10, n = to, times = 4000) => {
❷   const bar = (len, val, max) =>
      "#".repeat(Math.round((len * val) / max));

❸   const result = {};

    const compare = (a, b) => (a < b ? -1 : 1);

❹   let max = 0;
❺   for (let i = 1; i <= times; i++) {
❻     const arr = Array(n)
        .fill(0)
        .map((v, i) =>
          i < from || i > to ? i : String.fromCharCode(65 + i),
        );
❼     const x = fn(arr, from, to).join("-");
❽     result[x] = x in result ? result[x] + 1 : 1;
❾     max = Math.max(max, result[x]);
    }
```

```
❿ let count = 0;
  for (const [key, val] of Object.entries(result).sort(compare)) {
    count++;
    console.log(
      `${key}: ${String(val).padStart(5)} ${bar(50, val, max)}`,
    );
  }

  console.log("COUNT=", count);
};
```

The parameters for this logging are the shuffling function fn, the portion of the array to be shuffled (from, to), the size of the input array to use (n), and how many times to run the test ❶. An auxiliary function bar(...) ❷ draws a bar of pound characters: the maximum length is len for a val value that equals a maximum max value; smaller values get proportionally shorter bars. Use an object, result, to count how many times each permutation occurred ❸ (you could also use a JavaScript map). The max variable ❹ tracks the maximum number of times that any permutation appeared. You loop n times ❺, each time initializing an array to shuffle ❻. Use fn to shuffle the array, creating a string key out of the result ❼, and update the counts ❽ and maximum ❾ observed frequencies. The last step ❿ returns the results in tabular form, with permutations ordered alphabetically for clarity.

8.2 Random Three or Six

Tossing two coins produces four combinations. You can assign numbers to those combinations, and if a combination numbered 1 to 3 occurs, accept it; if you get a 4, redo the throw:

```
❶ const random01 = () => (randomBit() ? 0 : 1);

  const random3 = () => {
    let num = 0;
    do {
❷    num = 2 * random01() + random01();
❸  } while (num > 2);

❹  return ["3-hi", "2-md", "1-lo"][num];
  };
```

This assigns numbers to combinations of coins using the binary system, calling heads a 0 and tails a 1. Use the randomBit() function to produce bits ❶. You "throw" the die twice, and assign a number to the resulting combination ❷. Loop until you get a number from 0 to 2 ❸, and after getting it, map it to a result ❹.

To simulate a die, you'll need three coin tosses, but the logic is quite similar:

```
const randomDie = () => {
  let num = 0;
```

```
  do {
    num = 4 * random01() + 2 * random01() + random01();
  } while (num > 5);
  return num + 1;
};
```

Finally, simulating a 1 to 20 throw is trickier. You could use 5 bits, getting a result from 0 to 31 and discarding the 12 last values, but that would probably require several attempts; after all, a failure rate of 12 out of 32 is high. A better solution would be using 6 bits for a result of 0 to 63, discarding the last 4 (so, 0–59) and dividing the result by 3 to get a 0 to 19 result, with small odds (4 in 64) of needing a new attempt.

8.3 Not-So-Random Shuffling

If this algorithm produced uniform shuffles, you'd expect each value in the original array to appear the same number of times at each position of the shuffled array. For the initial value to end at the last position, you need all `randomBit()` calls to come up true, so it's far more likely that the initial value will end up not far from its starting position and only rarely appear in the last places.

8.4 Bad Swapping Shuffle

The problem is that this code doesn't generate every possible permutation with the same frequencies. You can see a hint of that by noticing that this code loops n times, choosing a random number with n possible values each time, so it has n^n ways to run. However, a permutation generation algorithm should run in $n!$ ways only. If you want to verify this by hand, try the algorithm for just three elements and simulate all 27 possible sequences of random numbers: from (0,0,0), which would shuffle ABC into CBA; (0,0,1), which shuffles ABC into BCA; and so on, up to (2,2,2), which produces CBA. Count how many times each possible permutation occurs, and you'll see that some are more favored than others. The algorithm doesn't produce an even distribution of shuffles.

8.5 Robson's Top?

The question is equivalent to finding out the largest factorial that can be calculated without losing precision. Using normal precision, this turns out to be 18; 19! is too long:

```
❶ for (let num = 1, fact = 1; fact < fact + 1; num++) {
    fact *= num;
    console.log(`${num}=${fact}`);
  }
```

You can test whether precision is lost by adding 1 and checking whether the result is changed; if you get the same result, it means that JavaScript hasn't enough digits to accommodate your large number ❶.

However, you could use `BigInt` values, and then you'd be able to work with much higher values—as long as the size of a factorial doesn't exceed the allowed memory. The following program will happily go past 19 until something crashes:

```
for (let num = 1n, fact = 1n; fact < fact + 1n; num++) {
  fact *= num;
  console.log(`${num}=${fact}`);
}
```

The only difference here is that it's using `BigInt` numbers: `1n` is such a number.

8.6 Sampling Testing

The solution is quite similar to what we used for shuffling, with a special detail related to whether you're doing sampling with or without repetition:

```
❶ const logResults = (fn, k, n = 10, times = 60000, noReps = true) => {
  const bar = (len, val, max) =>
    "#".repeat(Math.round((len * val) / max));

  const result = {};

  const compare = (a, b) => (a < b ? -1 : 1);

  let max = 0;
  for (let i = 1; i < times; i++) {
    const arr = Array(n)
      .fill(0)
      .map((v, i) => String.fromCharCode(65 + i));
❷   const x = noReps

        ? fn(arr, k).sort().join("-")
        : fn(arr, k).join("-");
    result[x] = x in result ? result[x] + 1 : 1;
    max = Math.max(max, result[x]);
  }

  let count = 0;
  for (const [key, val] of Object.entries(result).sort(compare)) {
    count++;
    console.log(
      `${key}: ${String(val).padStart(5)} ${bar(50, val, max)}`,
    );
  }

  console.log("COUNT=", count);
};
```

When sampling without repetition, sample B-C-A is the same as sample C-A-B, so you sort the elements to get a unique key A-B-C. However, when sampling with repetition, the two results are different. The added parameter `noReps` ❶ solves this; when counting the sample, you sort values (or not) depending on it ❷.

8.7 Single-Line Repeater

The following is a single statement but is shown here on several lines for clarity. You generate an array of the right size and use .map(...) to fill it with randomly chosen values:

```
const repeatedPick = (arr, k) =>
  Array(k)
    .fill(0)
    .map(() => arr[randomInt(0, arr.length)]);
```

The tricky part is that you need to use .fill(0) to put values in the array. If you don't do this, .map(...) won't do anything, because it skips uninitialized array locations.

8.8 Sort to Sample

The following logic does the work:

```
const sortingSample = (arr, k) => {
❶ const rand = arr.map((v) => ({ val: v, key: Math.random() }));

❷ for (let i = 0; i < k; i++) {
      let m = i;
      for (let j = i + 1; j < arr.length; j++) {
        if (rand[m].key > rand[j].key) {
          m = j;
        }
      }
      if (m !== i) {
        [rand[i], rand[m]] = [rand[m], rand[i]];
      }
    }

❸ return rand.slice(0, k).map((obj) => obj.val);
  };
```

This method of assigning random keys comes directly from the section "Shuffling by Sorting" on page 139 ❶. The logic that follows is a slight modification of the logic in "Selecting with Comparisons" on page 124 ❷, and the final code to leave just the original values is again from "Shuffling by Sorting" ❸.

8.9 Iterate, Don't Recurse

To understand why this works, consider what the calls to other(...) will be and with which arguments. Applying the conversion to the factorial function produces the well-known equivalent:

```
const factorial = (p) => {
  let result = 1;
  for (let i=1; i <= p; i++) {
    result = result * i
```

```
        }
    return result;
}
```

Applying the conversion to Floyd's algorithm is a bit trickier, because the function has two arguments (k and n), but since their difference is constant because they decrease in parallel, you can achieve the conversion. You'll need to rename some variables to avoid confusion; for example, we're already using i with another meaning in Floyd's code.

8.10 No Limits?

The check isn't needed; initially toSelect is not greater than toConsider. If they ever become equal, from that point onward, all elements will be chosen, because the test Math.random() < toSelect / toConsider will always succeed for all random values less than 1, and toSelect will eventually become 0.

Chapter 9

9.1 Searching Right

This is the code I used. The checkSearch(...) higher-order function takes a searching function to test and a boolean flag to indicate whether to use it with sorted or unsorted data. The actual files with data are called data32 and data_sorted_32:

```
const checkSearch = (fn, sorted = false) => {
❶ const data32 = sorted ? require("../data_sorted_32") : require("../
data32");

❷ const verify = (v, i, f) => {
  ❸ if (i !== f) {
      throw new Error(`Failure searching v=${v} i=${i} fn=${f}`);
    }
  ❹ if (i !== -1) {
      console.log("Searching v=", v, " i=", i);
    }
  };

❺ data32.forEach((v, i) => {
    const f = fn([...data32], v);
    verify(v, i, f);
  });

❻ const m1 = Math.min(...data32);
  const m2 = Math.max(...data32);
❼ for (let i = m1 - 3; i <= m2 + 3; i++) {
  ❽ if (!data32.includes(i)) {
    ❾ verify(i, -1, fn([...data32], i));
    }
  }
};
```

You use a sorted or unsorted set of data ❶ depending on the kind of algorithm. An auxiliary `verify(...)` function ❷ allows shorter code: the function tests whether the result matches what you expected ❸, throwing an error if not. For successful searches ❹, it displays the input and output. You try searching for every value in the input array ❺. You then find the minimum (`m1`) and maximum (`m2`) values of the array ❻ and then try all possible (invalid) searches ❼ from `m1` - 3 to `m2` + 3; whenever a value isn't included in the array ❽, you specifically try to find it, expecting to receive -1 as result ❾.

9.2 JavaScript's Own

The simplest solution would be `array.findIndex(x => x === key)`.

9.3 Infinite Search Levels?

If `levels` tends to infinity, `b` will always become 2, and that means you split the search area in half at each step. You've rediscovered binary search! Roughly speaking, `i` takes the place of `l` in the iterative binary search code, and `m` is the difference between `l` and `r`.

9.4 Exactly How Much?

You can calculate the average number of tests by finding the sum of $1 \times 1 + 2 \times 2 + 4 \times 3 + 8 \times 4$ and so on, up to 2^{n-1} found in n questions, and dividing by the total number of searches, $2^n - 1$. We worked this out in section "Analysis of Algorithms in Practice" on page 55, and the sum equals $(n + 1)2^n - (2^{n+1} - 1))$. Dividing, you get the average you wanted, which for large values of n is close to $n - 1$, so for any array length k, the answer is approximately $\log_2 k - 1$.

9.5 Three Tops Two?

When trying to decide in which third to keep searching, if the key is in the first third, you'll be able to decide by asking a single question, and if it's in the other two thirds, you'll need to ask a second question: on average, you need $1 \times 1/3 + 2 \times 2/3 = 5/3$ questions. Dividing the array by 3, you'll have $\log_3 n$ search steps, compared to $\log_2 n$ steps with binary search. Given that $\log_3 n$ is approximately $0.631 \log_2 n$, the performance of ternary search is about $5/3 \times 0.631$, which equals 1.052 times that of binary search, which is practically the same.

9.6 Binary First

The idea is if you find the key, instead of returning it, note the position and keep searching to the left:

```
const binaryFindFirst = (arr, key, l = 0, r = arr.length - 1) => {
❶ let result = -1;
  while (l <= r) {
    const m = (l + r) >> 1;
    if (arr[m] === key) {
❷    result = m;
```

```
❸ r = m - 1;
  } else if (arr[m] > key) {
    r = m - 1;
  } else {
    l = m + 1;
  }
}
❹ return result;
};
```

The code is the same as for binary search, with four differences. Initialize a result variable ❶ with the value that you'll return at the end ❹. When you find the key, update this variable ❷, but instead of returning, keep searching to the left ❸ just in case the key occurs again.

To find the last position of key in arr ❸, write l = m + 1 to keep searching but to the right.

9.7 Count Faster

Use the solutions to the previous problem to find the first and last positions of the key in the array. Of course, if the first search returns -1, the count is 0 and you don't have to do the second search.

9.8 Rotation Finding

You can use a variant of binary search for this, but it's not exactly the same; there's a careful detail to pay attention to:

```
const rotationFind = (arr) => {
❶ let l = 0;
  let r = arr.length - 1;
❷ while (arr[l] > arr[r]) {
❸   const m = (l + r) >> 1;
❹   if (arr[m] > arr[r]) {
      l = m + 1;
❺   } else {
      r = m;
    }
  }
❻ return l;
};
```

You set up l and r as for binary search ❶. You'll stop searching whenever you find that the value at l isn't greater than the value at r ❷, and while that isn't true, you'll search in a half of the array. As in binary search ❸, m is the middle of the array. If the value at m is greater than the value at r ❹, the rotation is in the right part of the array, and the minimum must be at least at m + 1. Otherwise, the value at l must be greater than the value at m ❺, and here you must be very careful because the place of the rotation could be m itself! So, when the rotation is on the left, you don't set r to m - 1, as in binary search, but rather to m. When you find that the value at l isn't greater than the value at r ❻, l is the position you're looking for.

9.9 Special First

No, you don't need to. Assume that arr[0] === key:

```
const exponentialSearch = (arr, key) => {
  const n = arr.length;
❶ let i = 1;
❷ while (i < n && key > arr[i]) {
    i = i << 1;
  }
❸ return binarySearch(arr, key, i >> 1, Math.min(i, n - 1));
};
```

In the logic, i starts at 1 ❶; the loop exits immediately ❷ because key (arr[0]) cannot be greater than arr[1]. The final binary search ❸ is done between 0 and 1 and succeeds.

Chapter 10

10.1 Iterating Through Lists

To find the size of a list, initialize a pointer to the first element and follow the next pointers until you get to the end, counting each node you visit:

```
const size = (list) => {
❶ let count = 0;
❷ for (let ptr = list; ptr !== null; ptr = ptr.next) {
    ❸ count++;
  }
  return count;
};
```

The count variable ❶ keeps a count of the elements. You'll do a loop, starting at the head and advancing until reaching the end ❷, and you'll update the count for each node ❸.

Similar code is used to find whether a given value is found in a list:

```
const find = (list, value) => {
❶ for (let ptr = list; ptr !== null; ptr = ptr.next) {
    if (ptr.value === value) {
    ❷ return true;
    }
  }
❸ return false;
};
```

The logic of going down the list is the same as for count(...) ❶, and for each element you test whether it matches the desired value. If so, you return true ❷, and if you reach the end of the list without a match, you return false ❸.

Adding an element is also a matter of advancing down the list until reaching either the end or the desired position:

```
const add = (list, position, value) => {
  if (position === 0) {
    list = { value, next: list };
  } else {
    let ptr;
    for (
      ptr = list;
      ptr.next !== null && position !== 1;
      ptr = ptr.next
    ) {
      position--;
    }
    ptr = { value, next: ptr.next };
  }
};
```

Finally, removing a given element also works with similar logic: go down the list until you reach the end of the list or the position you want to remove:

```
const remove = (list, position) => {
  if (!isEmpty(list)) {
    if (position === 0) {
      list.first = list.next;
    } else {
      let ptr;
      for (
        ptr = list;
        ptr.next !== null && position !== 1;
        ptr = ptr.next
      ) {
        position--;
      }
      if (ptr.next !== null) {
        ptr.next = ptr.next.next;
      }
    }
  }
  return list;
};
```

10.2 Going the Other Way

The idea is that you go through the list, pushing each value into a stack, and the final list will be reversed:

```
const reverse = (list) => {
❶ let newList = null;
❷ while (list !== null) {
```

```
❸ [list.next, newList, list] = [newList, list, list.next];
   }
❹ return newList;
};
```

Create the reversed list as a stack with newList as its pointer ❶, and use list to go element by element ❷ pushing it into the newList stack; you may want to draw a diagram of the pointer juggling that's going on here ❸. At the end, return the reversed list ❹. A question for you: Does this algorithm also work for null lists?

10.3 Joining Forces

The idea is to look for the last element of the first list and link it to the head of the second list:

```
const append = (list1, list2) => {
  if (list1 === null) {
❶ list1 = list2;
  } else {
❷ let ptr = list1;
    while (ptr.next !== null) {
❸ ptr = ptr.next;
    }
❹ ptr.next = list2;
  }
  return list1;
};
```

An interesting case is if the first list is empty ❶, the result of the operation is the second list. Otherwise, use ptr to go down the first list ❷, advancing while the end hasn't been reached ❸. When reaching the end ❹, just modify its next pointer to point at the second list.

10.4 Unloop the Loop

You don't need to store anything; just use two pointers. The idea is to advance down the list with two pointers, one a node at a time, and one at twice the speed. If the list doesn't have a loop, the second will reach the end and you're done. If the list has a loop, however, the two pointers will eventually meet (because the second moves faster than the first), and that means there's a loop:

```
const hasALoop = (list) => {
  if (list === null) {
❶ return false;
  } else {
❷ let ptr1 = list;
    let ptr2 = list.next;

❸ while (ptr2 !== null && ptr2 !== ptr1) {
❹ ptr1 = ptr1.next;
❺ ptr2 = ptr2.next ? ptr2.next.next : null;
    }
```

```
❻ return ptr2 === ptr1;
  }
};
```

If the list is empty ❶, then for certain there's no loop. Otherwise, ptr1 goes down the list a node at a time, and ptr2 advances two nodes at a time ❷. You'll keep working unless ptr2 reaches the end or ptr2 reaches ptr1 ❸. During the iteration, ptr1 advances one node ❹ and ptr2 advances twice ❺ unless it reached the end. At the end, if ptr2 had reached ptr1, there's a loop; otherwise, there isn't.

10.5 Arrays for Stacks

Given the .pop(...) and .push(...) methods, implementing a stack is straightforward:

```
❶ const newStack = () => [];

❷ const isEmpty = (stack) => stack.length === 0;

  const push = (stack, value) => {
❸ stack.push(value);
    return stack;
  };

  const pop = (stack) => {
❹ if (!isEmpty(stack)) {
      stack.pop();
    }
    return stack;
  };
❺ const top = (stack) =>
  isEmpty(stack) ? undefined : stack[stack.length - 1];
```

A new stack is just an empty array ❶, and to check for an empty stack, you just see whether the array's length is zero ❷. To push a new value ❸, use .push(...) and use .pop() ❹ to pop with a check for an empty stack. Finally, getting to the top of the stack is just a matter of looking at the last element of the array ❺.

10.6 Stack Printing

Given the implementations of size(...) and find(...) in the first question for this chapter, you shouldn't need further explanation for this:

```
const print = (list) => {
  for (let ptr = list; ptr !== null; ptr = ptr.next) {
    console.log(ptr.value);
  }
};
```

10.7 Height of a Stack

You could implement this in $O(n)$ time by using code similar to the code in the previous question, but a simpler solution is to add a height field to the stack definition initialized to zero and update that field appropriately when pushing or popping values. A stack would now be something more similar to a queue, in that an object is used instead of just a pointer:

```
const newStack = () => ({ first: null, height: 0 });
```

10.8 Maximum Stack

The idea is to push entries with two pieces of data: not only the value that's being pushed, but also the current maximum, which depends on the value being pushed and the previous maximum, which was at the top of the stack before the push. To find the minimum, you'd push three pieces of data: the value to be pushed, the maximum at that time, and the minimum at the same time. This allows knowing the maximum or minimum at any moment, in $O(1)$ time.

10.9 Queued Arrays

This operation essentially requires the same implementation as for stacks, except that for entering a queue, use .unshift(...) instead of push(...), so new values are added at the beginning of the array instead of at the end.

10.10 Queue Length

Trick question! Just apply the same logic as for question 10.9.

10.11 Queueing for Sorting

The following code does the work; it just rewrites the _radixSort(arr) function from the "Radix Sort" section on page 115, which sorts an input array:

```
const _radixSort = (arr) => {
  const ML = Math.max(...arr.map((x) => String(x).length));

  for (let i = 0, div = 1; i < ML; i++, div *= 10) {
    const buckets = Array(10)
      .fill(0)
❶    .map(() => ({ first: null, last: null }));

    arr.forEach((v) => {
❷    const digit = Math.floor(v / div) % 10;
❸    const newNode = { v, next: null };
      if (buckets[digit].first === null) {
        buckets[digit].first = newNode;
      } else {
        buckets[digit].last.next = newNode;
      }
```

```
      buckets[digit].last = newNode;
    });

    arr = [];
❹ buckets.forEach((b) => {
      for (let ptr = b.first; ptr; ptr = ptr.next) {
        arr.push(ptr.v);
      }
    });
  }

  return arr;
};
```

The differences in the code from that in Chapter 6 are highlighted.
Instead of creating arrays for each bucket, set up queues ❶. For each
value, after deciding which bucket it will go in ❷, make it enter the corresponding queue ❸. After distributing values into queues, go through
each of them ❹ to generate an array.

10.12 Stacked Queues

As suggested, the idea is to use two stacks: say IN and OUT. Push the
new value to the IN stack to enter the queue. Pop the OUT stack to exit
the queue, but if it was empty, first pop all values from the IN stack, one
at the time, and push them to OUT, before finally doing the pop. Every
value that enters the queue will go through two pushes (first to IN and
at a later time to OUT) and two pops (at some later time from IN and
eventually from OUT), so the amortized cost of the operation is $O(1)$.
Obviously, some exit operations (those that find an empty OUT stack)
will require more time.

10.13 Palindrome Detection

The idea is to split the string into separate characters and enter all the
letters into a dequeue. When you're done, repeatedly check whether the
front element is the same as the last element, and exit them both. If you
get down to zero elements (first is null) or just one (first equals last),
the string was a palindrome.

10.14 Circular Listing

The following logic does the job:

```
const print = (circ) => {
❶ if (!isEmpty(circ)) {
❷   let ptr = circ;
    do {
❸     console.log(ptr.value);
❹     ptr = ptr.next;
❺   } while (ptr !== circ);
  }
};
```

First check for an empty circular list; if so, you don't have to do anything ❶. Use ptr to go around the list ❷, printing each visited node ❸, advancing to the next ❹, and exiting the loop when reaching the initial node again ❺.

10.15 Joining Circles

Manipulating links isn't hard, but you have to be careful. Initially, you'll have this scenario:

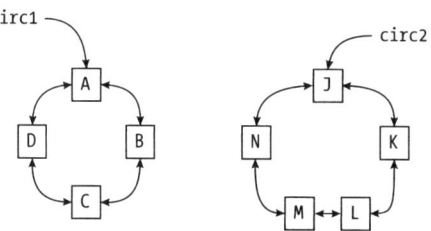

And you'll want to get to this situation:

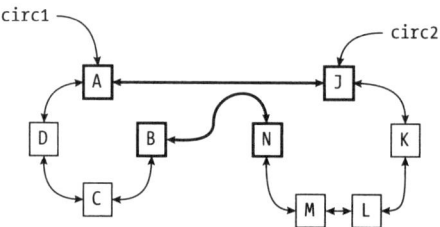

Assuming that both circ1 and circ2 are not null, you need four lines of code:

```
❶ circ2.prev.next = circ1.next;
❷ circ1.next.prev = circ2.prev;
❸ circ1.next = circ2;
❹ circ2.prev = circ1;
```

First, B follows N ❶. Set B so that N precedes it ❷. Similarly, J follows A ❸, and A precedes J ❹.

Chapter 11

11.1 Sentinels for Searches

If an ordered list includes a final +Infinity value, you can simplify the search because you know you'll never get past the end:

```
const find = (list, valueToFind) => {
  if (valueToFind < list.value) {
    return false;
  } else if (valueToFind === list.value) {
```

```
      return true;
  } else {
    // valueToRemove > list.value
    return find(list.next, valueToFind);
  }
};
```

Compare this code to the code in the section "Searching for a Value" on page 183; the first if is now simpler. (I agree that the speed gain may be minimal, but the technique is commonly used and worth knowing.)

11.2 More Sentinels?

An initial -Infinity value means that you'll never add a new value at the beginning of the list, so essentially that makes the pointer to the head a constant. This extra sentinel makes iterative code simpler; you can check it out yourself.

11.3 A Simpler Search?

You could change the function to return either null (if the value wasn't found) or a pointer to the value—which will just be the first in the list.

11.4 Re-skipping Lists

Start by deleting all pointers at levels other than the bottom one. Then, create level 1 by going through level 0 and choosing all elements at even positions; level 1 will end with about half the elements of level 0. Redo this process to create level 2 based on elements at even positions in level 1; then, create level 3 based on even positions in level 2, and so on. Stop when the topmost list has only one element.

11.5 Skip to an Index

You want to be able to produce, say, the 229th value of a list without having to go one by one past the previous 228 values. To sketch the solution, you'll define the "width" of a link as the number of values at the next level that are encompassed by the link. (In other words, how many nodes do you jump over by following the link?) You can create and update the widths when adding or removing values. All widths at the bottom level are 1. At any level, if a link goes from node A to node B, the width of the link is the sum of all the widths from A (inclusive) to B (exclusive).

Knowing these widths makes it easy to find any given position. You start at the topmost level, going horizontally as long as the sum of widths doesn't exceed the index you want. When this sum surpasses the index you want, you go down to the next level and keep going horizontally.

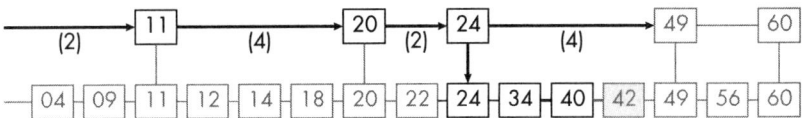

For instance, in the diagram provided, say you want to find the 11th element of the list (the widths of links are in parentheses below each link). At the first level, the first three links cover $2 + 4 + 2 = 8$ elements, so by following them in three steps, you're already at the 8th element of the list. The next link is 4, so that would go past the 11th, and you'd go down. The widths of the links there are all 1, so you advance a couple more times and find 40, the 11th value.

11.6 Simpler Filling

It would fill the array with a reference to the *same* list, instead of references to 100 distinct lists.

11.7 A Hashed Set

For searches and removals, there's no change whatsoever. For additions, the key is to keep searching until you either find the value (in which case you wouldn't add it again) or reach an empty space. For efficiency, should you find an available slot during the search, make a note of it, and instead of adding the new value at the empty space, put it into the available slot.

11.8 Wrong Seating

The key is to look at the problem in a different way. Instead of thinking that a person upon finding their seat occupied moves to some other random seat, imagine that *they* sit down and the previous occupant of the seat is the one who moves away. (In other words, the one who originally sat down in the wrong place.) By the time the 100th person enters, 98 people will be in their assigned seats, and the first person who entered the theater will be either in their seat or in the 100th person's seat—it depends on how the previous person moved with only two possible seats to choose: their seat or the 100th person's seat. (If they had chosen any other seat, they would have had to move again when the rightful occupant of the seat appeared.) The answer is 50 percent.

11.9 Progressive Resizing

The idea is to work with both tables at the same time, gradually removing values from the old one and inserting them in the new one. After you decide you need to resize, create a new, larger table, and from then on, all insertions will go to the new table. Whenever you want to search for a value, look in both tables; a value in the old table may have already been moved to the new table. (The same applies to removals.) Every time you do an operation (add, remove, or search), remove some values from the old table and insert them into the new one. When every value has been removed from the old table, just work with the new one.

Chapter 12

12.1 A Matter of Levels

The height of a tree would be the highest level of its nodes.

12.2 Breaking the Rules

Symbolic links (symlinks) may point to any file or directory, so they allow you to break the tree structure.

12.3 What's in a Name?

A *perfect tree* is complete and full. A *complete tree* isn't necessarily perfect (the bottom need not be complete), and it may not be full either, because a node may have only one child—for example, a tree with just two nodes. Finally, a *full tree* may be neither complete nor perfect; see Chapter 13 for an example of this.

12.4 A find() One-Liner

Using the ternary operator, you can make do as follows:

```
const find = (tree, keyToFind) =>
  !isEmpty(tree) &&
  (keyToFind === tree.key ||
    find(
      tree[keyToFind < tree.key ? "left" : "right"],
      keyToFind,
    ));
```

Because of space restrictions, this appears as several lines of text, but it's a single statement in any case.

12.5 Sizing a Tree

Given the definition that an empty tree has size 0 and that, otherwise, the size of a tree is 1 (for the root) plus the sizes of both subtrees, you can write a one-liner solution for this:

```
const {
  isEmpty,
} = require("../binary_search_tree.js");

const calcSize = (tree) =>
  isEmpty(tree)
    ? 0
    : 1 + getSize(tree.left) + getSize(tree.right);
```

12.6 Tall as a Tree

The height of a tree is the maximum length of a path from the root to a leaf. So, if you know the heights of both subtrees of the root, the height of the complete tree will be one more than the height of the highest subtree. You can program this very simply using recursion:

```
const { isEmpty } = require("../binary_search_tree.js");

const calcHeight = (tree) =>
  isEmpty(tree)
    ? 0
    : 1 + Math.max(getHeight(tree.left), getHeight(tree.right));
```

12.7 Copy a Tree

Recursion is the best solution: a copy of an empty tree is just an empty tree, and a copy of a nonempty tree is built out of the tree's root, plus copies of its left and right subtrees:

```
const { newNode, isEmpty } = require("../binary_search_tree.js");

const makeCopy = (tree) =>
  isEmpty(tree)
    ? tree
    : newNode(tree.key, makeCopy(tree.left), makeCopy(tree.right));
```

You can also build a copy another way, which should remind you of a postorder traversal:

```
const makeCopy2 = (tree) => {
❶ if (isEmpty(tree)) {
    return tree;
  } else {
❷  const newLeft = makeCopy2(tree.left);
❸  const newRight = makeCopy2(tree.right);
❹  return newNode(tree.key, newLeft, newRight);
  }
};
```

If the tree to copy is null ❶, nothing needs to be done. Otherwise, first make a copy of the left subtree ❷, then a copy of the right subtree ❸, and finally, build a tree out of the tree's key plus the two newly created trees ❹.

12.8 Do the Math

You need a postorder traversal for this, because before applying any operator, you need to know the value of its left and right subexpressions. You can do this with a function, since a whole class would be overkill:

```
const evaluate = (tree) => {
❶ if (!tree) {
    return 0;
❷ } else if (typeof tree.key === "number") {
    return tree.key;
❸ } else if (tree.key === "+") {
    return evaluate(tree.left) + evaluate(tree.right);
❹ } else if (tree.key === "-") {
    return evaluate(tree.left) - evaluate(tree.right);
❺ } else if (tree.key === "*") {
    return evaluate(tree.left) * evaluate(tree.right);
❻ } else if (tree.key === "/") {
    return evaluate(tree.left) / evaluate(tree.right);
❼ } else {
    throw new Error("Don't know what to do with ", tree.key);
  }
};
```

If the tree is empty ❶, you return 0, a reasonable value. Otherwise, if the root is a number ❷, just return that number, and if the root is an operator ❸❹❺❻, use recursion to evaluate both sides of the expression and return the calculated value. You also add a "catch-all" for any unexpected input ❼.

A simple example shows this at work:

```
const exampleInBook = {
  key: "*",
  left: {
    key: "+",
    left: { key: 2 },
    right: { key: 3 }
  },
  right: {
    key: 6
  }
};
```

This code returns 30, as expected. Can you figure out why I didn't include null pointers?

12.9 Making It Bad

You never want to have a node with two children, so the root must be either the minimum or the maximum of the set of keys. After that, the following key must also have a single child, so it must be either the minimum or the maximum of the remaining set of keys. If you follow this logic to the end, you have two options for the first key, times two options for the second key, times two options for the third key, and so on, up until the $(n - 1)$ key, after which a single option is left. The number of linear trees you can produce out of n keys is then 2^{n-1}.

12.10 Rebuild the Tree

Given the preorder and inorder lists, it's clear that the root of the tree must be the first value in the preorder. If you look for that value in the inorder list, all keys preceding it will come from the root's left subtree and all keys after it will be from the right subtree. Separate the preorder list in two, and you'll have the preorder and inorder listings for both subtrees; apply recursion, and you'll build the tree.

12.11 More Rebuilding?

Working with inorder and postorder would be possible; the only difference is that you'd find the root at the end of the postorder listing instead of at the beginning of the preorder listing. However, working with preorder and postorder isn't possible—and an example should suffice to show why. If I tell you that the preorder listing was "1, 2" and the postorder listing was "2, 1", two possible binary search trees produce those listings. Can you find them?

12.12 Equal Traversals

Preorder and inorder would be the same if there were no left trees, so the first answer is "trees with only right subtrees"; for inorder and postorder, the answer would similarly be "trees with only left subtrees." Finally, for preorder and postorder, the answer is "trees with no more than one key."

12.13 Sorting by Traversing

First, add all the keys into a binary search tree, create an empty array, and then do an inorder traversal providing a visit function that will push the key value into the array. (You'll use this technique later in question 12.26.)

12.14 Generic Order

The following code will do. Note the two recursive calls mixed with the three possible visit() calls:

```
const { isEmpty } = require("../binary_search_tree.js");

const anyOrder = (tree, order, visit = (x) => console.log(x)) => {
  if (!isEmpty(tree)) {
    order === "PRE" && visit(tree.key);
    anyOrder(tree.left, order, visit);
    order === "IN" && visit(tree.key);
    anyOrder(tree.right, order, visit);
    order === "POST" && visit(tree.key);
  }
};
```

12.15 No Recursion Traversal

Use a stack from Chapter 11. You could do specific solutions for each of the traversals, but let's go with a generic solution (actually implementing the anyOrder() function from earlier) to highlight how a stack makes avoiding recursion easy.

The idea is that you'll push into the stack the pending operations, which can be of two types: visit a key (type "K") or traverse a tree (type "T"). You'll push these operations, and the main code will be a loop that will pop an operation and do it, which may imply visiting or traversing, and the latter will cause more operations to be pushed in:

```
const anyOrder = (tree, order, visit = (x) => console.log(x)) => {
  let pending = newStack();
  let type = "";
❶ pending = push(pending, { tree, type: "T" });

❷ while (!isEmptyStack(pending)) {
❸   [pending, { tree, type }] = pop(pending);

❹   if (!isEmptyTree(tree)) {
❺     if (type === "K") {
```

```
        visit(tree.key);
❻ } else {
        if (order === "POST") {
          pending = push(pending, { tree, type: "K" });
        }
        pending = push(pending, { tree: tree.right, type: "T" });
        if (order === "IN") {
          pending = push(pending, { tree, type: "K" });
        }
        pending = push(pending, { tree: tree.left, type: "T" });
        if (order === "PRE") {
          pending = push(pending, { tree, type: "K" });
        }
      }
    }
  }
};
```

Start by creating a stack and pushing the tree you want to traverse ❶. While there are pending operations ❷, you'll pop the top one ❸, and if it doesn't point to an empty tree ❹, you'll execute whatever is needed. If the operation was a "K", just visit the node ❺, and if it was a "T" ❻, you'll have to push two operations (traverse the left and right subtrees) and a visit (for the root). The key point is to make sure to push them in backward order, so the operations will be popped in the right sequence; study this carefully. For instance, if you are doing a postorder traversal, you'll first push the root visit, then the right subtree traversal, and finally the left subtree traversal—and when you do those operations in reverse order, everything will come out right.

12.16 No Duplicates Allowed

Basically, you just have to check whether you have arrived at the value you were thinking of adding:

```
const add = (tree, keyToAdd) => {
  if (isEmpty(tree)) {
    return newNode(keyToAdd);
❶ } else if (keyToAdd === tree.key) {
    throw new Error("No duplicate keys allowed");
  } else {
❷   const side = keyToAdd < tree.key ? "left" : "right";
    tree[side] = add(tree[side], keyToAdd);
    return tree;
  }
};
```

You add a test for equality before continuing the search ❶, and the other minor change is that you don't test for "less-than-or-equal-to" ❷, because the key can never be equal.

12.17 Get and Delete

You can manage to get the minimum and remove it at the same time, if after finding the minimum value in the (not empty) tree, you copy its right subtree to the node; see the following diagram for an example:

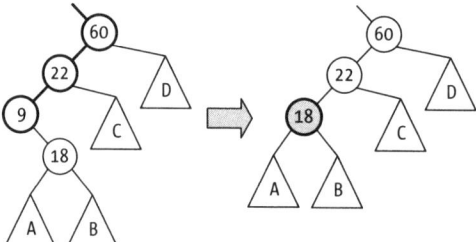

Here's how to implement this algorithm:

```
const _removeMinFromTree = (tree) => { // not empty tree assumed
  if (isEmpty(tree.left)) {
❶ return [tree.right, tree.key];
  } else {
    let min;
❷ [tree.left, min] = _removeMin(tree.left);
❸ return [tree, min];
  }
};
```

Assuming that the tree isn't empty, if you cannot go to the left, return the right tree and the node's key, and you're done ❶. Otherwise, recursively get and remove the minimum key from the left subtree ❷ and return the updated tree node and the found key ❸.

How do you use it? The change in remove() is small, affecting only one line: instead of first finding the minimum key and then removing it, make a single call to _removeMin():

```
const remove = (tree, keyToRemove) => {
  if (isEmpty(tree)) {
    // nothing to do
  } else if (keyToRemove < tree.key) {
    tree.left = remove(tree.left, keyToRemove);
  } else if (keyToRemove > tree.key) {
    tree.right = remove(tree.right, keyToRemove);
  } else if (isEmpty(tree.left) && isEmpty(tree.right)) {
    tree = null;
  } else if (isEmpty(tree.left)) {
    tree = tree.right;
  } else if (isEmpty(tree.right)) {
    tree = tree.left;
  } else {
    [tree.right, tree.key] = _removeMin(tree.right);
  }
  return tree;
};
```

12.18 AVL Worst

Suppose Hn is the number of nodes in the worst possible AVL tree of height n. The first few such trees are the following:

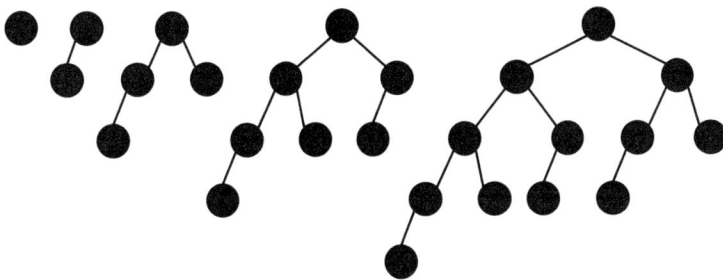

Build the next worst tree out of the previous two (plus a root), so H$_n$ equals H$_{n-1}$ + H$_{n-2}$ + 1: the sequence is 0, 1, 2, 4, 7, 12, 20, . . . , which is one less than the Fibonacci sequence 1, 2, 3, 5, 8, 13, 21, . . .

12.19 Singles Only

Given the structure restrictions of AVL trees, only a leaf can be a single child. Since every single child has one parent and there may be more (not single) nodes in the tree, single children cannot be more than 50 percent of all nodes.

12.20 Why One?

If the subtrees have sizes p and q, the tree's size is $p + q + 1$, so the fractions are $(p + 1) / (p + q + 2)$ and $(q + 1) / (p + q + 2)$, and adding the numerators together produces exactly $(p + q + 2)$, which is the denominator.

12.21 Easier Randomizing?

The developer is correct in that an ordered sequence of keys would become disordered, but a disordered sequence of keys would also exist that would become ordered after hashing, so while solving the problem of ordered additions, this technique wouldn't completely solve the worst-case problem.

12.22 Why Not Decrement?

If the key you want to delete isn't in the tree, you'd still (wrongly) decrement the node size.

12.23 Bad Splay?

If you only add keys, you'll get the same linear structure as for common binary search trees. However, after a few removals, the height of the tree is considerably lowered (and the tree becomes more "bushy"), while for common binary search trees, the shape would still be linear.

12.24 What Left Subtree?

It should be null. The minimum key in a tree has no left subtree; otherwise, it wouldn't be the minimum.

12.25 Code Transformation

When you splay the tree with reference to a key, you end with a tree that has at its root the closest value to the given key, so if the key is -Infinity (assuming keys are numeric; an empty string would do for alphabetic keys), you can deduce that _splay(tree, -Infinity) produces the same result as _splayMinimum(tree).

But you can go further. Take the _splay() code and assume keyToUp is -Infinity, a value that cannot be in the tree and that is smaller than all current keys. The highlighted parts in the code that follows can then be omitted, either because the result is already known or because it's unreachable code:

```
❶ const _splay = (tree, keyToUp) => {
❷ if (isEmpty(tree) || keyToUp === tree.key) {
    return tree;
  } else {
  ❸ const side = keyToUp < tree.key ? "left" : "right";
    if (isEmpty(tree[side])) {
      return tree;
  ❹ } else if (keyToUp === tree[side].key) {
      return _rotate(tree, side);
    } else {
    ❺ if (keyToUp <= tree[side].key === keyToUp <= tree.key) {
      ❻ tree[side][side] = _splay(tree[side][side], keyToUp);
        tree = _rotate(tree, side);
    ❼ } else {
        const other = side === "left" ? "right" : "left";
        tree[side][other] = _splay(tree[side][other], keyToUp);
        if (!isEmpty(tree[side][other])) {
          tree[side] = _rotate(tree[side], other);
        }
      }
      return isEmpty(tree[side]) ? tree : _rotate(tree, side);
    }
  }
};
```

The keyToUp parameter isn't needed ❶❻, since you are assuming its value. Tests for equality will always fail ❷❹. Since keyToUp is smaller than any key, the side variable will always end with "left" ❸, and the test at ❺ will always succeed, making some code ❼ unreachable.

After these simplifications, renaming _splay to _splayMin and changing tree[side] to tree.left leaves the following code:

```
const _splayMin = (tree) => {
  if (isEmpty(tree)) {
    return tree;
  } else {
    if (isEmpty(tree.left)) {
      return tree;
```

```
    } else {
      tree.left.left = _splayMin(tree.left.left);
      tree = _rotate(tree, "left");
      return isEmpty(tree.left) ? tree : _rotate(tree, "left");
    }
  }
};
```

Transforming this code to _splayMinimum() is easy now: join the two if statements into one by combining their tests, and you're done.

12.26 Full Rebalance

The idea is to get all the keys using the technique suggested in question 12.13 and produce a well-balanced tree, splitting the array in the middle. The key there will be the root for the balanced tree, the keys to its left will be used to produce the left subtree, and likewise for the keys to its right:

```
const {
  newBinaryTree,
  newNode,
  inOrder
} = require("../binary_search_tree.js");

❶ const _buildPerfect = (keys) => {
❷   if (keys.length === 0) {
      return newBinaryTree();
❸   } else {
❹     const m = Math.floor(keys.length / 2);
❺     return newNode(
        keys[m],
        _buildPerfect(keys.slice(0, m)),
        _buildPerfect(keys.slice(m + 1))
      );
    }
  };

❻ const restructure = (tree) => {
❼   const keys = [];
    inOrder(tree, (x) => keys.push(x));
❽   return _buildPerfect(keys);
  };
```

Start with the rebalancing code, _buildPerfect() ❶. Given an array of keys, if the array is empty ❷, a null tree is returned. Otherwise ❸, find the array's middle point ❹ and return a node as described earlier ❺; use recursion to build its balanced subtrees. The restructure() function ❻ is then quite short: generate an ordered list of keys using inOrder() ❼ and pass it to the _buildPerfect() function to produce the final output ❽.

Chapter 13

13.1 Missing Test?

No need; addChild() already does it.

13.2 Traversing General Trees

Implementation is not complex. If the tree is represented using arrays of children, preorder traversal for a nonempty tree entails visiting the root and then sequentially traversing each child. For a tree represented in a left-child, right-sibling style, the logic starts by visiting the root and then, starting at the first child, traversing it and moving to the next sibling until no more siblings are left. In both cases, postorder traversal first starts with the children and visits the root only after that.

13.3 Nonrecursive Visiting

The solution is quite similar to the breadth-first queue version:

```
depthFirstNonRecursive(visit = (x) => console.log(x)) {
❶ if (!isEmptyTree(tree)) {
  ❷ const s = new Stack();
    s = push(s, tree);
  ❸ while (isEmptyStack(s)) { **DZ missing ! operator**
      let t;
    ❹ [s, t] = s.pop();
      visit(t.key);
    ❺ [...t.childNodes].reverse().forEach((v) => { s = push(s, v); });
    }
  }
}
```

As in other traversals, if the tree is empty, you don't do anything ❶. Otherwise, you create a stack and push the tree's root into it ❷. Then, you do a loop. While the stack isn't empty ❸, pop the stack top ❹ and visit it, and then finish with this tricky detail: you must push all the children in *reverse* order (the rightmost first, the leftmost last), so the first child is visited first ❺. Be careful using reverse() because it modifies the array, so build a copy using destructuring.

13.4 Tree Equality

You could do some recursive logic to compare trees, but a simpler solution exists: use JSON.stringify() to produce string versions of both trees and compare them.

13.5 Measuring Trees

The code is similar to what was used for binary trees in Chapter 12:

```
const { Tree } = require("../tree.class.js");

class TreeWithMeasuring extends Tree {
  calcSize() {
    return this.isEmpty()
```

```
❶ ? 0
❷ : 1 + this._children.reduce((a, v) => a + v.calcSize(), 0);
}

calcHeight() {
  if (this.isEmpty()) {
❸ return 0;
  } else if (this._children.length === 0) {
❹ return 1;
  } else {
❺ return 1 + Math.max(...this._children.map((v) => v.calcHeight()));
  }
}
}
```

An empty tree has size 0 ❶; otherwise, its size is 1 (for the root itself) plus the sum of the sizes of all its subtrees ❷, which you can calculate using .reduce(). Then, for height, an empty tree has height 0 ❸, a leaf has height 1 ❹, and other trees have height 1 (for the root) plus the tallest height of any of its subtrees ❺.

13.6 Sharing More

In the implementation, since we borrowed only a single key from a sibling, we made do with simple code. To share half and half, you should set up an array with all the keys from the left sibling, plus the key of the parent, plus all the keys in the right sibling, and divide it. The key at the middle position will go into the parent, all the keys to its left to the left sibling, and all the keys to its right to the right sibling—and a similar procedure will apply to the pointers in the nodes.

13.7 Faster Node Searching

The binary search algorithm from Chapter 9 ends, upon a search failure, with the left index pointing at the link you should follow; check this out. Here's the code:

```
const _findIndex = (tree, key) => {
  let l = 0;
  let r = tree.keys.length - 1;

  while (l <= r) {
    const m = Math.floor((l + r) / 2);
    if (tree.keys[m] === key) {
      return m;
    } else if (tree.keys[m] > key) {
      r = m - 1;
    } else {
      l = m + 1;
    }
  }
  return l;
};
```

The algorithm is the standard one you already saw; the difference is that if the search fails, it returns the left pointer. As to the effect on the B-tree performance, the search in the node sped up by a constant factor (if the order of the B-tree is p, instead of p tests, we do log p tests), but the general order of algorithms remains $O(\log n)$.

13.8 Lowest Order

Nodes other than the root in a B-tree of order 2 should have one key and thus two children, so that implies a full binary tree since all leaves must be at the same level. So, yes, you already knew about B-trees of order 2!

13.9 Many Orders of Trees

By default, modules are singletons, which means that the code is imported only once, so all trees that you create would share the same ORDER variable. If you want to have different variables, instead of exporting an object with many properties (in module.exports), you need to export a function that, when invoked, returns the desired object. You can see some examples of this transformation in Iain Collins's "How (Not) to Create a Singleton in Node.js" at *https://medium.com/@iaincollins/ how-not-to-create-a-singleton-in-node-js-bd7fde5361f5*.

13.10 Safe to Delete?

When you get to delete a node with no right child, the root was red, implying it would have no left child. If the root had been black (the other possibility, considering the invariant), its left child would have been red, and you would have rotated *it* right, so it wouldn't then have an empty right child.

Chapter 14

14.1 Is It a Heap?

Simply loop through all the elements except the root and check that each is not greater than its parent. A first implementation could be:

```
function isHeap1(v) {
  for (let i = 1; i < v.length; i++) {
    if (v[i] > v[Math.floor((i - 1) / 2)]) {
      return false;
    }
  }
  return true;
}
```

You can also use .every() to shorten the code and make it more declarative:

```
function isHeap2(heap) {
  return heap.every(
    (v, i) => i === 0 || v <= heap[Math.floor((i - 1) / 2)],
  );
}
```

14.2 Making Do with Queues

If you assign monotonically ascending values to the elements you
enter in a priority queue, it will behave like a stack in a last-in, first-out
(LIFO) way. Similarly, if you assign monotonically descending values to
elements, the priority queue will emulate a common queue.

14.3 Max to Min

This is a trick question! Using Floyd's enhanced heap-building code,
you can transform *any* array into a min heap in linear time, so you obviously can convert a max heap into a min heap in this time.

14.4 Max or Min

Only three code changes are needed: one in the _bubbleUp() function
and two in the _sinkDown() function. Invert the current comparisons
from heap[a] > heap[b] to heap[a] < heap[b] and you're done.

14.5 Merge Away!

The needed algorithm works as follows. Create a min heap whose nodes
will be nodes from the lists. Initialize the heap by taking the first element of each list. Initialize an empty output list. Repeatedly, while the
heap isn't empty, pick the node corresponding to the root of the heap
and remove it from the heap. Add the selected node to the output list.
If the selected node had a next node, add it to the heap.

Assuming the nodes have key and next fields, the code (based on
the heap developed in this chapter, but inverting some comparisons to
produce a min heap as in question 14.4) could be as follows:

```
function merge_away(lists) {
  const heap = [];

  const add = (node) => {
    const _bubbleUp = (i) => {
      // Bubble up heap[i] comparing by heap[i].key
      // (you'll have to modify the bubbling code we
      // saw a bit for this).
    };

❶   if (node) {
      heap.push(node);
      _bubbleUp(heap.length - 1);
    }

    const remove = () => {
      const _sinkDown = (i, h) => {
        // sink down heap[i] comparing by heap[i].key
```

```
      };
    };

    const node = heap[0];
    heap[0] = heap[heap.length - 1];
    heap.pop();
    _sinkDown(0, heap.length);
    return node;
  };

❷ lists.forEach((list) => add(list));
❸ const first = { next: null };
  let last = first;
❹ while (heap.length > 0) {
    const node = remove();
  ❺ add(node.next);
  ❻ last.next = node;
    last = node;
    node.next = null;
  }
❼ return first.next;
}
```

The add() method pushes a node into the stack (unless it's null) and bubbles it up ❶. The complete logic requires setting up the heap with the first element of each list ❷ to then do the merging. Adding an empty initial value for the output list ❸ simplifies the code a bit; remember to skip this extra node when returning the merged list ❼. While there still are any nodes in the heap ❹ (meaning there still is some merging to do), remove the top element, add the next node of the corresponding list into the heap ❺, and add the removed element at the end of the list ❻; the dummy node avoided the need to test for an empty list. The final step just returns the list, without the extra initial node ❼.

You can test this code easily. The following outputs a few Fibonacci numbers starting at 1:

```
const list1 = {
  key: 2,
  next: { key: 3, next: { key: 8, next: null } },
};

const list2 = {
  key: 1,
  next: { key: 13, next: { key: 55, next: null } },
};

const list3 = null;

const list4 = { key: 21, next: null };

const list5 = { key: 5, next: { key: 34, next: null } };
```

```
const merged = merge_away([list1, list2, list3, list4, list5]);

let p = merged;
while (p) {
  console.log(p.key);
  p = p.next;
}
```

14.6 Searching a Heap

You could write a recursive function, but because the array is basically unordered, given any node in the heap, you'd have to search in both its left and right subtrees, so you will eventually have to look through the whole tree in an $O(n)$ procedure. It would be better simply to use `heap.find()`.

14.7 Removing from the Middle of a Heap

Changing a key in the heap is a bit similar to removing a key:

```
const removeMiddle = (heap, k) => {
❶ if (isEmpty(heap)) {
    throw new Error("Empty heap; cannot remove");
❷ } else if (k < 0 || k >= heap.length) {
    throw new Error("Not valid argument for removeMiddle");
  } else {
❸ [heap[k], heap[heap.length - 1]] = [heap[heap.length - 1], heap[k]];
    heap.pop();
❹ _bubbleUp(heap, k);
❺ _sinkDown(heap, k, heap.length);
  }
  return heap;
};
```

After a couple of checks to see whether the removal can be done ❶❷, moving the last value of the heap to the place of the removed value ❸ restores the structure property. The problem is that the value now at index k may not be placed correctly, violating the heap property. The simplest way to ensure that this is satisfied is to apply _bubbleUp() ❹ first and then sinkDown() ❺. At most, only one of those functions will do anything, and you'll end up with a fully compliant heap.

14.8 Faster Build

The changes to the `newHeap()` function are as follows:

```
const newHeap = (values = []) => {
  const newH = [];
❶ values.forEach((v) => newH.push(v));
  for (let i = Math.floor((newH.length - 1) / 2); i >= 0; i--) {
❷ _sinkDown(newH, i, newH.length);
  }
  return newH;
};
```

This starts with an empty array and copies the values (if any) into it ❶; then it uses _sinkDown() to build the heap ❷ that is returned.

14.9 Another Way of Looping

You could use forEach() in the following way (only the i parameter matters here):

```
v.forEach((_, i) => _bubbleUp(v, i));
```

14.10 Extra Looping?

It would work the same way (it would just be a tad slower), because the _sinkDown() procedure won't do anything for elements with no children.

14.11 Maximum Equality

It would be $O(n)$. Neither _sinkDown() nor _bubbleUp() would do any work, so they are $O(1)$, and there are n calls to those functions.

14.12 Unstable Heap?

A simple array works for both versions of the heapsort code: [1, 1]. The first 1 will end at the last place of the sorted array.

14.13 Trimmed Selection

If you find a value at level i of a heap, you know for sure that there are at least $(i-1)$ values greater than it, because of the heap property. Thus, if you are looking for the k greatest values of a heap, they cannot be at levels $(k+1)$ or beyond. If the heap has more than k levels, you can discard all beyond the kth level, and the selection process will be a bit faster. A heap with k complete levels has $2k-1$ nodes, so if the heap has more nodes than that, you can shorten it:

```
function selection(k, values) {
  const heap = [];

  const _sinkDown = // ...omitted here...

  // Build heap out of values.
  values.forEach((v) => heap.push(v));
  for (let i = Math.floor((heap.length - 1) / 2); i >= 0; i--) {
    _sinkDown(i, heap.length);
  }

  // Trim the heap, if possible.
  const maxPlace = 2 ** k - 1;
  if (heap.length > maxPlace) {
  ❶ heap.length = maxPlace;
  }

  // Do the selection.
  for (let i = heap.length - 1; i >= heap.length - k; i--) {
    [heap[i], heap[0]] = [heap[0], heap[i]];
```

```
    _sinkDown(0, i);
  }

  return heap.slice(heap.length - k);
}
```

You could use `.slice()` to shorten the `heap` array, but JavaScript allows you to modify its `.length` property directly ❶.

14.14 Is It a Treap?

You can produce an interesting recursive solution with functional aspects. What is a valid treap? If it's empty, it's obviously okay; otherwise, its children should be treaps too:

```
function isTreap(tr, valid = () => true) {
  return (
    tr === null ||
❶ (valid(tr) &&
    ❷ isTreap(
        tr.left,
        (t) => t.key <= tr.key && t.priority <= tr.priority,
      ) &&
    ❸ isTreap(
        tr.right,
        (t) => t.key >= tr.key && t.priority <= tr.priority,
      ))
  );
}
```

This checks a basic condition having to do with its key and priority compared to those of its parent ❶, but notice that the first time, for the root, a trivial validation is provided because obviously the root has no parent with which to compare! Then, this recursively checks that both children are also treaps, each fulfilling a specific different new condition: the left subtree should have smaller keys and priorities than its parent ❷, and the right subtree should have greater keys but lower priorities ❸.

14.15 Treap Splitting

Add the limit value to the treap with an `Infinity` priority: because of the high priority, this limit will become the new root of the treap, and because of the binary search tree structure, all keys smaller than the limit will be in the root's left subtree and greater keys will be in its right subtree, providing the desired partitioning.

14.16 Rejoining Two Treaps

To create a single treap out of two separate ones, create a dummy node with any random key and priority and then assign the first treap as its left subtree, assign the second treap as its right subtree, and finish by deleting the root dummy node.

14.17 Removing from a Treap

This would work the same way, but it would be just a bit slower, because it would do a series of if statements before going down tree[other].

14.18 Trees as Heaps

Using a balanced binary search tree would ensure logarithmic performance for all three operations, as you've seen earlier. However, you can do better by including a separate attribute for the maximum seen so far. You'd update it in $O(1)$ time after each addition by simply comparing the current maximum to the new key, and when removing the top, you could find the new maximum also in logarithmic time, so top() itself would become $O(1)$.

Chapter 15

15.1 Intuitive but Worse

That would be $O(m \log n)$, since there would be m insertions, each $O(\log n)$.

15.2 Sequential Cases

When inserting keys in descending order (the highest first), you get a badly shaped heap, but each add() takes constant time. Can you see why? And if you insert them in ascending order, you can achieve a full binary tree; see both cases shown next for keys from 1 to 7:

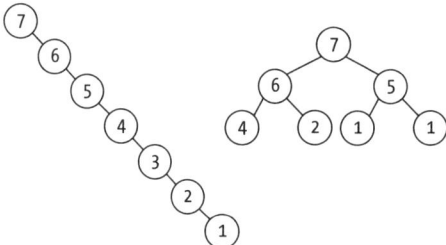

15.3 No Recursion Needed

You could just write the merge as follows; the bold lines show the changes:

```
const merge = (heap1, heap2) => {
  if (isEmpty(heap2)) {
    return heap1;
  } else if (isEmpty(heap1)) {
    return heap2;
  } else if (goesHigher(heap1.key, heap2.key)) {
    [heap1.left, heap1.right] = [merge(heap2, heap1.right), heap1.left];
    return heap1;
  } else {
    [heap2.left, heap2.right] = [merge(heap1, heap2.right), heap2.left];
    return heap2;
```

```
    }
};
```

The new lines are the same as in the previous case (when the first heap has the greatest key), except they exchange heap1 and heap2 throughout.

15.4 Change Needed

The change() method for a skew heap requires being able to remove a key and, after changing it, inserting it again. Given a reference to the node with the key, removing it requires a link to the parent (from which you must disconnect it), so an up pointer to a node's parent should be added. You could alternatively use bubbleUp(), but it would also need a link to the parent.

15.5 Just Adding

You initially have just a single tree with eight nodes. The following table shows which heaps you'd have at each stage and how many merges are needed after adding a value; cell items in bold represent merges. For example, adding the 9th value requires no merges, but adding the 10th produces two heaps of size 1, so a merge is needed to replace them with a heap of size 2. Similarly, adding the 11th value requires no merges, but the 12th needs two: first to merge two heaps of size 1 and then to merge two heaps of size 2.

Values	1	2	4	8	16
8	0	0	0	1	0
9	1	0	0	1	0
10	**0**	1	0	1	0
11	1	1	0	1	0
12	**0**	**0**	1	1	0
13	1	0	1	1	0
14	**0**	1	1	1	0
15	1	1	1	1	0
16	**0**	**0**	**0**	**0**	1

You had an initial heap with $n = 8$ values; after adding the other eight values, the total number of merges (in bold) was exactly eight, so on average, each addition required one merge. This is not a formal proof, of course, but the result can be proven mathematically for all binomial heaps.

15.6 Faster Binomial Top

You can add a variable like _heapTop in lazy binomial heaps to get the top value of the heap.

15.7 Easier Bubbling Up?

The problem is you are working with an addressable heap, and if you change keys around, old references to nodes won't be valid any longer and will point to different values.

15.8 Searching a Heap

This algorithm is obviously $O(n)$ and certainly *not* the kind of thing you do with heaps, but let's do it anyway. The only problem is to notice when to stop traversing the circular list of siblings:

```
❶ _findInTree(tree, keyToFind, stopT = null) {
❷ let node = null;
❸ if (tree && tree !== stopT) {
    ❹ if (tree.key === keyToFind) {
        node = tree;
    } else {
        node =
        ❺ this._findInTree(tree.down, keyToFind) ||
        ❻ this._findInTree(tree.right, keyToFind, stopT || tree);
    }
}
❼ return node;
}
```

This is a depth-first traversal. The stopT parameter remembers where you started in the list of siblings to avoid a loop ❶. Use node to store either a null value (if you don't find the key) or the found node otherwise ❷. If the tree you're looking at is neither null nor the stop point for the list ❸, check whether you found the key you were looking for at the root ❹; if so, save it to return it later ❼. If the root didn't match, search downward ❺, and if that returns a null, search to the right, providing the starting point as the stop value ❻. Using xxx || yyy is typical of JavaScript; if the value of the xxx expression isn't "falsy," it is returned; otherwise, the value of the yyy expression is returned.

15.9 Two in One

In _mergeA2B(), you could write:

```
if (high._down) {
  low.right = high.down;
  low.left = high.down.left;
  high.down.left.right = high.down.left = low;
}
```

In add(), you could join two assignments:

```
newTree.left = newTree.right = newTree;
```

In remove(), you'd do something similar:

```
bt.right = bt.left = bt;
```

And that would also apply in _separate():

```
node.left = node.right = node;
```

Is it worth it? Four lines saved versus the possibility of misreading or misunderstanding the code; it's your call!

Chapter 16

16.1 Maps for Tries

The needed changes with regard to object-based tries are as follows; I leave comments up to you, but the modified lines are in bold. Creating a trie requires the following:

```
const newNode = () => ({ links: new Map() });
```

Finding a key just changes the way you access links:

```
const _find = (trie, [first, ...rest]) => {
  if (isEmpty(trie)) {
    return null;
  } else if (first === EOW) {
    return isEmpty(trie.links.get(first))
      ? null
      : trie.links.get(first).data;
  } else {
    return _find(trie.links.get(first), rest);
  }
};
```

Make the same kind of change when adding a new key:

```
const _add = (trie, [first, ...rest], data) => {
  if (first) {
    if (isEmpty(trie)) {
      trie = newNode();
    }
    if (first === EOW) {
      trie.links.set(first, { data });
    } else {
      trie.links.set(first, _add(trie.links.get(first), rest, data));
    }
  }
  return trie;
};
```

And the same happens when removing a key:

```
const _remove = (trie, [first, ...rest]) => {
  if (isEmpty(trie)) {
    // nothing to do
  } else if (!first) {
    trie = null;
  } else {
    trie.links.set(first, _remove(trie.links.get(first), rest));
    if (isEmpty(trie.links.get(first))) {
      trie.links.delete(first);
      if (trie.links.size === 0) {
        trie = null;
      }
    }
  }
  return trie;
};
```

16.2 Ever Empty?

The answer is yes in case you attempt to add a key that was already in the trie.

16.3 Rotate Your Tries

Yes, you can apply rotations to a trie. You would work with the left and right links only and never affect the middle links.

16.4 Empty Middle?

True, unless you didn't want to store any data, in which case, the middle link of the EOW characters would be null.

16.5 Four-Letter Trie?

The longest path (the trie's height) would be from the root to the EOW for ZZZZ: 104 steps. The number of characters in the keys would be 4, and the alphabet is 26 letters, which means the height is the following: $4 \times 26 = 104$.

16.6 How Do They Look?

Both an array-based trie and an object-based trie would be a list of nodes, one for each letter. A radix tree would have a single node, with the ALGORITHM word in it. Finally, a ternary tree would be a vertical column of nodes, with a letter at each level.

Chapter 17

17.1 Where's the Path?

Add a next[i][j] matrix that tells you where to go if you are at i and want to get to j. Whenever you update dist[i][j] you'll also update next. Here's the updated algorithm:

```
const distances = (graph) => {
  const n = graph.length;

  const distance = [];
❶ const next = [];
  for (let i = 0; i < n; i++) {
    distance[i] = Array(n).fill(+Infinity);
❷ next[i] = Array(n).fill(null);
  }

  graph.forEach((r, i) => {
    distance[i][i] = 0;
❸ next[i][i] = i;
    r.forEach((c, j) => {
      if (c > 0) {
        distance[i][j] = graph[i][j];
      ❹ next[i][j] = j;
      }
    });
  });

  for (let k = 0; k < n; k++) {
    for (let i = 0; i < n; i++) {
      for (let j = 0; j < n; j++) {
        ❺ if (distance[i][j] > distance[i][k] + distance[k][j]) {
          distance[i][j] = distance[i][k] + distance[k][j];
          next[i][j] = next[i][k];
        }
      }
    }
  }

❻ return [distance, next];
};
```

Define the next matrix ❶ and fill it with null values as the default ❷. Mark that to go from a point to itself, the path obviously passes through that point ❸, and whenever there's an edge between two points, mark that too ❹. Whenever you find a better path between two points ❺, update not only dist but next as well. At the end, you must return the next matrix ❻, which you'll use in the following path-finding algorithm:

```
const path = (next, u, v) => {
❶ const sequence = [];
❷ if (next[u][v] !== null) {
  ❸ sequence.push(u);
  ❹ while (u !== v) {
    ❺ u = next[u][v];
    ❻ sequence.push(u);
  }
}
❼ return sequence;
};
```

Create a sequence array ❶ that will have all intermediate steps. If there is some way of going from the first point to the last ❷, push the initial point ❸, and if you haven't reached the destination ❹, advance to the next point ❺ and push it ❻. At the end, just return the sequence ❼ that includes all steps.

17.2 Stop Searching Sooner

Make sure to check whether there were any changes in any pass. The main loop of the algorithm changes as follows:

```
for (let i = 0; i < n - 1; i++) {
❶ let changes = false;
  edges.forEach((v) => {
    const w = v.dist;
    if (distance[v.from] + w < distance[v.to]) {
      distance[v.to] = distance[v.from] + w;
      previous[v.to] = v.from;
    ❷ changes = true;
    }
  });
❸ if (!changes) {
    break;
  }
}
```

At the beginning of each pass, set the changes variable to false ❶, but if you change any distance, set it to true ❷. After going through all the edges ❸, if there were no changes, you don't need to repeat the loops and can exit.

17.3 Just One Will Do

You have to modify the signature of the function to receive three arguments (graph, from, and the added to) and change the main loop to see whether the top of the heap is the destination to point; in that case, you stop:

```
while (heap.length && heap[0] !== to) {
```

17.4 The Wrong Way

The output is a topological sort of the original graph, but in reverse order: the first nodes that appear are those that were the last nodes in the previous graph.

17.5 Joining Sets Faster

There will be two changes. First, all the nodes in the forest will include a size attribute, initially set to 1:

```
const groups = Array(n)
  .fill(0)
  .map(() => ({ ptr: null, size: 1 }));
```

The other change appears when merging two sets into a single set. The main loop of the isConnected(...) code changes as follows:

```
for (let i = 0; i < n; i++) {
  for (let j = i + 1; j < n; j++) {
    if (graph[i][j]) {
      const pf = findParent(groups[i]);
      const pt = findParent(groups[j]);

      if (pf !== pt) {
        count--;
        if (pf.size < pt.size) {
          pt.size += pf.size;
          pf.ptr = pt;
        } else {
          pf.size += pt.size;
          pt.ptr = pf;
        }
      }
    }
  }
}
```

The code in bold marks the change. If you find two different roots, check which has the smallest size and link it to the other root. Don't forget to update its size to account for the added subset.

17.6 Take a Shortcut

The following modified findParent(...) routine creates the shortcuts as described; can you see how it works?

```
const findParent = (x) => {
  if (x.ptr !== null) {
    x.ptr = findParent(x.ptr);
    return x.ptr;
  } else {
    return x;
  }
};
```

17.7 A Spanning Tree for a Tree?

The exact same graph is produced: in a tree there is only one way to get from any point to another, so there are no alternatives for a different spanning tree.

17.8 A Heap of Edges

I'll leave this up to you, but I chose heapsort because of its assured performance.

Chapter 18

18.1 Getting Here

The front list isn't empty, which means someone exited the queue. A possible sequence could be as follows: X enters, then A enters, then B enters, then X exits (now the back list is empty and the front list consists of A and B), followed by C, D, and E entering in that order.

18.2 With Apologies to Abbott and Costello, Who's on Front?

The following logic solves the problem:

```
const front = (queue) => {
❶ if (isEmpty(queue)) {
    return undefined;
❷ } else if (queue.frontPart !== null) {
    return queue.frontPart.value;
❸ } else {
    let ptr = queue.backPart;
    while (ptr.next !== null) {
      ptr = ptr.next;
    }
    return ptr.value;
  }
};
```

If the queue is empty ❶, there's no front element; return undefined, throw an exception, or perform other similar actions. If the front part isn't empty ❷, its top element is simply the front of the queue. But if the front part is empty ❸, go through the back part until its end, because that's the front of the queue.

18.3 No Change Needed

There are many possibilities, but the simplest is to throw an exception if the key isn't found and wrap the tree search algorithm in a try...catch structure, so if the exception is thrown, just return the original tree.

18.4 A New Minimum

Remember that you apply this function to find the minimum value of a nonempty binary search tree:

```
const minKey = (tree) =>
  isEmpty(tree.left) ? tree.key : minKey(tree.left);
```

If the root has no left child, the root's value is the minimum. Otherwise, the minimum value of the tree will be the minimum of the root's left child, because all values in that subtree are smaller than the root's value.

BIBLIOGRAPHY

Aguilar, Luis Joyanes. *Fundamentos de Programación: Algoritmos, estructuras de datos y objetos.* 4th ed. Madrid: McGraw-Hill, 2008.

Aravinth, Anto. *Beginning Functional JavaScript.* Berkeley, CA: Apress, 2017.

Atencio, Luis. *Functional Programming in JavaScript.* New York: Manning Publications, 2016.

Bae, Sammie. *JavaScript Data Structures and Algorithms.* Berkeley, CA: Apress, 2019.

Baldwin, Douglas, and Greg Scragg. *Algorithms and Data Structures: The Science of Computing.* Boston: Charles River Media, 2004.

Bird, Richard, and Philip Wadler. *Introduction to Functional Programming.* Hoboken, NJ: Prentice Hall International, 1988.

Braithwaite, Reginald. *JavaScript Allongé.* 6th ed. *https://leanpub.com/javascriptallongesix/read.*

Brass, Peter. *Advanced Data Structures.* Cambridge, UK: Cambridge University Press, 2008.

Cormen, Thomas, Charles Leiserson, Ronald Rivest, and Clifford Stein. *Introduction to Algorithms.* 3rd ed. Cambridge, MA: MIT Press, 2009.

Dale, Nell. *C++ Plus Data Structures*. 3rd ed. Burlington, VT: Jones and Bartlett Publishers, 2003.

Drozdek, Adam. *Data Structures and Algorithms in Java*. 2nd ed. Boston: Thomson Learning, 2005.

Fogus, Michael. *Functional JavaScript*. Sebastopol, CA: O'Reilly Media, 2013.

Goldman, Sally, and Kenneth Goldman. *A Practical Guide to Data Structures and Algorithms Using Java*. Boca Raton, FL: Chapman & Hall/CRC Press, 2008.

Groner, Loiane. *Learning JavaScript Data Structures and Algorithms*. 3rd ed. Birmingham, UK: Packt Publishing, 2018.

Harmes, Ross, and Dustin Díaz. *Pro JavaScript Design Patterns*. Berkeley, CA: Apress, 2008.

Jansen, Remo. *Hands-On Functional Programming with TypeScript*. Birmingham, UK: Packt Publishing, 2019.

Karumanchi, Narasimha. *Data Structures and Algorithms Made Easy*. Hyderabad, Telangana, India: CareerMonk Publications, 2010.

Kereki, Federico. *Mastering JavaScript Functional Programming*. 3rd ed. Birmingham, UK: Packt Publishing, 2023.

Khot, Atul, and Raju Kumar Mishra. *Learning Functional Data Structures and Algorithms*. Birmingham, UK: Packt Publishing, 2017.

Knuth, Donald. *The Art of Computer Programming, Volume 1: Fundamental Algorithms*. 3rd ed. Boston: Addison-Wesley, 1997.

———. *The Art of Computer Programming, Volume 2: Seminumerical Algorithms*. 3rd ed. Boston: Addison-Wesley, 1997.

———. *The Art of Computer Programming, Volume 3: Sorting and Searching*. 2nd ed. Boston: Addison-Wesley, 1998.

Lonsdorf, Brian (Dr. Boolean). *Professor Frisby's Mostly Adequate Guide to Functional Programming*. https://github.com/MostlyAdequate/mostly-adequate-guide.

Mantyla, Dan. *Functional Programming in JavaScript*. Birmingham, UK: Packt Publishing, 2015.

Masood, Adnan. *Learning F# Functional Data Structures and Algorithms*. Birmingham, UK: Packt Publishing, 2015.

McMillan, Michael. *Data Structures and Algorithms with JavaScript*. Sebastopol, CA: O'Reilly Media, 2014.

Mehlhorn, Kurt, and Peter Sanders. *Algorithms and Data Structures: The Basic Toolbox*. Berlin: Springer, 2008.

Morin, Pat. *Open Data Structures*. Edmonton, AB: AU Press, 2013.

Mukkamala, Kashyap. *Hands-On Data Structures and Algorithms with JavaScript.* Birmingham, UK: Packt Publishing, 2018.

Okasaki, Chris. *Purely Functional Data Structures.* Cambridge, UK: Cambridge University Press, 1998.

Parker, Alan. *Algorithms and Data Structures in C++.* Boca Raton, FL: CRC Press, 1993.

Sedgewick, Robert. *Algorithms in C.* Boston: Addison-Wesley, 1990.

Sedgewick, Robert, and Philippe Flajolet. *An Introduction to the Analysis of Algorithms.* 2nd ed. Boston: Addison-Wesley, 2013.

Sedgewick, Robert, and Kevin Wayne. *Algorithms.* 4th ed. Boston: Addison-Wesley, 2011.

Simpson, Kyle. *Functional-Light JavaScript. https://github.com/getify/Functional-Light-JS.*

Thareja, Reema. *Data Structures Using C.* 2nd ed. New Delhi: Oxford University Press, 2014.

Wengrow, Jay. *A Common-Sense Guide to Data Structures and Algorithms.* Raleigh, NC: The Pragmatic Programmer, 2017.

Wirth, Niklaus. *Algorithms + Data Structures = Programs.* Englewood Cliffs, NJ: Prentice-Hall, 1976.

INDEX

trinary heaps, 340
TypeScript, 18–19

V

Visual Studio Code (VSC), 4, 13–15, 18

W

Waterman, Alan, 153
WebAssembly (WASM), 185

weight-bounded balance (BB[α]) trees, 255–261, 281
    adding keys to, 257
    creating, 256–257
    finding by rank, 260–261
    fixing balance, 257–259
    performance, 261
    removing keys from, 257
Williams, John W. J., 327–329
worst-case performance of algorithms, 54

RESOURCES

Visit *https://nostarch.com/data-structures-and-algorithms-javascript* for errata and more information.

More no-nonsense books from **NO STARCH PRESS**

THE NATURE OF CODE
Simulating Natural Systems with JavaScript
BY DANIEL SHIFFMAN
640 PP., $39.99
ISBN 978-1-7185-0370-0

ELOQUENT JAVASCRIPT, 4TH EDITION
BY MARIJN HAVERBEKE
456 PP., $49.99
ISBN 978-1-7185-0410-3

DATA STRUCTURES THE FUN WAY
An Amusing Adventure with Coffee-Filled Examples
BY JEREMY KUBICA
304 PP., $39.99
ISBN 978-1-7185-0260-4

THE RECURSIVE BOOK OF RECURSION
Ace the Coding Interview with Python and JavaScript
BY AL SWEIGART
328 PP., $39.99
ISBN 978-1-7185-0202-4

ALGORITHMIC THINKING, 2ND EDITION
Learn Algorithms to Level Up Your Coding Skills
BY DANIEL ZINGARO
480 PP., $49.99
ISBN 978-1-7185-0322-9

GRAPH ALGORITHMS THE FUN WAY
Powerful Algorithms Decoded, Not Oversimplified
BY JEREMY KUBICA
416 PP., $59.99
ISBN 978-1-7185-0386-1

PHONE:
800.420.7240 OR
415.863.9900

EMAIL:
SALES@NOSTARCH.COM

WEB:
WWW.NOSTARCH.COM

Never before has the world relied so heavily on the Internet to stay connected and informed. That makes the Electronic Frontier Foundation's mission—to ensure that technology supports freedom, justice, and innovation for all people—more urgent than ever.

For over 30 years, EFF has fought for tech users through activism, in the courts, and by developing software to overcome obstacles to your privacy, security, and free expression. This dedication empowers all of us through darkness. With your help we can navigate toward a brighter digital future.